The Behavioural Biology of Zoo Animals

"Zoo animals" as a population are a diverse array of species from all around the globe. When managed in captivity, it is important that key aspects of natural ecology are factored into animal care, as well as considerations relating to welfare, life history, and behavioural needs. *The Behavioural Biology of Zoo Animals* is the first book on captive animal behaviour and how this applies to welfare.

This book enables all aspects of zoo husbandry and management (nutrition, enclosure design, handling and training, enrichment, population management) to be based on a sound knowledge of the species, its evolutionary history, and its natural history. Chapters from expert authors cover a vast range of taxa, from primates and elephants, to marine mammals and freshwater fish, to reptiles, birds, and invertebrates. The final part looks to the future, considering animal health and wellbeing, the visitor experience, and future visions for zoos and aquariums.

For on-the-ground practitioners as well as students of zoo biology, animal science, and welfare, this book provides an explanation of key areas of behavioural biology that are important to fulfilling the aims of the modern zoo (conservation, education, research, and recreation). It explains how evidence from the wild can be implemented into captive care to support the wider aims of the zoo, shedding light on the evidence-based approaches applied to zoo biology and animal management.

Chapter 3 is available to download Open Access on the www.taylorfrancis.com website.

The Behavioural Biology of Zoo Animals

Edited by
Paul Rose

First edition published 2023
by CRC Press
6000 Broken Sound Parkway NW, Suite 300, Boca Raton, FL 33487-2742

and by CRC Press
4 Park Square, Milton Park, Abingdon, Oxon, OX14 4RN

CRC Press is an imprint of Taylor & Francis Group, LLC

Library of Congress Cataloging-in-Publication Data
Names: Rose, Paul (Zoologist), editor.
Title: The behavioural biology of zoo animals / [edited by] Paul Rose.
Description: First edition. | Boca Raton : CRC Press, 2023. | Includes
bibliographical references and index. |
Summary: "This is the first book on captive animal behaviour and how this applies to welfare. The book enables all aspects of zoo husbandry and management (nutrition, enclosure design, handling and training, enrichment, population management) to be based on a sound knowledge of the species, its evolutionary and natural history. Chapters from expert authors cover a vast range of taxa, from primates and elephants to marine mammals and freshwater fish, to reptiles, birds and invertebrates. It explains how evidence from the wild can be implemented into captive care to support the wider aims of the zoo (conservation, education, research and recreation)"-- Provided by publisher.
Identifiers: LCCN 2022032368 (print) | LCCN 2022032369 (ebook) | ISBN 9781032077192 (hardback) | ISBN 9781032077161 (paperback) | ISBN 9781003208471 (ebook)
Subjects: LCSH: Zoo animals--Behavior. | Animal welfare.
Classification: LCC QL77.5 .B44 2023 (print) | LCC QL77.5 (ebook) | DDC 590.73--dc23/eng/20220804
LC record available at https://lccn.loc.gov/2022032368
LC ebook record available at https://lccn.loc.gov/2022032369

ISBN: 978-1-032-07719-2 (hbk)
ISBN: 978-1-032-07716-1 (pbk)
ISBN: 978-1-003-20847-1 (ebk)

DOI: 10.1201/9781003208471

Typeset in Minion
by SPi Technologies India Pvt Ltd (Straive)

To my parents

Contents

Foreword

Many of us enjoy a trip to the zoo. Of course, we love wildlife documentaries as well, stunning footage of animals finding food, looking for mates, bringing up offspring, and generally getting on with their lives in the wild. But nothing compares with the experience of seeing an animal close up in the flesh; smelling it, hearing it, getting a real appreciation of the animal as a wonderful living thing. For most of us, this is the nearest we get to close encounters with "wild" animals.

Astonishingly, more people go to zoos every year than to football matches, or in some countries to their major tourist attractions, with an estimated 700 million visitors annually to zoos and aquariums worldwide. This provides strong support for the rationale behind part of the zoos' conservation mission, which involves education about animals and their conservation, and awareness-raising about the threats those animals face in the wild. In this sense we can regard the animals housed in zoos as ambassadors, representing those in the wild, and promoting familiarity with and caring about those species which need conservation help. But they are more than that. They are also "insurance" populations, a reservoir of genes and individuals who can supplement dwindling wild populations if and when they are needed.

I have deliberately put the word "wild" in inverted commas when referring to zoo-housed animals as "wild" animals in the paragraph above. This is because in some respects many of the animals housed in zoos are not strictly wild at all, but are zoo-born, often from parents who are also zoo-born. Zoos aim to achieve their conservation mission without taking animals from the wild unless the requirements of conservation make it really necessary. All of this creates a potential dilemma for zoos. Their animals are no longer living in the environment that they have evolved to survive in, and the zoo must consequently provide a captive environment which, while not exactly duplicating the wild, nevertheless provides the animal with the behavioural opportunities we might expect it to have in the wild. This is essential to ensure that those animals have the best possible welfare, but is also necessary to guard against the risk that during generations of captive breeding, the animals in zoos might eventually lose some of the behaviours that help them survive in the wild, and which characterise their species. At the same time, however, the animals housed in zoos have to be comfortable with the unavoidable consequences of life in captivity, which include being faced with those 700 million human visitors every year, but also having their lives, their feeding schedules, their mating opportunities, and other aspects of their lives being managed for them.

The way of dealing with this dilemma is through evidence-based animal management. The evidence comes from scientific research, which gives us a greater understanding of how those animals behave in the wild and how they respond to captivity. When we implement that understanding through changes to management, the research can tell us how effective our interventions are. The sheer growth of zoo-based research over the past two or three decades would probably come as a surprise to anyone not familiar with this field. Hundreds of peer-reviewed papers on zoo science are published every year, and keeping up to date with them can be a daunting task. This book is our guide through this literature. We are also very aware that, like many other sciences, zoo biology suffers from a taxon bias, a tendency for researchers to concentrate on a small number of high-profile orders or species of animal. In the zoo world, these tend to be the primates (and particularly the great apes), elephants, and large carnivores, and it is sometimes said that these need the most attention because of their sentience, awareness, and complex needs. But it is the case that the frogs, the insects, the birds, the lizards, and all those other taxa are just as deserving of attention as the big charismatics like pandas, tigers, and gorillas; and one of the strengths of this volume is that it devotes chapters to these usually under-represented taxa as well.

The chapters in this book help to guide us through this literature, and point out how the knowledge gained in these studies can be used effectively to ensure that the animals in zoos live the best quality lives in the best living conditions we can provide, and how this contributes to achieving the conservation goals of the modern zoo.

Geoff Hosey

Preface

Welcome to *The Behavioural Biology of Zoo Animals* (BBZA). This book covers behavioural ecology information and evidence essential to the development of biologically relevant animal husbandry and care regimes for a range of zoo and aquarium-housed species. The overall goal of BBZA is to provide the zookeeper or aquarist, curator, or living collections manager, research student or academic, conservation biologist, or animal welfare officer (among other important zoo roles and personnel too) with information on how zoo species live in a wild state and therefore how they should live when housed in the zoo. This book evaluates and discusses overarching theory, providing specific examples and case studies throughout; it directs and instructs the reader to important sources of information to build their knowledge base of what zoo and aquarium animals need when designing and implementing husbandry and housing.

The book contains 26 chapters, split across three sections. The first section of the book is introductory, explaining and evaluating key background information on what we mean by behavioural biology, how we can collect evidence on behaviour to help zoo management, what methods are at our disposal for research in the zoo, and how population management can be based around animal behaviour information. These introductory chapters aim to show what evidence is available, why it matters, and how it can be collected and ultimately applied.

The second section of the book is the largest, and provides an overview of the behavioural biology of specific taxonomic groups and species of animals often found in zoos and aquariums. While it would be impossible to cover every single species of wild animal currently housed in zoos and aquariums, the breadth of taxa covered by this book shows the wide impact that researching and then applying evidence can have on animal care and welfare. Each taxonomic chapter provides examples of species' ecology and adaptations placed within the context of the adaptive benefits of behaviour patterns, and how these must be provided for in the zoo. Of course, natural behaviour is only one way of assessing the suitability and relevance of zoo husbandry, and the welfare states of the individual animals experiencing that husbandry, but it is an important one. Behaviour evolves to provide advantages to animals within ecosystems. If zoo and aquarium-housed animals are to uphold the key aims of the modern zoo they must be provided with an environment that allows them to gain adaptive benefits from behaviour patterns. The zoo's educational aims are best supported when animals behave in a biologically relevant manner; animals with greater behavioural flexibility and resilience make for better subjects to use for conservation work; research outputs are more widely generalisable and robust when science is conducted on animals that are not having to cope with impoverished or irrelevant environmental conditions; and visitors will be more engaged and feel positive in their views on the zoo, and experiences of their visit, if they are observing animals that they perceive to be contented within their enclosures.

The third and final section of the book poses questions "for the future" and aims to illuminate the relevance of behavioural biology to zoo education programmes (ensuring that material is delivered in a correct, equitable and accessible manner), to native species (their conservation and management within zoo grounds), to animal health and veterinary care (behaviour can be an excellent indicator of health status and good health can be promoted by species-appropriate husbandry and housing), to animal welfare (measurement and assessment of welfare take into account the relevance of animal behaviour patterns) and to zoo professionals (how zoo keepers use evidence, gather evidence, and are the guardians of correct and appropriate husbandry and care regimes for the species they are responsible for).

The living collections housed by zoos and aquariums are their most valuable and most impactful assets. The living collection needs to be at the forefront of the zoo's expenditure and investment as a visitor's experiences of these wild animals is that which leaves the most lasting impression. Zoos need to continue their efforts in shaping collection plans and working with field conservation initiatives to support wild-based conservation with in-zoo efforts. Conservation action is best fulfilled with a collaborative approach across the zoo and the wild. Zoos need to justify the species that are kept to ensure there is an added value to the presence of said species in the zoo. Behavioural biology underpins these goals and aspirations. To inspire the next generation of conservation biologist, environmental protectors, and animal behaviour experts, zoos and aquariums must ensure that husbandry and housing are continually reviewed and evaluated, building more capacity for conservation and human behaviour change. In a rapidly changing world of global climatic instability and impoverished biodiversity, the role of zoos and aquariums in championing the natural world and ensuring all generations care for it is never greater.

I am grateful to all authors for their wisdom, knowledge, experience, and insight into the different topics that BBZA covers. I hope that this book is a useful and informative volume for all those working with zoo- and aquarium-housed species, regardless of the level of education or stage of career, and that it can help guide the future direction of zoo animal management for the better.

Editor

Dr Paul Rose gained his PhD in 2018, investigating the behaviour and welfare of captive flamingos using a variety of methodological approaches including social networks analysis. He is a lecturer in the Psychology Department at the University of Exeter where he specialises in animal behaviour. Paul also works for the Wildfowl & Wetlands Trust (WWT) where he manages the Animal Welfare & Ethics Committee and is a research associate for WWT's captive animal research programme. Paul is also a lecturer on the zoo animal management courses for University Centre Sparsholt. As a member of Defra's Zoos Expert Committee, Paul writes and reviews policy pertaining to the running and licencing of zoos and aquariums. Paul is also co-chair of the IUCN Flamingo Specialist Group and is a member of the IUCN SSC Giraffe & Okapi Specialist Group. He is the co-chair of the BIAZA Research Committee and a member of the BIAZA Bird Working Group steering committee. Paul completed his first piece of in-zoo research in 2002 and has been professionally involved in zoo animal behaviour and welfare in an academic and practitioner capacity since 2006. His research predominantly focuses on behaviour and welfare, and evidence for zoo husbandry. He is a Senior Fellow of the Higher Education Academy and a Fellow of the Zoological Society of London.

Contributors

John E. Andrews
AZA Population Management Center at Lincoln Park Zoo
Chicago, Illinois, USA

Jonathan Beilby
Chester Zoo
The North of England Zoological Society
Chester, UK

Louise Bell
University Centre Myerscough
Preston, UK

James E. Brereton
University Centre Sparsholt
Winchester, UK

Jack Boultwood
Wild Planet Trust
Paignton, UK

María Díez-León
Animal Welfare Science and Ethics, Department of Pathobiology and Population Sciences, Royal Veterinary College
University of London
Hatfield, UK

Matthew Fiddes
CJ Hall Veterinary Surgery
London, UK

Ricardo Lemos de Figueiredo
School of Biosciences, College of Life and Environmental Sciences
University of Birmingham
Edgbaston, UK

Marianne Freeman
University Centre Sparsholt
Winchester, UK

Phillip J. Greenwell
Lieu dit Salce
Saint Georges, France

Jessica Harley
Knowsley Safari
Prescot, UK

Animal Behaviour & Welfare Research Group, Department of Biological Sciences
University of Chester
Chester, UK

Linda Henry
SeaWorld San Diego
San Diego, California, USA

Ian Hickey
Chester Zoo
The North of England Zoological Society
Chester, UK

Brent A. Huffman
Toronto Zoo
Toronto, Ontario, Canada

Kerry A. Hunt
University Centre Sparsholt
Winchester, UK

Louise Jakobsen
Browse Poster (Registered Charity in England & Wales no. 1178456)
London, UK

Georgia C.A. Jones
Department of Life and Environmental Sciences
Bournemouth University
Poole, UK

Joanna Klass
Animal Management Department
Woodland Park Zoo
Washington, USA

Jack Lewton
Department of Life Sciences
Imperial College London
Silwood Park, UK

Beau-Jensen McCubbin
Natural History Museum
London, UK

Christopher J. Michaels
Zoological Society of London
London Zoo, Outer Circle
London, UK

Andrew Mooney
Department of Zoology
Trinity College Dublin
Dublin, Ireland

Steve Nash
Wild Planet Trust
Paignton, UK

Zoe Newnham
Marwell Wildlife
Hampshire, UK

Michelle O'Brien
Wildfowl & Wetlands Trust
Slimbridge Wetlands Centre
Slimbridge, UK

Lisa M. Riley
Centre for Animal Welfare
University of Winchester
Winchester, UK

Paul Rose
Centre for Research in Animal Behaviour, Psychology, Washington Singer Labs
University of Exeter
Exeter, UK

Wildfowl & Wetlands Trust
Slimbridge Wetlands Centre
Slimbridge, UK

Lewis Rowden
Zoological Society of London
London Zoo, Outer Circle
London, UK

Jake Scales
Meade Barn
Tadley, UK

Chloe Stevens
Animals in Science Department, RSPCA
Wilberforce Way, Horsham
West Sussex, UK

Christopher D. Sturdy
City College Norwich (Easton Campus)
Norwich, UK

Lisa Ward
Wild Planet Trust
Paignton, Devon, UK

Samantha Ward
School of Animal, Rural and Environmental Sciences
Nottingham Trent University
Brakenhurst, UK

Michael Weiss
Washington Singer Labs
University of Exeter
Exeter, UK

Center for Whale Research
Washington, USA

David J. Wright
Earlham Institute
Norfolk, UK

PART I

Setting the scene

DOI: 10.1201/9781003208471-1

1

Introduction to the behavioural biology of the zoo

PAUL ROSE
University of Exeter, Exeter, UK
WWT, Slimbridge Wetland Centre, Slimbridge, UK

1.1 INTRODUCTION TO THE BEHAVIOURAL BIOLOGY OF THE ZOO

Zoos and aquariums (hereafter "zoos") have markedly evolved in their outlook, philosophy, and standards of animal care since the inception of the world's first scientific zoo by the Zoological Society of London in 1826. Consideration of population viability and sustainability, animal welfare standards, and the effects of human–animal interactions have been given more credence as zoos aim to add more value to their living collections (in terms of conservation, education, and scientific research outputs). Added value comes from displaying animals in naturalistic environments that promote the performance of natural behaviours and is supported by husbandry and management techniques that are species-specific and grounded in evidence. Evidence for species-specific husbandry should be obtained by a thorough understanding of the biology of the animal, and the subsequent integration of such information in the development and application of animal care routines for that species. This approach is then supported and further refined by data collection on the animal's responses to their care and on the implementation and interpretation of husbandry regimes by the zoos housing that species.

Key information on the wild animal, which can be gathered by in-person fieldwork or from the available literature and then further evaluated as required, concerns a species' behavioural ecology, the evolution of specific behaviour patterns that provide fitness benefits, behavioural adaptations to its environment and daily activity budgets that dictate the amount of time (and therefore energy, and therefore motivational drive) that individuals of the species partition to specific behaviours.

Behaviour comprises all the observable responses that an individual provides to a stimulus from its external or internal environment (Barnard, 1983). Behaviour is measured as states (long-duration activities that take up a large proportion of the

DOI: 10.1201/9781003208471-2

individual's circadian cycle, such as foraging) or events (instantaneous to short-duration acts, such as social interaction) (Bateson & Martin, 2021) and data on these different behavioural categories allows for the creation of time-activity budgets or rates of occurrence that can then be used to infer welfare states, or responses to the environment, or suitability of husbandry and care regimes (Rose & Riley, 2021). Animal welfare is defined as the state of the individual as it attempts to cope with its environment (Broom, 1986) and encompasses both physiological and psychological measures. Observation of behaviour allows for inferences of animal welfare state-important information for the re-assessment and re-alignment of husbandry and management practices.

As such, the purpose of this text is to explain the importance of the evidence-based approach from an animal behaviour perspective. It aims to evaluate and examine why the care of wild species housed under managed conditions should be based on their evolutionary ecology and natural history. It analyses key, underpinning principles of how zoo populations are managed, how collection plans integrate with population viability and sustainability, and why animal behaviour knowledge increases the chances of successful conservation action. To inform the next generation of zoo scientists and to enthuse those who are already established in the field, this book critiques fundamental research questions that should be answered to further the evidence-based approach, and details suitable methodologies for data collection and data analysis pertaining to behavioural biology research questions in the zoo. To provide a solid basis for the taxa-specific chapters (that integrate knowledge of animal behaviour with the practicalities of animal care in the zoo) research approaches, design and evidence gathering techniques, and relevant analytical procedures are examined to enable the use of repeatable methods and the generation of valid results that zoo professionals can have in confidence in when reviewing and rationalising animal care.

1.2 WHAT IS BEHAVIOURAL BIOLOGY?

Behavioural biology relates to the natural biology, ecology, and activity patterns of the species in its wild state. Characteristics of the species' behaviour are moulded by selection pressures that provide fitness benefits to individuals that perform them, and therefore the population overall is adapted to the prevailing environment and is able to survive, thrive, and breed within its specific niche. Species-specific husbandry is enhanced when it is based on information on evolutionary history and ecological specificity, combined with data on daily activity patterns and motivational needs, in conjunction with a sound understanding of anatomical and physiological adaptations. All of these features or traits of a species, and how they perform behaviour under natural conditions, provide insight into behavioural development, responses to the immediate environment, and the welfare state in captivity (Latham & Mason, 2004).

By observing and measuring behaviour in a standardised manner, using ethological tools to guide interpretation of what animals do and why, we gain an insight into the importance of specific behaviour patterns to a species' function and role within its ecosystem. One of the most widely known and most commonly used ways of asking questions of behaviour are the "Four Questions" devised by pioneering ethologist Professor Niko Tinbergen (Tinbergen, 1963). The Four Questions centre around two proximate questions, which explain the initial happenings of the behaviour at the individual level, and two ultimate questions that define why a species has evolved these proximate systems and how they advance survival and reproduction (Figure 1.1). The proximate questions are concerned with how an individual animal is able to perform specific behaviours and are influenced mainly by the animal's endocrine and sensory systems. These proximate questions are mechanism or causation (how the individual is able to commence the performance of a behaviour) and ontogeny or development (how the behaviour can be shaped over time by individual experience or learning). The two ultimate questions of function (how the behaviour allows the species to solve a survival or reproductive challenge) and evolution (the history of evolutionary change for that behaviour and why it is expressed in the manner it is in the modern animal) explain why the species has evolved structures and adaptations to enhance the survivability and reproductive success of the species. The ultimate questions enable the evolutionary history of behaviour to be compared across species to see what is similar between, and what is unique to, specific taxa.

	Cause	Origin
Proximate (How the individual's structures function)	*Causation* How do flamingos know when to display? - Signals from feather colour of flock mates. - Nutritional requirements are met = sufficient energy for display. - Number of birds in breeding condition can be judged. - Weather conditions assessed as optimal for breeding.	*Development* What allows a flamingo to refine its courtship display? - Observation of adult birds during juvenile stage. - Practising display movements. - Experience gained over time to make display actions more impactful.
Ultimate (Why the species has evolved such structures)	*Adaptation* What is the function of the behaviour in this species? - To select a partner for breeding. - To provide the highest change of successful nesting by encouraging the largest number of flamingos to breed at once. - To synchronise breeding during optimal environmental conditions.	*Evolution* What is the route of selection on this behaviour? - Defining mate choice and sexual selection in this species. - Birds with the highest diversity of display movements are most attractive. - Birds that nested as a group were more likely to raise chicks successfully.

Figure 1.1 An example of Tinbergen's Four Questions applied to the group courtship display performed by flamingos (*Phoenicopteriformes*). Understanding the causation and development of the behaviour at the individual level allows zoos to provide the right conditions (nutrition, space, social group) for birds to commence display, and consideration of the evolutionary adaptation of the behaviour is important for population management and future sustainability.

1.2.1 Learning from the past and developing the future

Using behavioural biology to understand zoo animal welfare and husbandry needs is not a new undertaking. Previous research into the prevalence of abnormal repetitive behaviour (stereotypic pacing) in members of the order Carnivora identified that important aspects of behavioural biology (notably home range size) predicted the likelihood that a species would perform pacing in the zoo (Clubb & Mason, 2007), and potentially more likely to suffer from poorer welfare. In this instance, behavioural biology informs enclosure size and space provided, as well as long-term collection planning based on the suitability of the species to "do well" in the zoo and enhance the zoo's mission. Similarly, Ryder and Feistner (1995) review several examples of conservation success (breeding programmes and reintroduction initiatives) when captive management of cheetahs *(Acinonyx jubatus)*, golden lion tamarins *(Leontopithecus rosalia)* and giant pandas *(Ailuropoda melanoleuca)* has been based around important aspects of behavioural biology. In this instance, management of breeding pairs around mate choice and sexual selection, how individuals communicate, and their means of sensory perception and coordination, climbing, and locomotory behavioural diversity.

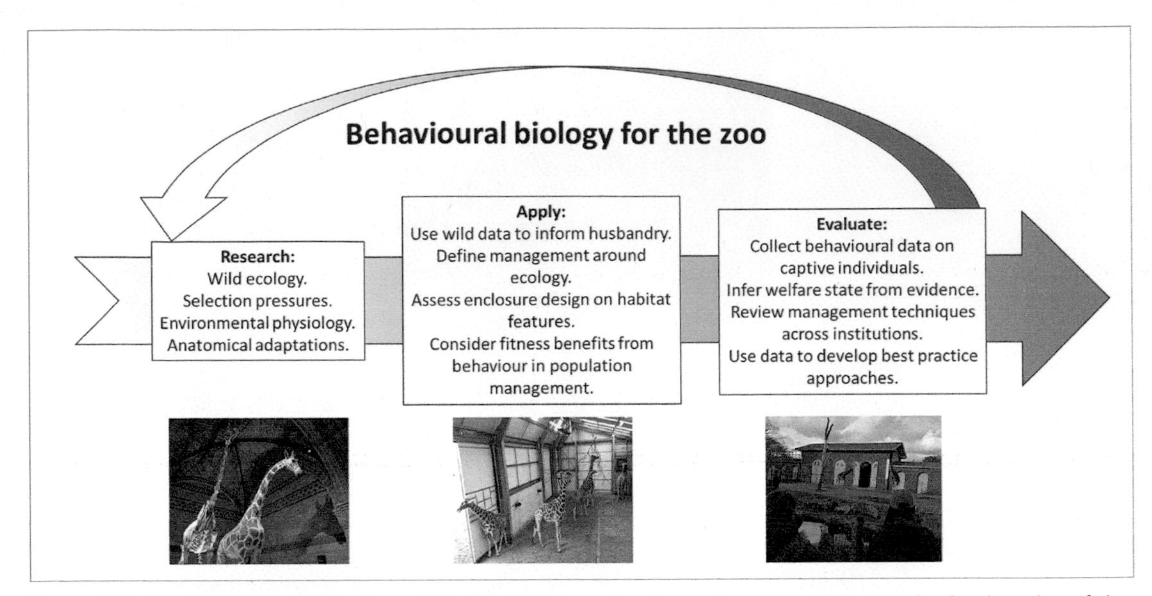

Figure 1.2 Integration of behavioural biology information from the wild animal to the husbandry of the zoo and how a continual process of re-evaluation ensures that zoo husbandry remains relevant.

Husbandry should be advanced and continuously re-evaluated alongside of our knowledge of the animal's behaviour from the wild. Such wild data, combined with evidence from in-zoo research, provides a strong foundation for the attainment of positive welfare states and viable population goals for all individuals housed under human care (Figure 1.2). As we expand our understanding of how animal's view and interact with the world around them (through different means of sensory perception), how they communicate intra- and interspecifically, how they develop their behaviour patterns across the course of their lifetimes, and how new information about their evolutionary relationships and evolutionary history further explains their adaptive traits, we can advance and adapt the management strategies used for such species housed under human care.

From early, pioneering studies, such as the works of Charles Henry Turner on invertebrate communication, social behaviour, and experiential learning (Abramson, 2009) that broke new ground in how we consider the organisation and expression of behaviour. To the re-purposing of human research methods for the collection and interpretation of animal behavioural, health and welfare data, such as those provided by social network analysis (Rose & Croft, 2015). And to the application of new technologies that provide even deeper insights into how animals utilise their habitats, capturing information previously hidden from the in-person observer (Smith & Pinter-Wollman, 2021) and further explaining the adaptive benefits animals gain from behavioural performance (Brandl et al., 2021). This evolution in the methods of animal behaviour data collection has allowed a deeper understanding of species' behaviour patterns that ultimately benefit captive individuals who can be provided with a more species-specific, more ecologically relevant environment in the zoo.

1.3 WHY THE WILD IS IMPORTANT

Wild animals do not live in boundless freedom. They are constrained by many biotic and abiotic factors within the habitats where they dwell. For example, predation and disease, resource competition, food availability and distribution, social group dynamics, and interactions within and between species, climatic changes and weather, and seasonal, temporal, and spatial variation. To cope with these pressures, and to stand the best chance of surviving to breed successfully, species adapt their behaviours according to their needs within the environment. Natural selection imposes the route of change on the population, with behavioural characteristics becoming fixed within a population if they convey adaptive benefits to the individuals that possess them. When wild species are housed in the managed environment of the zoo, it is important to consider the evolutionary pathway of their behaviours to ensure that selection pressures (that will differ in the artificial setting of the zoo compared to a wild functioning habitat) do not change behavioural performance and heritability of behavioural characteristics adversely. For example, variation in individual animal behavioural syndrome ("personality") or differing levels of behavioural motivation that is beneficial for the wild, but determinantal to the zoo, may be lost if individuals that are more behaviourally flexible breed readily in captivity, and therefore disproportionately influence the behavioural phenotypes of future generations. Comparison of species' ecological differences and evaluating behavioural traits that may predict why species thrive or do not cope well with the zoo environment (Mason, 2010) emphasises that wild data are essential for moving forwards with collection planning, husbandry practices, and assessment of animal welfare states.

Even when species appear similar in morphology and activity, and can be sympatric in the wild, important ecological differences and behavioural specialisms impact on their suitability for the captive environment. For example, Figure 1.3 illustrates the very similar appearance of two species of coral reef fish, the Moorish idol (*Zanclus cornutus*) and the pennant coralfish (*Heniochus acuminatus*) but this outward resemblance does not mean that both species have the same requirements in the aquarium. Pennant coralfish are easy to manage and thrive in the aquarium; recreation of the functional aspects of their habitat is relatively easy and as this fish naturally consumes a varied diet, captive nutrition is easier to provide. Moorish idols are considered extremely challenging to maintain over the long term in aquaria; they range over wide areas of coral reefs and have a specialised diet that mainly consists of algae and sponges. Their spatial needs correspond with this continual grazing foraging strategy. The ecological strategy and behavioural adaptations of the Moorish idol means replication of the functional aspects of its habitat in the aquarium can be near to impossible.

Fascinatingly, studies of the Moorish idol by none other than Niko Tinbergen's ethological counterpart, Professor Konrad Lorenz, highlight both the difficulties of housing this species in a managed environment and illuminate why behaviour at every level needs consideration when species are brought from the wild to the zoo (Howlett, 1999). Lorenz was intrigued by coral reef fish and was planning on writing a book on their ethology at the time of his death in 1989. He constructed an enormous aquarium for his Moorish idols (using the money he had won from his 1973 Nobel Prize), stocking it with living rock and providing exact water chemistry and circulation as per the ocean.

Figure 1.3 Species with a similar overall appearance and overlapping wild distribution but totally different captive requirements. Left, the Moorish idol that is very challenging to maintain under human care, contrasted to the easier-to-care-for pennant coralfish, on the right.

Even with this care and detail, Lorenz was not at first successful in keeping any Moorish idols alive. A subsequent attempt and tweaks to the care of the fish was effective, and Lorenz was able to document the varied and complex social behaviour of the fish, which has been likened by other authors to the fission-fusion systems seen in many primate troops. Lorenz noted that the fish displayed preferences for specific individuals within social groups that form part of the overall population, and that the fish used colour as a form of communication. These social traits and use of colours are now documented in other species of coral reef fish, including the butterflyfish family (*Chaetodontidae*) that the pennant coralfish is a member of. This social signalling may be the reason why these two unrelated species look the same in appearance, even though their ecological requirements are vastly different.

This example clearly shows the importance of data collection on the wild animal to enhance our understanding of its evolutionary biology and life history strategy so that we know why it acts in the way it does, and what it needs from its environment to be successful. Such information can then guide captive care and the behavioural biologist of the zoo must consider what fitness benefits occur due to the performance of specific behaviours, and ultimately provide an environment that enables such processes to continue as normal, even though the species is under ex-situ management.

1.4 APPLYING BEHAVIOURAL BIOLOGY TO THE ZOO

The challenge for the zoo is to provide an environment that enables a wild animal to experience environmental fluctuation and challenge in a way that promotes the performance of advantageous behaviours that deliver fitness benefits. As Rees (2011) states "*A management goal of zoos should be to provide an environment in which the animal, if it could know, would not know that it was in captivity.*" And when zoos house challenging or "problematic" species, in terms of performance of abnormal behaviour or heightened likelihood of poorer welfare, researchers state that a zero-tolerance approach to such un-natural behaviours is the only ethical thing to do (Mason et al., 2007). Behavioural biology is essential to achieving both these aims. It provides the evidence that we can use to recreate the functional and ecologically relevant aspects of a species' habitat (Robinson, 1998; Brereton & Rose, 2022) so that reliance on abnormal behaviour is not the way that individual animals cope with the situation they find themselves in. Applying behavioural biology to the zoo means that essential and core animal husbandry activities are evidence-based, species appropriate, and in keeping current views on how living collections should be displayed.

Examples of how zoo and aquarium animal care should be based on behavioural biology are provided below:

- Environmental enrichment: Promotion of important appetitive behaviours that provide an outlet for key motivational states. Prolonging the time spent on behavioural states to equate time-activity patterns to ecological norms.
- Enclosure design: Functional substitution (i.e., creation of key parts of a habitat or ecosystem that a specific species interacts with) provides opportunities for a diverse array of behavioural performance and can build in resilience and behavioural flexibility into the activity patterns of zoo-housed species.
- Nutrition: Dietary provision and feeding practices, based on anatomy, physiology, and foraging strategy, promotes naturalistic time budgets, keeping animals active and healthy and reducing food waste and any costs associated with poor health (e.g., from veterinary interventions).
- Veterinary issues and animal health: Behaviour that promotes fitness benefits enables animals to survive for longer. Increased beneficial activity, reduced lethargy, and improved behavioural diversity that enhances physical and psychological health, reduces the likelihood of veterinary interventions in some cases. Costs are saved, staff time is conserved, and animal health is promoted holistically.
- Breeding: Further consideration of mate choice in conservation programmes enhances the chance of a successful pairing. Embedding knowledge of sexual selection into population management increases viability and reproductive output, allowing animals more choice and control over their breeding partners (within the constraints of the recommendations and programme for breeding as per each individual species).

- Reintroduction: Conservation of key adaptive behaviours (e.g., optimal foraging, anti-predatory, social interactions) is essential if ex situ-bred individuals are to successfully repopulate areas of their wild ranges as per conservation action plans. Such adaptive behaviours can be conserved by the use of species-specific environmental enrichment and ecologically relevant enclosure design that promotes a full range of appetitive (the "doing" is more important than the final result) and consummatory (where the animal reaches a goal and is satiated) behaviours.
- Preparing for the wild: Knowledge on social groups, and the benefits of experience or social learning, as well as training individual animals to interact with resources that they are likely to encounter can be based on evidence from wild research. Preparation to enhance behavioural diversity enables individuals in conservation breeding programmes to be fully equipped (behaviourally) for the dynamic nature of wild habitats.
- Social interactions: Measurement and assessment of social choices, social structure and social dynamics can be compared between the zoo and the wild to ensure that the adaptive benefits of sociality are not lost when a species is managed in the zoo, and likewise, that any differences to the social choices seen in captive individuals (compared to the wild) are not detrimental to long-term behavioural profiles.
- Assessment of animal welfare: Understanding natural behaviour and, perhaps most importantly, behaviours that are strongly motivated in performance and those that have evolved key fitness benefits should be the basis for animal welfare assessment. It is important to remember that not all behaviours will be performed in captivity in the same way as in the wild (and vice versa), therefore the context of the behaviour's performance within the prevailing environment of the individual needs to be judged in a replicable and valid manner.
- Training: Many zoo animals, from all taxonomic groups, are now trained to take part in their own husbandry and care routines. Successful training is based on knowledge of the behavioural capabilities of the species being worked with; therefore, the outputs from a training programme (both for the animal and the trainer) are enhanced when behavioural biology is a core part of the initiative.
- Education and value to visitors: Promoting species to visitors from an engagement and educational perspective is one of the key aims of the modern zoo. Added value, to the animals themselves and to the visitor's trip to the zoo, comes from opportunities to perform (the animal)/observe (the visitor) natural behaviour in a manner that explains the natural history of the species being housed. Conservation importance and advocacy for human behaviour change (both for that species in the wild, or for habitats closer to home in the visitors' local area) can be more impactful when the display of the animal (as the original "hook" for caring about conservation) is appropriate.

1.5 CONCLUSIONS

This introductory chapter has highlighted the importance of "the wild" to the zoo; why all zoo professionals, from keepers to curators to managers and scientists, need a thorough understanding of the ecology of the species whose care they are directly involved with. Without evidence from the wild, and without data collection on animal behaviour, husbandry standards will not progress nor be species appropriate. The following chapters of this book define, evaluate, and explain:

- How we can collect information on behavioural biology and apply it to zoo and aquarium animal care.
- Why we need to consider the added value that behavioural biology brings to population management, to collection planning, and to conservation action.
- How we measure and infer animal welfare states.
- What different species need from the zoo to ensure that they thrive, reproduce, develop, and behave in a naturalist manner.

This book is by no means an exhaustive text on zoo animal behaviour; instead, it aims to distil the most essential elements of behavioural science to key, broader concepts of zoo animal management. Concepts that are currently explored extensively in the literature and horizon scans for the future-what do we need to know, what data should we collect, and why? Finally, this book evaluates a selected number of taxonomic groups in terms of their ecological characteristics and needs in the zoo, and in doing so it aims to apply fundamental behavioural biology science to animal husbandry practice and the advancement of the evidence-based approach that is the foundation for excellent animal welfare, conservation success, and population viability.

REFERENCES

Abramson, C.I. (2009). A study in inspiration: Charles Henry Turner (1867–1923) and the investigation of insect behavior. *Annual Review of Entomology, 54*(1), 343–359.

Barnard, C.J. (1983). *Animal behaviour*. Boston, USA: Springer.

Bateson, M., & Martin, P. (2021). *Measuring behaviour: An introductory guide* (4th ed.). Cambridge, UK: Cambridge University Press.

Brandl, H.B., Griffith, S.C., Farine, D. R., & Schuett, W. (2021). Wild zebra finches that nest synchronously have long-term stable social ties. *Journal of Animal Ecology, 90*(1), 76–86.

Brereton, J., & Rose, P.E. (2022). An evaluation of the role of 'biological evidence' in zoo and aquarium enrichment practices. *Animal Welfare, 31*(1), 13–26.

Broom, D.M. (1986). Indicators of poor welfare. *British Veterinary Journal, 142*(6), 524–526.

Clubb, R., & Mason, G.J. (2007). Natural behavioural biology as a risk factor in carnivore welfare: How analysing species differences could help zoos improve enclosures. *Applied Animal Behaviour Science, 102*(3), 303–328.

Howlett, R. (1999). Monkey business in the aquarium. *Nature, 397*(6716), 211–211.

Latham, N., & Mason, G. (2004). From house mouse to mouse house: The behavioural biology of free-living Mus musculus and its implications in the laboratory. *Applied Animal Behaviour Science, 86*(3), 261–289.

Mason, G.J. (2010). Species differences in responses to captivity: Stress, welfare and the comparative method. *Trends in Ecology & Evolution, 25*(12), 713–721.

Mason, G.J., Clubb, R., Latham, N., & Vickery, S. (2007). Why and how should we use environmental enrichment to tackle stereotypic behaviour? *Applied Animal Behaviour Science, 102*(3–4), 163–188.

Rees, P.A. (2011). *An introduction to zoo biology and management*. Oxford, UK: John Wiley & Sons.

Robinson, M.H. (1998). Enriching the lives of zoo animals, and their welfare: Where research can be fundamental. *Animal Welfare, 7*(2), 151–175.

Rose, P.E., & Croft, D.P. (2015). The potential of social network analysis as a tool for the management of zoo animals. *Animal Welfare, 24*(2), 123–138.

Rose, P.E., & Riley, L.M. (2021). Conducting behavioural research in the zoo: A guide to ten important methods, concepts and theories. *Journal of Zoological and Botanical Gardens, 2*(3), 421–444.

Ryder, O.A., & Feistner, A.T.C. (1995). Research in zoos: A growth area in conservation. *Biodiversity & Conservation, 4*(6), 671–677.

Smith, J.E., & Pinter-Wollman, N. (2021). Observing the unwatchable: Integrating automated sensing, naturalistic observations and animal social network analysis in the age of big data. *Journal of Animal Ecology, 90*(1), 62–75.

Tinbergen, N. (1963). On aims and methods of ethology. *Zeitschrift für Tierpsychologie, 20*(4), 410–433.

2

Behavioural biology in animal collection planning and conservation

JESSICA HARLEY
Knowsley Safari, Prescot, UK
University of Chester, Chester, UK

2.1 INTRODUCTION TO ANIMAL COLLECTION PLANNING

Historically zoos and aquariums (hereinafter "zoos") housed and managed animals based on competition (between institutions), personal preference, and the availability of animals (Diebold & Hutchins, 1991). In 1986, the Ark paradigm was developed to make the best use of zoo space through global collaboration to ensure viable and sustainable populations of animals for the conservation of the species (Soulé, Gilpin, Conway, & Foose, 1986). A few years later, Hutchins, Willis, and Wiese (1995) called for more efficiency, economy and effectiveness in strategic collection planning and suggested that every species held in zoos should not solely be held based on status in the wild or future reintroduction (Ark paradigm) but on the species immediate potential to contribute to conservation through public education, research, and fundraising. These concepts have driven animal collection planning for the last few decades with the primary focus to avoid loss of genetic diversity within a set period, i.e., the retention of 90% of a species' genetic diversity for the next 200 years (Soulé et al., 1986). However, maintaining sustainable zoo animal populations has not been without its challenges. Che-Castaldo, Gray, Rodriguez-Clark, Schad Eebes, and Faust (2021) investigated the viability of cooperative breeding programmes of over 400 vertebrate species to determine if zoos could mitigate demographic and genetic declines in zoo populations, and the results were mixed. Their findings supported previous sustainability studies (Lees & Wilcken, 2009), concluding that many zoo populations do not meet indicators of long-term sustainability, i.e., they found most populations had fewer than 200 individuals and descended from <20 founders. However, the study also found that cooperative management can prevent the degradation of genetic health as most populations did not decline in genetic diversity, and some even increased.

Today, collection planning includes further actions to bridge the in situ and ex-situ conservation

DOI: 10.1201/9781003208471-3

gap and includes the development of management strategies and conservation actions for free-living and captive populations, which are produced collaboratively, resulting in a single conservation plan (Byers, Lees, Wilcken, & Schwitzer, 2013). The International Union for the Conservation of Nature (IUCN) Species Survival Commission (SSC) Conservation Planning Specialist Group (CPSG) describes this as the "One Plan Approach" (OPA). This integrated species conservation framework is central to the World Association of Zoos and Aquariums (WAZA) Conservation Strategy of 2015. To facilitate an OPA, the CPSG, in collaboration with regional zoo associations, developed the Integrated Collection Assessment and Planning (ICAP), with the first report delivered in 2016 for Canids and Hyaenids (Traylor-Holzer, Leus, & Byers, 2018). The purpose of the ICAP is to feed into Regional Collection Plans (RCPs) with Taxon Advisory Groups (TAG) assessing ICAP ex-situ recommendations to help facilitate collection planning, conservation messaging, and research for implementation at the regional and institutional level. Both the European Association of Zoos and Aquariums (EAZA) and the American Association of Zoos and Aquariums (AZA) TAGs used the ICAP outcomes for informing RCPs (Traylor-Holzer et al., 2018).

2.2 THE REGIONAL AND INSTITUTIONAL COLLECTION PLAN

The WAZA Global Species Management Programmes (GSPM) work in partnership with regional management programmes (e.g., AZA, EAZA), and implementation of these programmes occur at the regional and institutional level. For example, EAZA's RCP evaluates which conservation and/or non-conservation roles are appropriate for taxon and includes the development of Long-Term Management Plans (LTMPs) (as applicable) to examine in detail which genetic and demographic goals are best connected to the roles and circumstances of the taxon. Presently, EAZA manages Ex-Situ Programmes (EEP) for over 400 different species. The day-to-day activities of the EEP coordinator and their respective Species Committee manage the studbook, develop breeding and non-breeding recommendations, and address holders' needs in line with the objectives of the population. In North America, the AZA manage more than 500 Species Survival Plan® (SSP) programmes with TAGs overseeing the day-to-day activities while AZA scientific advisors assist the SSP in developing Breeding and Transfer Plans (BTP). The BTPs are based on studbook analysis of the genetic and demographic status and recommendations for breeding and transfers and made based on their findings to ensure maximum genetic diversity while minimising mean kinship.

Professional zoo industry bodies promote the use of Individual Collection Plans (ICPs). Accredited zoos (e.g., EAZA, AZA) must have ICPs as part of the accreditation process. The ICP typically considers the zoo's Conservation, Education, and Research (CER) objectives and provides a formal overview of the role of each species within the collection. Collection planning is predominantly led by the animal management team with input from the CER departments and then shared with all relevant senior management. The ICP includes the status and details of the steps required to progress toward those defined objectives; an example of what an ICP should look like is provided in Table 2.1.

When an institution assesses a new species for inclusion or whether to continue with the species as part of the collection plan, ideally, they should follow an assessment criterion to inform decision-making. Several considerations should be included in the process, beginning with the conservation status of the species and consideration of the institution's bespoke conservation target, e.g., the goal of 50% of the population IUCN Red List species listed as Vulnerable (VU), Endangered (EN), or Critically Endangered (CR). Although there are no mandatory criteria for an institution, the respective regional membership organisation provides guidelines for zoos. These guidelines provide a comprehensive list of considerations that should be included when assessing ICPs. The core components for assessing an ICP include conservation, animal welfare, sustainability, visitor experience, and available financial resources (Table 2.2).

Institutional collection plans often have competing factors that vary between zoos depending on their mission and business-critical factors. For example, for a zoo, the drivers of the institutional collection plan could be conservation and visitor experience, and the controls (or factors that challenge the drivers) are the animal's welfare (current and prospective), sustainability (which includes the ex-situ (Ark) population size and current RCP recommendations, as well as the available financial resources that will support housing the species.

Table 2.1 Example of the information which should be included in an Institutional Collection Plan (ICP) for each species currently part of the collection or considered for future acquisition.

Common name	**Amur leopard**
Scientific name	*Panthera pardus orientalis*
Current number individuals and sex ratio	2.0.0
Planned number individuals and sex ratio	1.1.0
Conservation value	High IUCN Red List CR planned reintroduction
Education value	High Strong public interest, important conservation messaging. Key to successful fundraising for in situ education activities to reduce human-animal conflict and improve HAR in the Far East
Ex-situ research value	Medium Peer reviewed publication on efficacy of GnRH agonist to mitigate intraspecific aggression in male leopards. Future husbandry research planned
EAZA RCP recommendations	Insurance population: Breed
Planned actions for individual collection	Create new breeding pair when aged sibling males are naturally phased out
Justification of action (include welfare consideration)	EEP recommendation for future breeding pair Staff suitably trained, purpose-built leopard breeding facility with multiple houses and exterior enclosures for management of breeding individual and offspring

The welfare component includes outcomes of any animal welfare assessment protocols, i.e., effective reviews of biological and ecological considerations to quantify the suitability of the proposed or existing species, environment (surrounding and conditions), and the animal's adaptive abilities to ensure the animal's needs are met. Each zoo must have a clear process for the ICP and determine the priority for the drivers and control that exist within the individual organisation, simultaneously ensuring they complement the regional professional zoo body's requirements and those of the wider zoo industry, an example is provided in Figure 2.1.

2.3 KNOWLEDGE OF BEHAVIOURAL BIOLOGY IS ESSENTIAL IN ANIMAL COLLECTION PLANNING

Central to the success of a zoo's conservation objectives is providing optimum care, management, and welfare for zoo-housed animals. Essential to achieving these standards of care is understanding an animal's natural biology, ecology, and activity patterns to inform housing and husbandry practices. An important role of regional ex-situ population management programmes is the development of Best-Practice Guidelines (BPG) or Animal Care Manuals (ACM). These guidelines aim to provide zoo practitioners with recommendations to enhance animal care based on the best available evidence centred on the animal's natural history and ex-situ management recommendations. Despite new documents being added regularly, there remains a paucity of guidelines across regional membership organisations. For example, EAZA members zoos collectively care for more than 7,800 different species yet fewer than 60 species-specific BPG are currently available, and numbers are similar for AZA ACM. The lack of these guidelines highlights the vital role at the institutional level to review current management practices against scientific literature and incorporate these findings into their animal management and the collection planning processes. For instance, a more refined understanding of an

Table 2.2 Examples of institutional collection planning considerations.

Collection planning considerations	Institutional goals
Conservation	
IUCN Red List, ZSL's EDGE, local conservation designations	Achievement of 50% VU, E, or CR
Zoo's conservation planning and programme	To foster links between zoos animal collection and active conservation activities – extends out to conservation partners and any stakeholders
Animal welfare	
Welfare assessment scores – the tool should be based on the four Physical/ Functional domains of the Five Domains Model (Health, Nutrition, Environment, Behaviour) or Welfare Quality® principles (good housing, good feeding, good health, and appropriate behaviour)	Enables the zoo to identify strengths and weaknesses, prioritisation of capital, or any mitigation actions necessary to facilitate positive welfare experiences. Allows identification of species where issues may not be rectifiable and when phasing out of species is required
Ecological and behavioural considerations, natural history, best-practice guidelines	Enables zoo to reflect on how suited or adaptive an animal is to the challenging environment provided
Sustainable ex-situ populations	
Ex-situ Management Programme	Supports RCP recommendations and the EAZA LTMPs for sustainable populations-balanced with the welfare needs of the animal
Ark (population size)	Support maintenance of genetic diversity for viable and sustainable zoo populations
Visitor experience	
Species engagement	Will drive visitation, visitor appeal, and perception
Exhibit engagement	Will drive guest experience and the zoo's educational and conservation messaging
Educational engagement and messaging – can we effectively deliver messaging?	The species provide a key link to the zoo's vision, mission, and cultural drivers
Financial resources	
Capital expenditures – what is available for new builds or redevelopments?	Funds available for enclosure development – e.g., animal welfare needs, visitor experience, and staff efficiency and their welfare
Running costs – do we have the cash flow?	Funds are available for yearly cost per animal and the population held
Staff – how many people do we need to manage this species?	We have the staff or the ability to source requirement of the number and experience of staff to manage species

Adapted from EAZA and BIAZA Institutional Collection Planning guidelines and published individual ICP from Chester Zoo, Twycross Zoo East Midland Zoological Society and Knowsley Safari's Animal Collection Assessment Tool (ACAT).

animal's behaviour may be elucidated with novel research methods. The giraffe (*Giraffa camelopardalis*) was first thought to have no social structure until 2000 when van der Jeugd and Prins (2000) reported more stable groups were found in areas with high giraffe density, which was suggestive of the occurrence of resource defence polygyny. A few years later social network analysis (SNA) (Croft,

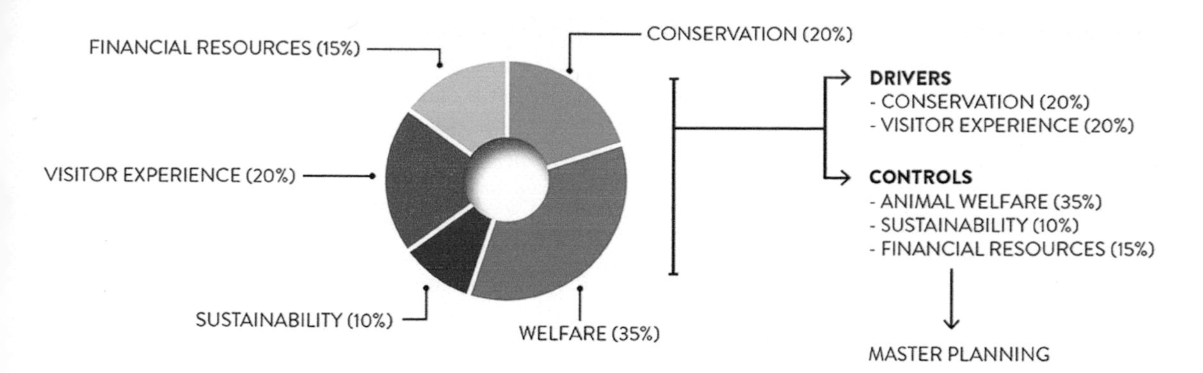

Figure 2.1 ACAT criterion Knowsley Safari. In this example, Knowsley Safari recognises the drivers (conservation and visitor experience) are an important factor (40%) but welfare priorities (35%), financial costs to maintain the species including development (15%) and sustainability of the population (10%) are the principles (controls) dictating whether species fit in with the long-term ICP for the collection. These individual drivers and controls vary between zoos and in their importance in line with the zoo's business and mission-critical drivers.

Krause, & James, 2004) emerged as a tool to better understand animals' social structure, and a resurgence of interest in giraffe behaviour occurred. Today, we better understand giraffe sociality, which has highlighted the significance of non-random relationships and the strength of the female-female bonds. This, in turn, informs practice and highlights important components in population management, such as maintaining females in stable herds (Lewton & Rose, 2020). Additionally, SNA should be used in other aspects of collection planning, such as recommendations for animal transfers or the internal needs of the institution regarding population management. Using research methods such as SNA can positively impact individual welfare and mitigate any group breakdown, which may result from moving the wrong individuals (e.g., socially significant).

An animal's behavioural biology must be intrinsically linked to its provision of housing and husbandry and considered in the planning process. For example, an animal's natural habitat and behavioural characteristics may influence an animal's response to visitors. Let's consider cryptic species or those from closed habitats (e.g., wooded areas/forests). These species may be more affected by visitors than those from open habitats (grasslands). This may relate to evolutionary exposure to humans (Queiroz & Young, 2018). When assessing future acquisitions in collection planning, a zoo with high visitor footfall could use these findings to inform decision-making. Considerations should include animal welfare and whether the zoo has a quiet area suitable with limited disturbance to house the species. Next, the zoo may consider their visitor experience and whether reducing visitor viewing opportunities meets the visitor engagement criterion for the zoo. Additionally, the ICP process must include a review of the institution's financial resources to ensure funding is available for modifications or a new build that may mitigate any potential negative visitor effect, e.g., additional retreat spaces, soundproofing, and the addition of vegetation to provide sufficient opportunities for the animal to camouflage themselves. Another situation where considering an animal's behavioural biology may occur is when an institution's conservation goal includes a focus on housing conservation dependent species (e.g., 50% of species that are either VU, E, or CR). For example, a zoo currently houses plains zebra (*Equus quagga*) and are considering phasing out plains zebras which are listed as Least Concern (LC) to be involved in the ex-situ management programme for the endangered Grévy's zebra (*E. grevyi*). Although husbandry requirements may appear similar across equids a deeper understanding of the species, behavioural biology is required to meet their specific needs. Equid species are highly social, but they do not all form the same type of social groups, which include either defence polygyny (harem-forming) or resource defence polygyny (territorial, fission-fusion) (Klingel, 1974). Plains and mountain zebra (*E. zebra*) are harem-forming, and herds will contain one or more adult males and one or more adult females with subadults of both sexes that have not yet dispersed. Male Grévy's zebra are territorial and may be alone or with females of varying reproductive states. Females are rarely alone, but the groups they form are temporary (Boyd, Scorolli, Nowzari, & Bouskila, 2016). Grévy stallions may not tolerate year-round contact with females, and the existing plains zebra enclosure earmarked for the species may not be sufficient (Schook, Powell, & Zimmermann, 2016). Thus, the new acquisition would necessitate a larger or separate enclosure, in which case the ICP process would include a review of available land for development and the financial implications of modified housing for Grévy's zebra to meet their conservation objectives. An enhanced understanding of an animal's natural history enables the care provider to design appropriate housing, offer biologically relevant substrates and furnishings, and ensure appropriate social opportunities are available to meet an animal's behavioural needs. Knowledge of an animal's behavioural biology and its application in their management is a vital component of animal collection planning and conservation.

BOX 2.1: Breeding success and behavioural biology

The cheetah (*Acinonyx jubatus*) provides an excellent example of how a thorough understanding of an animal's behavioural biology can contribute to its sustainability and welfare under managed care. Cheetahs are wide-ranging carnivores and persist in low density across their range, thus requiring expansive, connected habitat for their survival (Durant, Mitchell, Ipavec, & Groom, 2015). The cheetah is listed as vulnerable by the IUCN Red List of Threatened Species and the population trend is decreasing, with ~7000 individuals occupying grassland and savannah habitats in eastern and southern Africa. Small populations can also be found in North Africa and Iran. The social organisation is atypical for felids. Males may be solitary or social, living in stable male coalitions. Females tend to be nonterritorial and are solitary or with dependent young (Caro, 1994). Females are highly promiscuous, and studies have indicated high levels of mixed paternity in litters (Gottelli, Wang, Bashir, & Durant, 2007).

Population threats include habitat loss, human-animal conflict due to livestock farming and persecution, vehicular and rail accidents, entrapment in hunting/bushmeat snares and the illegal pet trade (Durant et al., 2015). Their near absence of genetic diversity further compromises the diminishing cheetah population, a consequence of a bottleneck event dating ~6000 to 20,000 years ago (O'Brien, Wildt, Goldman, Merril, & Bush, 1983). Excessive inbreeding can cause an increase in the deleterious recessive alleles, culminating in decreased fitness of a population, which is referred to as inbreeding depression. Cheetahs exhibit morphological and physiological signs of inbreeding effects with abnormal sperm and low motility (Culver, Driscoll, Eizirik, & Spong, 2010). Thus, managed breeding programmes are vital for this species to ensure a viable and sustainable population.

Captive cheetahs have had a history of inferior reproduction compared to their wild counterparts. Although housed in zoological collections since 1829, the successful breeding of cheetahs before the 1970s was extremely rare (Marker et al., 2018). Today, zoos are more consistently breeding cheetahs with success. Although there are still barriers to overcome, one of the most significant breakthroughs for successful reproductive performance has been the zoo's consideration of the cheetah's natural history to inform husbandry and housing practices. Over the last few decades, the design of enclosures and management of each animal has changed to reflect wild cheetah behaviour and ecology. Thanks to a few pioneering institutions and several scientific studies to identify factors influencing the breeding success of cheetahs, today's exhibits are designed to permit easy movement of cheetahs between enclosures. Females are housed separately as even subtle agonistic behaviour between females may suppress oestrous cyclicity (Brown, 2011). Males are generally housed in coalitions (Figure 2.2) out of the visual and olfactory range of females. In addition, facilities provide multiple individuals for breeding to better mimic natural mate choice.

Figure 2.2 A male coalition of Northern Africa cheetah (*Acinonyx j. soemmeringii*) comprised of males from two different litters (two from each litter). The photo is from the Sheikh Butti Maktoum's Wildlife Centre which was the first collection to breed cheetahs in the Middle East. Between 1995 and 2009 they successfully bred 98 cheetahs (S. McKeown, personal communication, 26th January 2022). Photo courtesy of Sean McKeown.

2.4 CONCLUSION

- Zoos no longer base animal collection planning on competition, personal preference, and the availability of animals. Instead, collection planning focuses on ensuring viable and sustainable populations of animals to conserve the species.
- Collection planning includes the OPA, and the development of management strategies and conservation actions for free-living and captive populations are produced collaboratively, resulting in a single conservation plan.
- To facilitate an OPA, the IUCN SSC CPSG, in collaboration with regional zoo associations, develop (taxon) reports known as Integrated Collection Assessment and Planning (ICAP).
- GSPM, work in partnership with regional management programmes, with the implementation of the programmes occurring at the regional and institutional level.
- An important role of regional ex-situ population management programmes is the development of BPGs. These guidelines aim to provide zoo practitioners with recommendations to enhance animal care based on the best available evidence and centred on the animal's natural history.
- Accredited zoos (AZA, EAZA etc.) must have a written ICP. Criteria for the development and assessment of a zoo's animal collection plan should include conservation, animal welfare, sustainability of ex-situ population, visitor experience, and the zoo's financial resources.
- An animal's behavioural biology must be fundamentally tied to its care provision.

- Enhanced understanding of an animal's natural history enables the care provider to design appropriate housing, offer biologically relevant substrates and furnishings, and ensure appropriate social opportunities are available to meet an animal's behavioural needs.

ACKNOWLEDGEMENTS

Special thanks to Jonathan Cracknell at Knowsley Safari for providing invaluable help and comments on this chapter and to Sean McKeown from Fota Wildlife Park for supplying a number of wonderful cheetah photos to choose from.

REFERENCES

Boyd, L., Scorolli, A.L., Nowzari, H.A. & Bouskila, A.M. (2016). Social organization of wild equids. In J.L. Ransom, & P. Kaczensky (Eds.), *Wild equids: Ecology, management, and conservation* (pp. 7–22). Baltimore, MD: The Johns Hopkins University Press.

Brown, J.L. (2011). Female reproductive cycles of wild female felids. *Animal Reproduction Science, 124*(3–4), 155–162.

Byers, O., Lees, C., Wilcken, J. & Schwitzer, C. (2013). The one plan approach: The philosophy and implementation of CBSG's approach to integrated species conservation planning. *WAZA Magazine, 14*, 2–5.

Caro, T.M. (1994). *Cheetahs of the Serengeti Plains: Group living in an asocial species*. Chicago, USA: University of Chicago Press.

Che-Castaldo, J., Gray, S.M., Rodriguez-Clark, K.M., Schad Eebes, K. & Faust, L.J. (2021). Expected demographic and genetic declines not found in most zoo and aquarium populations. *Frontiers in Ecology and the Environment, 19*(8), 435–442.

Croft, D.P., Krause, J. & James, R. (2004). Social networks in the guppy (*Poecilia reticulata*). *Proceedings of the Royal Society of London. Series B: Biological Sciences, 271*(suppl.6), S516–S519.

Culver, M., Driscoll, C., Eizirik, E. & Spong, G. (2010). Genetic applications in wild felids. In D.W. Macdonald & A.J. Loveridge (Eds.), *Biology and conservation of wild felids* (pp. 107–124). Oxford, UK: Oxford University Press.

Diebold, E. & Hutchins, M. (1991). Zoo bird collection planning: A challenge for the 1990s. In *Proceedings American Association of Zoological Parks and Aquariums Regional Conferences* (pp. 244–252). Wheeling: American Association of Zoological Parks and Aquariums.

Durant, S., Mitchell, N., Ipavec, A. & Groom, R. (2015). Acinonyx jubatus. *The IUCN Red List of Threatened Species 2015*, pp. e.T219A50649567. https://doi.org/10.2305/IUCN.UK.2015-4.RLTS.T219A50649567.en. Accessed on 26 December 2021.

Gottelli, D., Wang, J., Bashir, S. & Durant, S.M. (2007). Genetic analysis reveals promiscuity among female cheetahs. *Proceedings of the Royal Society of London Series B: Biological Sciences, 274*, 1993–2001.

Hutchins, M., Willis, K. & Wiese, R.J. (1995). Strategic collection planning: Theory and practice. *Zoo Biology, 14*(1), 5–25.

Klingel, H. (1974). A comparison of the social behaviour of the Equidae. In V. Geist & F. Walther (Eds.), *The behaviour of ungulates and its relation to management* (pp. 124–132). Morges, Switzerland: IUCN.

Lees, C.M. & Wilcken, J. (2009). Sustaining the Ark: The challenges faced by zoos in maintaining viable populations. *International Zoo Yearbook, 43*(1), 6–18.

Lewton, J. & Rose, P.E. (2020). Evaluating the social structure of captive Rothschild's giraffes (*Giraffa camelopardalis rothschildi*): Relevance to animal management and animal welfare. *Journal of Applied Animal Welfare Science, 23*(2), 178–192.

Marker, L., Vannelli, K.M., Gusset, M., Versteege, L., Meeks, K.D., Wielebnowski, N.C., Louwman, J.W., Louwman, H. & Lackey, L.B. (2018). History of cheetahs in zoos and demographic trends through managed captive breeding programs. In P.J. Nyhus, L. Marker, L.K. Boast & A. Schmidt-Küntzel (Eds.), *Cheetahs: Biology & conservation (Biodiversity of the World: Conservation from genes to landscapes)* (pp. 309–321). London, UK: Academic Press.

O'Brien, S.J., Wildt, D.E., Goldman, D., Merril, C.R. & Bush, M. (1983). The cheetah is depauperate in genetic variation. *Science, 221*(4609), 459–462.

Queiroz, M.B. & Young, R.J. (2018). The different physical and behavioural characteristics of zoo mammals that influence their response to visitors. *Animals, 8*(8), 139.

Schook, M.W., Powell, D.M. & Zimmermann, W. (2016). Wild equid captive breeding and management. In J. Ransom & P. Kaczensky (Eds.), *Wild equids. Ecology, management, and conservation* (pp. 149–163). Baltimore: The Johns Hopkins University Press.

Soulé, M., Gilpin, M., Conway, W. & Foose, T. (1986). The millenium ark: How long a voyage, how many staterooms, how many passengers? *Zoo Biology*, *5*(2), 101–113.

Traylor-Holzer, K., Leus, K. & Byers, O. (2018). Integrating ex situ management options as part of a one plan approach to species conservation. In B. Minteer, J. Maienschein & J. Collins (Eds.), *The ark and beyond: The evolution of zoo and aquarium conservation* (pp. 129–141). Chicago, USA: University of Chicago Press.

van der Jeugd, H.P. & Prins, H.H. (2000). Movements and group structure of giraffe (Giraffa camelopardalis) in Lake Manyara National Park, Tanzania. *Journal of Zoology*, *251*(1), 15–21.

3

Behavioural biology, conservation genomics, and population viability (Open Access)

DAVID J. WRIGHT
Earlham Institute, Norfolk, UK

3.1 WHY CONSIDER POPULATION VIABILITY IN A BEHAVIOURAL BIOLOGY TEXTBOOK?

A better question might be: "what is the aim of conservation biology?" We can consider the central purpose of conservation efforts to be the maintenance of the evolutionary potential of species and ecosystems. Evolutionary potential is being able to continue evolving, to be viable and to be sustainable. This important concept was highlighted in a landmark work by Sir Otto Frankel in which he argued that protecting the ability of populations to continue to adapt to future challenges should be the key objective for conservationists (Frankel, 1974). This fundamental idea applies to populations both inside and outside of captivity. DNA is the principal molecular mechanism by which information is inherited over generations and we can think of it as a vessel of this evolutionary potential. Inevitably then, we must consider the genetic diversity carried within the genomes of captive populations if we are to fully realise the benefits, and mitigate the risks, of captive breeding as a conservation tool to maintain evolutionary potential. Indeed, genetic diversity has long been identified as one of the IUCN's three core global conservation priorities along with species and ecosystem diversity (McNeely, Miller, Reid, Mittermeier, & Werner, 1990). In this introductory chapter, it is not possible to provide an exhaustive coverage of conservation genomics, nor is that appropriate. There are excellent texts that delve into this vast subject in the detail and depth it requires (e.g., Allendorf et al., 2021). Instead, the aim here is to discuss some of the broad concepts to provide evolutionary context when considering the management of behavioural biology in zoo populations. Previous chapters have outlined the many ways in which zoo practitioners can and must use evolutionary ecology to better understand the

DOI: 10.1201/9781003208471-4

behaviours of species in their care, and how collection planning can and should accommodate them to maximise conservation outcomes. Here, we start to think about the implications of these topics on an evolutionary scale; about the impact of captive management on species at the genetic level, and the considerations needed in managing species for evolutionary potential.

A logical place to start is by considering the phenotypic traits of species in our care. All traits are the result of complex interactions between heritable factors, such as DNA or epigenetics (biochemical modifications that change gene expression without altering DNA), and environmental factors (Bošković & Rando, 2018; Falconer & Mackay, 1996). Therefore, even behaviours observed in our captive populations are under at least some degree of genetic or epigenetic influence. For example, meerkats (*Suricata suricatta*) exhibit intergenerational transmission of hormone-mediated aggressive behaviours that maintain their despotic cooperative breeding system (Drea et al., 2021). However, for most organisms, the genetic and epigenetic basis for behaviour is not well understood (Plomin, DeFries, & McClearn, 2008). This is largely because most phenotypic traits are the product of complex networks of genes and gene regulators that interact with each other and numerous environmental influences to varying degrees in space and in time (Falconer & Mackay, 1996). Identifying specific trait-associated genes and their effects on traits such as behaviours or disease risk is therefore challenging. Despite this, investigation of complex genetic traits can provide very valuable opportunities for conservation action. The plight of the Tasmanian devil (*Sarcophilus harrisii*) is an excellent example. Devil populations have faced catastrophic decline due to a highly fatal facial tumour disease that spreads through the direct aggressive behaviour of conspecifics. Significant research effort to uncover the evolutionary genomics of this transmissible cancer in both wild and captive populations has provided vital insight into susceptibility, resistance, and genetic rescue opportunities. These data have directly informed conservation action plans for the species (Hamede et al., 2021). However, the resources available for conservation are often very limited and we must also understand the general evolutionary principles of captive population management, to derive maximum benefit. We can, in part, achieve this by considering genome diversity and the factors that influence it at population and species levels.

3.2 UNDERSTANDING THE IMPACTS OF CAPTIVITY ON GENOMIC DIVERSITY

A core aim of captive breeding is keeping populations as evolutionarily close to their wild counterparts as possible. This is easier said than done. Captive populations are often held at sizes far smaller than would be observed in the wild, and therefore their response to evolutionary forces is much different, making the task of managing their diversity complex. One such force, genetic drift, is the stochastic change in allele (gene variant) frequencies between generations due to random sampling of gametes and the fact all populations are finite in size. As alleles can be randomly fixed (present at that genetic locus in all individuals) or lost completely from a population, drift erodes diversity over time. Drift is so influential in the evolutionary trajectory of populations that effective population size (Section 3.3) was developed to more accurately describe populations genetically (Wright, 1931) and is a core parameter in conservation biology. In the face of such stochasticity, the initial diversity of a captive population is very important. However, many captive populations experience founder effects, a type of genetic drift caused by using few individuals (founders) to establish a new population from an established larger population. This can lead to both genotypic and phenotypic differences in the subsequent population and loss of genomic diversity (Mayr, 1999). Captive populations may have also experienced population bottlenecks, perhaps from threats in the wild, again leading to the loss of genomic diversity. As heterozygosity (see Section 3.3) of loci across the genome decreases, populations can lose heterozygote advantage, the fitness benefit of possessing diversity at particular genomic loci (Charlesworth & Willis, 2009). Further to this, the probability of inheriting related (identical by descent) copies of an allele from both parents increases. This is more commonly known as inbreeding. The smaller the population, the greater the likelihood of related alleles meeting in an individual and, without intervention small, isolated populations will become increasingly inbred over time. Inbreeding can result in inbreeding depression, the reduction in the fitness of inbred individuals. It is predominantly caused by the expression of deleterious mutations (Charlesworth & Willis, 2009) and is detrimental to population viability through impacts such as reduced offspring survival or increased disease susceptibility. Captive populations have played an important role in understanding the impacts of inbreeding depression, including both opportunistic observations of zoo populations (e.g., Ballou & Ralls, 1982) and experimental studies (e.g., Lacy, Alaks, & Walsh, 2013). Mitigating effects associated with small population size form the basis of much of captive population genetic management.

Captive populations are also subject to the effects of natural selection. Adaptation to captivity, where populations evolve in response to selection pressures of the captive environment, is well documented in zoo populations (Frankham, 2008). The fitness of individuals in captivity improves as the population adapts, but this is at an overwhelming detriment to fitness in the wild. Adaptation to captivity impacts many diverse physiological and behavioural traits in different species (Frankham, 2008). A study in white-footed mice (*Peromyscus leucopus*) demonstrated rapid adaptation in behavioural and reproductive phenotypes of the mice under different experimental management regimes. Interestingly, it showed evidence of evolution even in management regimes intended to minimise adaptation to captivity and the impacts of drift (Lacy et al., 2013). We can also learn much about adaptation to captivity by studying domestication. Perhaps the most famous domestication experiment has been conducted with foxes (*Vulpes vulpes*). Beginning in the 1950s, foxes were subjected to selection solely for tameness, but developed many other traits also associated with domestic dogs (*Canis lupus familiaris*) including floppy ears, tail-wagging, and coat colour variation. The work importantly demonstrated that the behavioural traits of tameness were genetically linked to many other phenotypes known as 'domestication syndrome' and suggests that wholescale evolutionary changes in the regulatory networks of gene expression must have occurred (Trut, Oskina, & Kharlamova, 2009). This example demonstrates some impacts of extreme selection pressure, but the magnitude of selection differences between captivity and the wild is not the only consideration. How long populations are subject to selection differences is also important. In steelhead trout (*Oncorhynchus mykiss*), inherited differences in gene expression across hundreds of genes were discovered after just a single generation in captivity (Christie, Marine, Fox, French, & Blouin, 2016). This demonstrates that captive breeding can lead to detectable population changes incredibly quickly.

So, with evolutionary changes poised to rapidly alter captive populations, are they all doomed to fail? It is important to consider that the relationship between diversity and sustainability is not straightforward (Estoup et al., 2016; Frankham, Ballou, & Briscoe, 2002; Mable, 2019). Some species seem to persist with relatively low diversity, such as the Mauritius kestrel (*Falco punctatus*) whose population successfully recovered from just a single breeding pair (Groombridge, Jones, Bruford, & Nichols, 2000). Others, like the cane toad (*Rhinella marina*), can even become hugely successful invasive species despite reduced diversity (Selechnik et al., 2019). Entire evolutionary lineages appear to maintain long-term viability with generally low genomic diversity, such as the rhinoceros (*Rhinocerotidae*) family (Liu et al., 2021). Whilst our understanding of how such cases occur is still developing, they emphasise the importance of considering the evolutionary ecology of the species we manage and provide some welcome optimism for the future of endangered species harbouring low genomic diversity. Encouragingly, populations reintroduced to the wild can also recover fitness, provided they have at least the minimum remaining fitness to successfully establish in the wild and sufficient remaining diversity for wild selection pressures to act upon (Frankham, 2008). However, maintaining existing genomic diversity and fitness remains a core objective for conservationists, as this will generally provide species with the best possible chances of achieving long-term viability.

3.3 METRICS OF VIABILITY IN CAPTIVE POPULATIONS

We know that genomic diversity is important for the evolutionary potential of captive populations, but how do we quantify it and decide what is enough to be viable? As science and technology have progressed, many different metrics have been developed to aid captive species management. The thresholding and quantification of diversity has improved as evidence has accrued from an increasing number of species, management plans, and conservation histories. Examples of some key metrics that are widely utilised and encountered in management plans include:

- Effective population size (N_e): the size of an idealised model population that experiences the same amount of drift, or exhibits the same mean inbreeding rate, as the actual census population size (N_c) in question. N_e is usually much smaller than N_c.
- Gene diversity (expected heterozygosity, H_E): the probability that two alleles of a locus randomly sampled from a population do not share a common ancestor, assuming random mating and Hardy-Weinberg equilibrium (a testable model of allele frequency behaviour).
- Founder genome equivalent (FGEs): how many founders (i.e., wild individuals) would equal the same amount of gene diversity as exists in the living captive population.
- Inbreeding coefficient (F): the probability that two alleles at a locus in an individual share a common ancestor (identical by descent).
- Kinship coefficient (K): the probability that an allele in a locus in individual 1 is identical by descent to an allele randomly sampled in that same locus in individual 2.

Metrics like these are often assessed together to create a snapshot of the genetic state of a population. Tools specifically designed for aiding zoo biologists leverage these metrics to manage populations and studbooks are available, including the well-established PMx (Lacy, Ballou, & Pollak, 2012). How best to manage existing diversity in a population will depend on a range of factors such as whether a pedigree exists and is reliable, the current diversity and demography of the population or metapopulation in both captivity and the wild, and predictions of how quickly the existing diversity is expected to be lost.

Best practice in addressing genetic and epigenetic impacts on captive populations has been the subject of decades of research and scientific debate (e.g., Frankham, 1995; Margan et al., 1998; Ralls et al., 2018). Since their development in the late 1980s, many conservation management plans have loosely followed the 'Ark Paradigm'; a benchmark of retaining 90% gene diversity for *ca.* 100–200 years, as a general guide (Seal, Foose, & Ellis, 1994; Soulé, Gilpin, Conway, & Foose, 1986). However, this may not be appropriate for species with long generation times such as tortoises (Testudinidae) for example, where specifying a number of generations (e.g., 10) may be more biologically relevant. Of course, some captive populations may fall below these preferred diversity thresholds. In such cases, zoos will generally aim to mitigate further damage by implementing management actions aimed at slowing the rate in which the remaining diversity is lost. How much initial gene diversity there may be is dependent upon both the evolutionary history and conservation status of the population or species (Section 3.2). This highlights the importance of a taxon-specific approach to assessment and management and many species indeed benefit from species-specific management plans, actively curated by zoological collections. Global organisations such as the IUCN and governing bodies of zoos and aquaria publish and regularly update best practice guidelines for captive breeding programmes and genetic management in captivity. Additionally, experts in the field also produce comprehensive general management guidelines for small populations (e.g., Ballou et al., 2012). Collectively, this provides a wealth of continually evolving resources for promoting the evolutionary potential of captive populations whilst underlining the central role of evolutionary biology in furthering species conservation.

3.4 FUTURE CONSIDERATIONS IN THE VIABILITY OF CAPTIVE POPULATIONS

The challenge for zoo conservationists is mitigating the impacts of captivity on diversity and viability whilst also maximising the health and behavioural welfare of the species being managed. For example, the obligate cooperative breeding system of meerkats is maintained in the wild by aggressive behaviours such as eviction and infanticide. How might we best conserve this complex trait and ensure genomic representation in the captive population for long-term viability, alongside welfare considerations? Much of this book deals with the important task of accommodating behavioural welfare in captive populations, and in this chapter we have discussed why this needs to be accomplished with evolutionary biology insight. A recent innovation in captive population management is the One Plan Approach to conservation (Byers, Lees, Wilcken, & Schwitzer, 2013). This method integrates evidence-based management of captive population genomic diversity with the broad range of multidisciplinary considerations necessary for successful species conservation, including behavioural welfare, collection planning, and in situ conservation action (see Chapter 2). This evidence-based management approach inherently requires an evolutionary perspective and further highlights the central role evolutionary biology plays in all aspects of species conservation.

The advancement of technology continues to provide novel tools and techniques for conservation biology (Segelbacher et al., 2021). This will inevitably lead to quicker, easier, and cheaper assaying of populations and enclosures, using increasingly non-invasive, more degraded, and smaller samples. The result is a rapidly increasing quantity and quality of data for decision-making processes (see Chapters 2, 4, and 5). Bio-banking and cryopreservation are important examples of the benefits of technology advancement for endangered species conservation. Initiatives around the globe such as the Frozen Ark Consortium, Frozen Zoo (San Diego Zoo Wildlife Alliance), CryoArks and EAZA (European Association of Zoos and Aquaria) BioBank harvest and store samples such as gametes, blood, cell lines, and embryos. Many consortiums operate not just within the zoo community but also involve museums and academic institutions in coordinated efforts to catalogue and conserve diversity. As technology improves, it becomes more viable to make use of these archives. For example, the black-footed ferret (*Mustela nigripes*) is an ex-situ conservation success story. Thought to be extinct, a small relic population was discovered, captive-bred, and successfully reintroduced across North America. However, due to the extreme bottleneck the species endured, it harbours low genomic diversity. Advancement in cloning technology has enabled researchers to clone a female black-footed ferret from DNA samples taken from the original founding population. Importantly, this individual carries genome diversity lost from the current population and so offers an opportunity for the genetic rescue of the population using ancestral diversity (U.S. Fish & Wildlife Service, 2021). This is a fascinating example of the possibilities for species conservation, albeit with important ethical considerations (Sandler, Moses, & Wisely, 2021). Another exciting aspect of technological innovation is that it will make captive populations increasingly valuable resources not only for conservation, but in furthering our understanding of fundamental biology, as new approaches and investigative opportunities emerge. Consider indirect genetic effects, for example. As discussed at the start of this chapter, the phenotype of an organism results from the interaction between inherited and environmental factors. Indirect genetic effects are a type of environmental influence in which the expression of the genotype of one individual influences the phenotype of another individual (Wolf, Brodie-III, Cheverud, Moore, & Wade, 1998). Captive populations, with their intensive and global management plans, could provide ample opportunities to further our understanding of such evolutionary mechanisms. It is easy to imagine how zoo populations would benefit from research into genomic and phenotypic consequences of conspecific interaction and social setting and how this could inform animal welfare in captivity.

3.5 CONCLUSIONS

Maintaining evolutionary potential is a core objective of conservation biology. Genomic diversity is important to this long-term viability as it underpins the very mechanisms of evolutionary change. Without diversity, populations experience a range of effects which increase their extinction risk. Our understanding of how best to maintain diversity for species undergoing ex-situ conservation management continues to evolve as the science and technology behind captive management improves. Here, we have briefly discussed how evolution underpins all aspects of the species in our care and how it needs to inform our ideas and methods for supporting the expression of behaviours. A deeper understanding of the often-complex behavioural biology of species managed in captivity is increasingly important. In terms of the evolutionary trajectory and viability of populations undergoing intensive captive management, it is likely of greater significance than we currently appreciate.

ACKNOWLEDGEMENTS

Sincere thanks to L. Wright, W. Nash, E. Bell and R. Shaw for their insightful reviews. This work was supported by the Biotechnology and Biological Sciences Research Council (BBSRC), part of UK Research and Innovation, through the Core Capability Grant BB/CCG1720/1 at the Earlham Institute.

REFERENCES

Allendorf, F.W., Funk, W.C., Aitken, S.N., Byrne, M., Luikart, G., & Antunes, A. (2021). *Conservation and the genomics of populations* (3rd Revised edition). Oxford, UK: Oxford University Press.

Ballou, J.D., Lees, C., Faust, L., Long, S., Lynch, C., Bingaman Lackey, L., & Foose, T.J. (2012). Demographic and genetic management of captive populations. In D.G. Kleiman, K.V. Thompson, & C. Kirk Baer (Eds.), *Wild mammals in captivity: principles and techniques for zoo management* (2nd edition), pp. 219–252. Chicago, USA: University of Chicago Press.

Ballou, J.D., & Ralls, K. (1982). Inbreeding and juvenile mortality in small populations of ungulates: A detailed analysis. *Biological Conservation, 24*(4), 239–272.

Bošković, A., & Rando, O.J. (2018). Transgenerational epigenetic inheritance. *Annual Review of Genetics, 52*, 21–41.

Byers, O., Lees, C., Wilcken, J., & Schwitzer, C. (2013). The one plan approach: the philosophy and implementation of CBSG's approach to integrated species conservation planning. *WAZA Magazine, 14*, 2–5.

Charlesworth, D., & Willis, J.H. (2009). The genetics of inbreeding depression. *Nature Reviews. Genetics, 10*(11), 783–796.

Christie, M.R., Marine, M.L., Fox, S.E., French, R.A., & Blouin, M.S. (2016). A single generation of domestication heritably alters the expression of hundreds of genes. *Nature Communications, 7*, 10676.

Drea, C.M., Davies, C.S., Greene, L.K., Mitchell, J., Blondel, D.V., Shearer, C.L., Feldblum, J.T., Dimac-Stohl, K.A., Smyth-Kabay, K.N., & Clutton-Brock, T.H. (2021). An intergenerational androgenic mechanism of female intrasexual competition in the cooperatively breeding meerkat. *Nature Communications, 12*(1), 7332.

Estoup, A., Ravigné, V., Hufbauer, R., Vitalis, R., Gautier, M., & Facon, B. (2016). Is there a genetic paradox of biological invasion? *Annual Review of Ecology, Evolution, and Systematics, 47*(1), 51–72.

Falconer, D.S., & Mackay, T.C.F. (1996). *Introduction to quantitative genetics* (4th ed.). London, UK: Longman.

Frankel, O.H. (1974). Genetic conservation: our evolutionary responsibility. *Genetics, 78*(1), 53–65.

Frankham, R. (1995). Conservation Genetics. *Annual Review of Genetics, 29*, 305–327.

Frankham, R. (2008). Genetic adaptation to captivity in species conservation programs. *Molecular Ecology, 17*(1), 325–333.

Frankham, R., Ballou, J.D., & Briscoe, D.A. (2002). *Introduction to conservation genetics.* Cambridge, UK: Cambridge University Press.

Groombridge, J.J., Jones, C.G., Bruford, M.W., & Nichols, R.A. (2000). "Ghost" alleles of the Mauritius kestrel. *Nature, 403*(6770), 616.

Hamede, R., Madsen, T., McCallum, H., Storfer, A., Hohenlohe, P.A., Siddle, H., Kaufman, J., Giraudeau, M., Jones, M., Thomas, F., & Ujvari, B. (2021). Darwin, the devil, and the management of transmissible cancers. *Conservation Biology: The Journal of the Society for Conservation Biology, 35*(2), 748–751.

Lacy, R.C., Alaks, G., & Walsh, A. (2013). Evolution of *Peromyscus leucopus* mice in response to a captive environment. *PloS One, 8*(8), e72452.

Lacy, R.C., Ballou, J.D. & Pollak, J.P. (2012). PMx: software package for demographic and genetic analysis and management of pedigreed populations. *Methods in Ecology and Evolution/ British Ecological Society, 3*(2), 433–437.

Liu, S., Westbury, M.V., Dussex, N., Mitchell, K.J., Sinding, M.-H.S., Heintzman, P.D., Duchêne, D.A., Kapp, J.D., von Seth, J., Heiniger, H., Sánchez-Barreiro, F., Margaryan, A., André-Olsen, R., De Cahsan, B., Meng, G., Yang, C., Chen, L., van der Valk, T., Moodley, Y., Rookmaaker, K., Bruford, M.W., Ryder, O., Steiner, C., Bruins-van Sonsbeek, L.G.R., Vartanyan, S., Guo, C., Cooper, A., Kosintsev, P., Kirillova, I., Lister, A.M., Marques-Bonet, T., Gopalakrishnan, S., Dunn, R.R., Lorenzen, E.D., Shapiro, B., Zhang, G., Antoine, P., Dalén, L., & Gilbert, M.T.P. (2021). Ancient and modern genomes unravel the evolutionary history of the rhinoceros family. *Cell, 184*(19), 4874–4885.e16.

Mable, B.K. (2019). Conservation of adaptive potential and functional diversity: integrating old and new approaches. *Conservation Genetics, 20*(1), 89–100.

Margan, S.H., Nurthen, R.K., Montgomery, M.E., Woodworth, L.M., Lowe, E.H., Briscoe, D.A., & Frankham, R. (1998). Single large or several small? population fragmentation in the captive management of endangered species. *Zoo Biology, 17*(6), 467–480.

Mayr, E. (1999). *Systematics and the origin of species, from the viewpoint of a zoologist.* Cambridge, USA: Harvard University Press.

McNeely, J.A., Miller, K.R., Reid, W.V., Mittermeier, R.A., & Werner, T.B. (1990). *Conserving the world's biological diversity.* Gland, Switzerland; Washington, DC, USA: IUCN; WRI, CI, WWF-US, and the World Bank.

Plomin, R., DeFries, J.C. & McClearn, G.E. (2008). *Behavioral genetics.* London, UK: Macmillan.

Ralls, K., Ballou, J.D., Dudash, M.R., Eldridge, M.D.B., Fenster, C.B., Lacy, R.C., Sunnucks, P., & Frankham, R. (2018). Call for a paradigm shift in the genetic management of fragmented populations. *Conservation Letters, 11*(2), e12412.

Sandler, R.L., Moses, L., & Wisely, S.M. (2021). An ethical analysis of cloning for genetic rescue: Case study of the black-footed ferret. *Biological Conservation, 257,* 109118.

Seal, U.S., Foose, T.J., & Ellis, S. (1994). Conservation Assessment and Management Plans (CAMPs) and Global Captive Action Plans (GCAPs). In P.J.S. Olney, G.M. Mace, & A.T.C. Feistner (Eds.), *Creative conservation: interactive management of wild and captive animals* (pp. 312–325). Dordrecht, Netherlands: Springer.

Segelbacher, G., Bosse, M., Burger, P., Galbusera, P., Godoy, J.A., Helsen, P., Hvilsom, C., Iacolina, L., Kahric, A., Manfrin, C., Nonic, M., Thizy, D., Tsvetkov, I., Veličković, N., Vilà, C., Wisely, S.M., & Buzan, E. (2021). New developments in the field of genomic technologies and their relevance to conservation management. *Conservation Genetics.* https://doi.org/10.1007/s10592-021-01415-5.

Selechnik, D., Richardson, M.F., Shine, R., DeVore, J.L., Ducatez, S., & Rollins, L.A. (2019). Increased adaptive variation despite reduced overall genetic diversity in a rapidly adapting invader. *Frontiers in Genetics, 10,* 1221.

Soulé, M., Gilpin, M., Conway, W., & Foose, T. (1986). The millenium ark: how long a voyage, how many staterooms, how many passengers? *Zoo Biology, 5*(2), 101–113.

Trut, L., Oskina, I., & Kharlamova, A. (2009). Animal evolution during domestication: the domesticated fox as a model. *BioEssays: News and Reviews in Molecular, Cellular and Developmental Biology, 31*(3), 349–360.

U.S. Fish & Wildlife Service. (2021, February 18). Innovative genetic research boosts black-footed ferret conservation efforts by USFWS and partners. https://www.fws.gov/mountain-prairie/pressrel/2021/02182021-USFWS-and-Partners-Innovative-Genetic-Cloning-Research-Black-footed-Ferret-Conservation.php

Wolf, J.B., Brodie-III, E.D., Cheverud, J.M., Moore, A.J., & Wade, M.J. (1998). Evolutionary consequences of indirect genetic effects. *Trends in Ecology & Evolution, 13*(2), 64–69.

Wright, S. (1931). Evolution in mendelian populations. *Genetics, 16*(2), 97–159.

4

Behavioural biology, applied zoo science, and research

RICARDO LEMOS DE FIGUEIREDO
University of Birmingham, Edgbaston, UK

MARÍA DÍEZ-LEÓN
University of London, Hatfield, UK

4.1 INTRODUCTION

As part of their mandate and mission, zoos conduct research for a variety of reasons – from informing their own evidence-based husbandry to contributing to the conservation of species in the wild – and thus the type of research, including behavioural research, across zoological institutions is as varied as the fields of study. Behavioural research carried out by zoos and related facilities, such as designated conservation breeding units, is not only performed ex-situ, but increasingly zoos lead behavioural research in situ (e.g., on reintroduced individuals and also on behavioural aspects of free-living animals such as foraging behaviours, personality, etc.).

Zoo-based research is a useful tool for the behavioural scientist, as the close access to human-habituated animals enables the answering of questions on their behavioural adaptations and abilities that could otherwise be difficult to achieve. These advances in knowledge may have direct applications to zoo management, animal welfare and/or conservation (applied science) but often their primary aim is simply to increase our understanding of the natural world (fundamental science; Figure 4.1). Examples of the latter include contributions to cognition in the form of detecting motor lateralisation in zoo-housed wolves (*Canis lupus*) by observing potential paw preferences during manipulation (Regaiolli, Mancini, Vallortigara, & Spiezio, 2021), or improving our understanding of self-control in several species of zoo-housed parrots (Psittaciformes) where both individual and species differ in performance in a "delayed gratification" task (Brucks et al., 2021). Findings like these, and the questions answered by fundamental science,

DOI: 10.1201/9781003208471-5

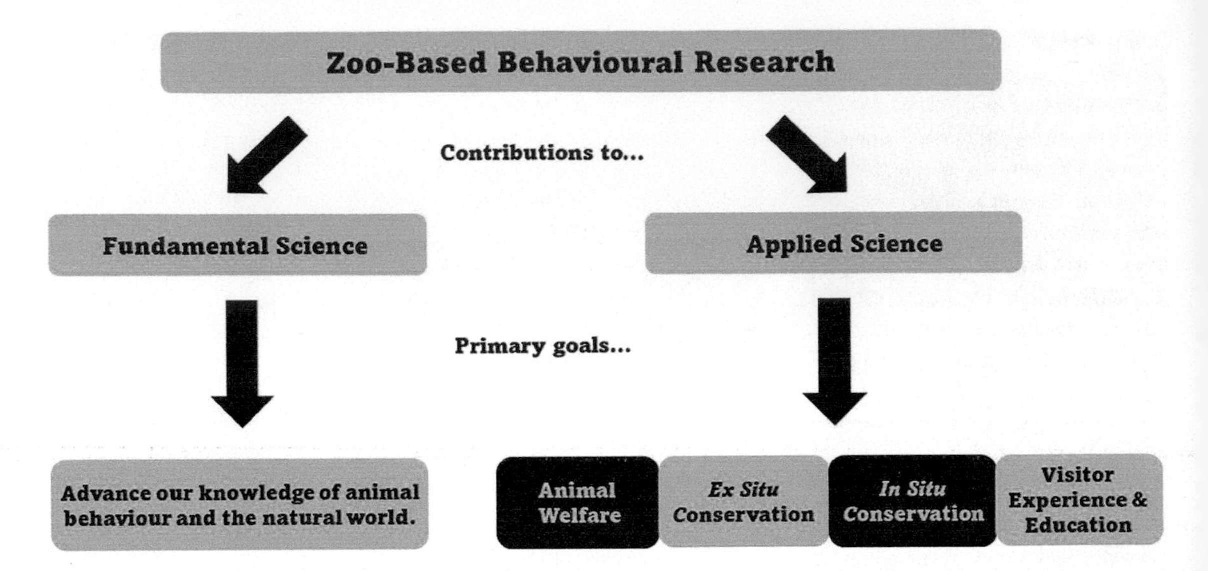

Figure 4.1 Diagram highlighting the contributions of zoo-based behavioural research to science.

may not have direct or immediate applications to zoo practitioners in the short term, but that does not make them any less important. Increasing our knowledge of the natural world is in itself a tool for improving animal care and conservation action and should be encouraged as much as possible.

For the remainder of this chapter, we will primarily focus on highlighting areas where applied behavioural research is integral to the mission (and vision) of zoos: animal welfare, ex-situ and in-situ conservation, and visitor experience, engagement, and education.

4.2 BEHAVIOURAL RESEARCH AT THE ZOO: ANIMAL WELFARE

The assessment of animal welfare (primarily understood as subjective states, see Chapter 21) in zoos is a complex and challenging task. Behaviour indicators in their various expressions feature prominently in zoo (and other) welfare frameworks. For example, one of the more recent ones, the "Five Domains" framework, includes the use of behavioural indicators that not only evidence absence of negative welfare states, but also indicate positive ones (Mellor, 2016). Further, in their official welfare strategy, the World Association of Zoos and Aquariums (WAZA) calls for their members to apply behavioural science in the management, husbandry, and welfare assessment of their animals. More specific welfare applications of behavioural research include the assessment of enclosure design, the effectiveness of enrichment strategies (Figure 4.2), the effects of visitor and keeper–animal interactions, the food preferences of animals and how to best feed them. Chapter 21 provides a detailed review of how the study of behaviour contributes to uphold animal welfare standards and design animal welfare assessment tools; here we introduce but a few examples on how behavioural research can help infer welfare state and contribute to its improvement in zoo animals.

Perhaps the most well-studied of these are abnormal behaviours, and more specifically, abnormal repetitive behaviours (e.g., the pacing behaviour common in some zoo-held carnivore and ungulate species, among others). These complex and multi-factorial behaviours are generally linked to poor welfare states (though individual personality and other factors might also play a role when displayed within the same conditions). While not unique to zoo settings, zoo research has been fundamental in both initiating the study of these behaviours, as well as advancing our understanding of their welfare implications. Indeed, some the first detailed descriptions of the development of these behaviours come from zoo observations of the thwarting of social behaviour in okapis (*Okapia johnstoni*) and dingos (*Canis lupus dingo*) (Meyer-Holzapfel, 1968), as do some of the more advanced multi-species comparisons to identify risk factors (Kroshko et al., 2016), or meta-analyses assessing

Figure 4.2 Gorilla interacting with enrichment device at Hellabrunn Zoo (Germany). Behavioural research can help assess the impacts of enrichment such as this on the activity and welfare of zoo-housed animals. Photo: Pixaby.com

the effectiveness of environmental enrichment in preventing, reducing, or abolishing these behaviours (Swaisgood & Shepherdson, 2006). While research on understanding these behaviours is likely to continue (as, in addition to the potential welfare concerns of both the animals performing and not performing them, visitors dislike seeing animals displaying these behaviours), research is becoming more refined, both in methodology (e.g., by use of detailed video analysis, Cless, Voss-Hoynes, Ritzman, & Lukas, 2015) and understanding the different causal basis of different forms of behaviours displayed, as these might have different implications for welfare and therefore treatment. Along with refining our understanding of these behaviours as welfare indicators, zoo research is borrowing tools from human psychology and behavioural ecology to develop at least three new potential behavioural welfare indicators: behavioural diversity (Miller, Gallup, Vogel, Vicario, & Clark, 2012), strength of social networks (Rose & Croft, 2020), and boredom (Burn, 2016).

Behaviour can also be used to complement research in other areas that are relevant to welfare and zoo husbandry and design, and there is a vast literature available on zoo behavioural research in practice, helping to inform better zoo management and husbandry practices and consequently contributing to improved animal welfare. For example, food preference studies can help uncover diet suitability, as in the case of zoo-housed lesser anteaters (*Tamandua tetradactyla*) observed to spend more time feeding on live termites than ants, likely due to their differential nutritional and digestibility values (Zárate, Mufari, Abalos Luna, Villarreal, & Busso, 2022). Studies on human effects on the welfare of zoo animals rely heavily on behavioural work. For instance, Stoinski, Jaicks, and Drayton (2012) found that some individuals within a group of zoo-housed gorillas (*Gorilla gorilla gorilla*) responded to greater crowd sizes with increased aggression. Similarly, behavioural observations are often key to validating personality studies in zoo-held animals, which are commonly carried out by surveying experienced keepers (e.g., Haage, Maran, Bergvall, Elmhagen, & Angerbjörn, 2017), an issue we turn to below, as personality not only has welfare implications but its importance to successful conservation breeding is increasingly acknowledged.

4.3 BEHAVIOURAL RESEARCH AT THE ZOO: EX-SITU CONSERVATION BREEDING

Ex-situ conservation efforts, such as conservation breeding (i.e., captive breeding for reintroduction), are an integral part of species conservation only likely to become more important as the number of species under threat increases. A recently published study reported that ex-situ measures were taken in over half of the mammal and bird species whose extinction was prevented by conservation (Bolam et al., 2021). Conservation breeding by zoos and aquariums, specifically, helped in the recovery of over a quarter of the vertebrate species whose conservation status had improved by 2015, as recognised by the International Union for the Conservation of Nature (IUCN). However, reintroduction programmes are notoriously challenging and have a high failure rate (Bubac, Johnson, Fox, & Cullingham, 2019). Captive-bred animals are less likely to survive once reintroduced than translocated wild-born individuals (Jule, Leaver, & Lea, 2008), as natural behaviours tend to get lost or modified with generations in captivity (McPhee, 2004). Similarly, breeding animals in captivity in the first place can be problematic due to either loss of reproductive behaviour or other behavioural issues (Martin-Wintle, Wintle, Díez-León, Swaisgood, & Asa, 2019).

Behavioural research is a key approach to addressing and mitigating these problems. As the IUCN recommends in their reintroduction guidelines, captive-bred individuals selected for reintroduction programmes must display "appropriate behaviours" in comparison to the original or remaining wild populations. Only by scientifically assessing the behaviour of the animals can we determine what natural behaviours are missing or under-represented and assess their suitability for reintroduction in the wild. Reports of behavioural research as a tool for ex-situ conservation can be found in the literature. For example, behavioural assessment of courtship behaviours can be fundamental to improve breeding success in ex-situ programmes (Martin-Wintle et al., 2019). Similarly, behavioural assessment of personality has been used as predictor of reintroduction success in several species (e.g., Haage et al., 2017), and in some cases used to select reintroduction candidates: a team of researchers in Brazil studied the behavioural responses of three captive maned wolves (*Chrysocyon brachyurus*) to foraging, intraspecific and interspecific stimuli, in order to select one individual for reintroduction to the wild (Coelho, Schetini de Azevedo, & Young, 2012). While which personality-types fare better after reintroduction can be species-specific and even season-specific, these studies yield findings with important applications to the management of the conservation breeding programme.

Many reintroduction programmes – such as those for Iberian lynx (*Lynx pardinus*), golden-lion tamarin (*Leontopithecus rosalia*), European mink (*Mustela lutreola*), or the black-footed ferret (*Mustela nigripes*), to name a few – partially owe their success to behavioural research and management. For example, captive Iberian lynx and European mink in the Iberian ex-situ programmes are monitored 24/7 via video-surveillance and both qualitative and quantitative behavioural data are recorded to determine whether individuals exhibit appropriate foraging and social behaviours, as well as appropriate avoidance responses towards humans. Behavioural monitoring provides the conservationists in the project with answers to these and other questions, allowing them to assess and improve animal care and to determine which individuals are suitable for release into the wild. On top of assisting with the selection and preparation of animals for reintroduction, behavioural research is also an important tool for the post-release stage. Stoinski, Beck, Bloomsmith, and Maple (2003) found deficient skills in foraging and locomotion in captive-bred individuals in comparison with their wild-born counterparts, highlighting the importance of monitoring the behaviour of the animals once released, so adjustments can be made to improve their chances of survival.

Behavioural research is also used to evaluate and optimise environmental enrichment and training, strategies that are not only used to improve animal welfare in zoos but also to enable wild-type behaviours and prepare captive animals for life in the wild. Reading, Miller, and Shepherdson (2013) reviewed the literature for evidence of the impacts of environmental enrichment on the success of reintroduction programmes; they found examples across species of enrichment improving reintroduction success by enabling the development of behaviours that are crucial for survival in the wild (i.e., locomotion, predator avoidance, foraging, social interactions, and physical fitness).

4.4 BEHAVIOURAL RESEARCH AT THE ZOO: IN-SITU CONSERVATION

The in situ (i.e., in their wild, natural habitat) study of populations and individual animals is vital for our understanding of animal biology, behaviour included, with important applications to species conservation. It is, however, very challenging, as researchers have to track, find, monitor, and/or observe animals in the vast and very complex environments they live in, with many species being particularly cryptic, small, fast, and/or difficult to follow over large distances. Advances in technology and the development of new methodologies help researchers in this task, and zoo-based animals and research can be instrumental in the fine-tuning of this process. Easier access to animals in the zoo environment allows methodologies, technology, and equipment to be tested and refined before it is used on wild animals. Furthermore, behavioural researchers often develop ethograms, test research protocols, and practice their data collection on zoo-housed animals prior to heading to the field, ensuring a higher degree of preparation. On top of facilitating research, trialling methods on zoo-housed individuals may also reduce impacts on animal welfare and reduce ethical conflicts. As zoo animals tend to be more habituated to humans, and are often trained for handling and medical procedures, testing methods and equipment on them first will reduce the need for invasive procedures on wild animals, whose welfare is more likely to be negatively impacted by them.

The following are just a few examples to highlight how important these applications can be. Firstly, the Humboldt penguins (*Spheniscus humboldti*) at Parc Zoologique de Paris are helping researchers from partner institutions to test remote-controlled robots, developed for use in the study of the behaviour of their wild counterparts. By assessing the behavioural response of the zoo-housed penguins, they are refining the design of the robot so it can more closely approach wild penguins (Parc Zoologique de Paris, 2021). Similarly, researchers piloted in zoo-housed red pandas (*Ailurus fulgens*) the use of radio collars aimed to be fitted to their wild counterparts: behavioural – as well as other – signs of stress in the zoo pandas were used to highlight the need to further refine the design of the radio collars to minimise impact when fitted in wild individuals (van de Bunte, Weerman, & Hoff, 2021). In another study, a team of researchers tested and calibrated an accelerometer for research on free-living Bewick's swans (*Cygnus columbianus bewickii*), using zoo-housed individuals. The authors highlighted how useful it was to have access to the captive swans, as direct observations on free-living individuals are difficult to obtain and they achieved high accuracy in the accelerometer's calibration (Nuijten, Prins, Lammers, & Mager, 2020). Edwards, Bungard, and Griffiths (2022) also perfectly demonstrate how in-situ and ex-situ research can be mutually beneficial: using environmental data collected in the field, they designed climate-controlled chambers for captive golden mantella frogs (*Mantella aurantiaca*), recreating the environment of their natural habitat. These researchers then studied the mantella's daily activity in these replicated environments, and their findings have applications to field research by informing researchers about the best time of day to capture or count individuals for conservation and research purposes.

4.5 BEHAVIOURAL RESEARCH AT THE ZOO: VISITOR EXPERIENCE, ENGAGEMENT & EDUCATION

Zoos and aquariums are popular visitor attractions with over 700 million visitors every year, as estimated by WAZA. Visitors are one of the main stakeholders in the zoo industry with admissions being responsible for the majority of its income. Zoos must therefore guarantee their income by drawing visitors to their gates, providing them with a recreational day and ensuring their satisfaction. On top of fun and recreation, modern zoos must also fulfil their pedagogic role by providing their vast numbers of visitors with an engaging and educational environment. Achieving an optimal balance of visitor fun, engagement, and education is not a simple task, and even less so when also considering animal welfare and ethics, which modern zoos must always do. Conducting research is integral to achieving this balance.

In order to better understand their visitors' experiences and learning outcomes, zoos are often assisted by social scientists, who scientifically assess what visitors are getting out of zoo visits. As animals are the main attraction of zoological institutions, integrating animal behaviour research and the social sciences allows for a better understanding of visitor–animal dynamics. This multidisciplinary approach to visitor studies has been used several times, with examples easily found in the literature. Back in the late 1990s, Altman (1998) recorded visitor conversations while they were observing different bear species (Ursidae) at the zoo, and simultaneously collected data on the activity of the animals they were watching. The results suggested that visitors' attention and their conversations were influenced by the activity the bears were engaged in, with "animated activities," as described by the author, eliciting more interest in the visitor. A few years later, Anderson, Kelling, Pressley-Keough, Bloomsmith, and Maple (2003) looked at zoo visitor perceptions in relation to Asian small-clawed otter (*Aonyx cinerea*) activity and training events and found that visitors perceived the animals more positively when they were more active. The author also noticed, however, that visitors rated otters as "good pets" more often when these were active, suggesting that zoos could inadvertently encourage exotic pet ownership. More recent studies looked into the relationship between animal behaviour and visitor emotions at the zoo, with results suggesting that active behaviours elicited more emotional responses in the visitors, which in turn were linked to increased conservation mindedness (Powell & Bullock, 2015). These are just a few of several studies highlighting the impacts of animal behaviour on visitors, directly or indirectly affecting their recreational and educational experiences, their perceptions of animals, and of conservation.

Applying animal behaviour research to visitor studies at zoos and aquariums is therefore an effective way to assess the visitor experience more accurately. It also provides zoo managers and practitioners with evidence to guide the decisions that will lead towards satisfied and happy visitors that should be more environmentally aware, better educated, and feeling more connected to nature.

4.6 CONCLUSIONS AND FUTURE AVENUES FOR ZOO-BASED BEHAVIOURAL RESEARCH

We finish by highlighting ways in which behavioural research can overcome some of the common challenges of zoo-based research that will be expanded upon in the following chapter, as well as advocating for partnerships between zoological institutions and research and academic institutions to advance applied behaviour research at the zoo.

- Increase use of multi-zoo studies:
  - As advocated by (for example) BIAZA's research guidelines (search for these on the BIAZA website, biaza.org.uk), these types of studies have the advantage of increasing sample size (particularly for species held at very low numbers in single zoos, e.g., solitary megafauna) and providing informative, detailed behavioural data.
- Increase the use of automated forms of behavioural recording:
  - Observing behaviour is time-consuming and while multi-zoo studies are important to have representative samples, they are logistically more laborious and time-consuming to set up, and require relatively more funding (either to allow the researcher to travel to observe behaviour, or to set up video equipment that allows for remote capture). Research has already started and is likely to continue to bring in automation of behavioural recording and/or incorporate elements of citizen science.
- Increase use of behavioural databases and detailed surveys:
  - As with multi-zoo studies, these allow for large sample sizes to be considered (potentially being able to sample the whole population), are relatively inexpensive – beyond the researchers' time to compile and analyse the data – and can cover multiple taxa in a more efficient way that relying on cameras or live observation, but can often lack depth and detail.
- Increase use of meta-analyses and databases of case studies:
  - Case studies are crucial for evidence-based husbandry and in addition to informing husbandry at individual institutions, they

can become the basis for meta-analyses and/or allow for the compilation of the type of databases mentioned above while avoiding some of their caveats. In addition, they represent a unique way to train future zoo behaviour researchers: they remain a popular project choice among students, who have the time for data collection and analyses that zoo personnel might otherwise lack.

- Partnerships and cross-fertilisations between non-zoo and zoo researchers and institutions:
  - Zoos provide a unique opportunity to conduct translational research and overcome some of the limitations that laboratory researchers find in standardised, non-naturalistic settings.
- Mixed-methods approach, involving the integration of behavioural research with other disciplines:
  - Behavioural research can be used to validate findings provided by other fields of study (e.g., physiology and the use of hormonal analysis as stress indicators).
  - Multi-disciplinary approaches can allow for refinement of behavioural methods by for example increasing understanding of causal factors underpinning behaviour or their fitness consequences.

Zoo-based behavioural research is therefore a key component of the mission of the modern zoo, as it contributes to the advancement of animal behaviour science while also generating and providing evidence with applications to, among others, animal welfare, ex-situ and in-situ conservation, and visitor experience and education. Future behavioural research at the zoo should aim to become even more collaborative and multi-disciplinary to overcome some of the barriers imposed by the idiosyncrasies of the zoo setting.

REFERENCES

Altman, J.D. (1998). Animal activity and visitor learning at the zoo. *Anthrozoös, 11*(1), 12–21.

Anderson, U.S., Kelling, A.S., Pressley-Keough, R., Bloomsmith, M.A., & Maple, T.L. (2003). Enhancing the zoo visitor's experience by public animal training and oral interpretation at an otter exhibit. *Environment and Behavior, 35*(6), 826–841.

Bolam, F.C., Mair, L., Angelico, M., Brooks, T.M., Burgman, M., Hermes, C., Hoffmann, M., Martin, R.W., McGowan, P.J.K., Rodrigues, A.S.L., Rondinini, C., Westrip, J.R.S., Wheatley, H., Bedolla-Guzmán, Y., Calzada, J., Child, M.F., Cranswick, P.A., Dickman, C.R., Fessl, B., … Butchart, S.H.M. (2021). How many bird and mammal extinctions has recent conservation action prevented? *Conservation Letters, 14*(1), 1–11.

Brucks, D., Petelle, M., Baldoni, C., Krasheninnikova, A., Rovegno, E., & von Bayern, A.M.P. (2021). Intra- and interspecific variation in self-control capacities of parrots in a delay of gratification task. *Animal Cognition*, https://doi.org/10.1007/s10071-021-01565-6

Bubac, C.M., Johnson, A.C., Fox, J.A., & Cullingham, C.I. (2019). Conservation translocations and post-release monitoring: Identifying trends in failures, biases, and challenges from around the world. *Biological Conservation, 238*, 108239.

Burn, C. (2016). Bestial boredom: A biological perspective on animal boredom and suggestions for its scientific investigation. *Animal Behaviour, 130*, 141–151.

Cless, I.T., Voss-Hoynes, H.A., Ritzman, R.E., & Lukas, K.E. (2015). Defining pacing quantitatively: A comparison of gait characteristics between pacing and non-repetitive locomotion in zoo-housed polar bears. *Applied Animal Behaviour Science, 169*, 78–85.

Coelho, C.M., Schetini de Azevedo, C., & Young, R.J. (2012). Behavioral responses of maned wolves (*Chrysocyon brachyurus*, Canidae) to different categories of environmental enrichment stimuli and their implications for successful reintroduction. *Zoo Biology, 31*(4), 453–469.

Edwards, W.M., Bungard, M.J., & Griffiths, R.A. (2022). Daily activity profile of the golden mantella in the "Froggotron"—a replicated behavioral monitoring system for amphibians. *Zoo Biology, 41*(1), 3–9.

Haage, M., Maran, T., Bergvall, U.A., Elmhagen, B., & Angerbjörn, A. (2017). The influence of spatiotemporal conditions and personality on survival in reintroductions – evolutionary implications. *Oecologia, 183*, 45–56.

Jule, K.R., Leaver, L.A., & Lea, S.E.G. (2008). The effects of captive experience on reintroduction survival in carnivores: A review and analysis. *Biological Conservation, 141*(2), 355–363.

Kroshko, J., Clubb, R., Harper, L., Mellor, E., Mohenreschlager, A., & Mason, G. (2016). Stereotypic route tracing in captive Carnivora is predicted by species-typical home range sizes and hunting styles. *Animal Behaviour, 117*, 197–209.

Martin-Wintle, M., Wintle, N. Díez-León, M., Swaisgood, R., & Asa, C. (2019). Improving the sustainability of captive populations with free mate choice. *Zoo Biology, 38*, 119–132.

McPhee, M.E. (2004). Generations in captivity increases behavioral variance: Considerations for captive breeding and reintroduction programs. *Biological Conservation, 115*(1), 71–77.

Mellor, D.J. (2016). Updating animal welfare thinking: Moving beyond the "five freedoms" towards "a life worth living." *Animals, 6*(3), 21.

Meyer-Holzapfel, M. (1968). Abnormal behaviour in zoo animals. In Fox, M.W. (Ed.), *Abnormal behaviour in animals*, Saunders, London, UK.

Miller, M.L., Gallup, A.C., Vogel, A.R., Vicario, S.M., & Clark, A.B. (2012). Evidence for contagious behaviors in budgerigars *(Melopsittacus undulatus)*: An observational study of yawning and stretching. *Behavioural Processes, 89*(3), 264–270.

Nuijten, R., Prins, E., Lammers, J., & Mager, C. (2020). Calibrating tri-axial accelerometers for remote behavioural observations in Bewick's swans. *Journal of Zoo and Aquarium Research, 8*(4), 231–238.

Parc Zoologique de Paris. (2021). Collaboration avec le CNRS et le centre scientifique de Monaco: Une nouvelle étape pour nos manchots. https://www.parczoologiquedeparis.fr/fr/actualites/collaboration-avec-le-cnrs-et-le-centre-scientifique-de-monaco-une-nouvelle-etape-pour-nos-manchots-3449

Powell, D.M., & Bullock, E.V.W. (2015). Evaluation of factors affecting emotional responses in zoo visitors and the impact of emotion on conservation mindedness evaluation of factors affecting emotional responses in zoo visitors and the impact of emotion on conservation mindedness. *Anthrozoös, 27*(3), 389–405. https://doi.org/10.2752/175303714X13903827488042

Reading, R.P., Miller, B., & Shepherdson, D. (2013). The value of enrichment to reintroduction success. *Zoo Biology, 32*(3), 332–341.

Regaiolli, B., Mancini, L., Vallortigara, G., & Spiezio, C. (2021). Paw preference in wolves (*Canis lupus*): A preliminary study using manipulative tasks. *Laterality, 26*(1–2), 130–143.

Rose, P.E., & Croft, D. (2020). Evaluating the social networks of four flocks of captive flamingos over a five-year period: Temporal, environmental, group and health influences on assortment. *Behavioural Processes, 175*, 104–118.

Stoinski, T.S., Beck, B.B., Bloomsmith, M.A., & Maple, T.L. (2003). A behavioral comparison of captive-born, reintroduced golden lion tamarins and their wild-born offspring. *Behaviour, 140*(2), 137–160.

Stoinski, T.S., Jaicks, H.F., & Drayton, L.A. (2012). Visitor effects on the behavior of captive western lowland gorillas: The importance of individual differences in examining welfare. *Zoo Biology, 31*(5), 586–599.

Swaisgood, R., & Shepherdson, D. (2006). Environmental enrichment as a strategy for mitigating stereotypies in zoo animals: A literature review and meta-analysis. In Mason, G. & Rushen, J. (Eds.), *Stereotypic animal behaviour: Fundamentals and applications to welfare* (2nd edition). CABI, Cambridge, UK.

van de Bunte, W., Weerman, J., & Hoff, A.R. (2021). Potential effects of GPS collars on the behaviour of two red pandas (*Ailurus fulgens*) in Rotterdam Zoo. *PLoS One, 16*(6), e0252456.

Zárate, V., Mufari, J.R., Abalos Luna, L.G., Villarreal, D.P., & Busso, J.M. (2022). Assessment of feeding behavior of the zoo-housed lesser anteater *(Tamandua tetradactyla)* and nutritional values of natural prey. *Journal of Zoological and Botanical Gardens, 3*(1), 19–31.

5

Behavioural biology methods and data collection in the zoo

JACK LEWTON
Imperial College London, Silwood Park, UK

SAMANTHA WARD
Nottingham Trent University, Brackenhurst, UK

5.1 DESIGNING BEHAVIOURAL STUDIES IN THE ZOO

5.1.1 Getting started

Starting a new behavioural research project can sometimes be quite daunting. Thinking of a new research question that is useful and contributes to the world of zoo research is possible and probably easier than you think. The key idea is to fill a knowledge gap; whether that be for a specific zoo, i.e., a case study style project such as "how does scent enrichment impact on the behaviour of two Sumatran tigers *(Panthera tigris sondaica)*?" or something a little larger and involve a multi-zoo approach, e.g., "Does animal personality impact on primate behavioural plasticity?" Here are three ways that might help to get you thinking of an appropriate project:

i) Head to the zoo and watch the animals (with permission from the zoo of course. Any formal research project conducted in a zoo needs to be submitted to zoo staff beforehand for review and scrutiny before data collection commences). What questions do you ask yourself while you watch them? Your critical mind will start to ponder "Why are they doing that?" or "How do the animals know to do that?"
ii) Read academic journal articles. You will find that within the discussion section, authors will provide justification for their results based on previous research. Usually, they will add something like "further research is needed to investigate this in more detail."
iii) Contact your target zoo or relevant university researchers to see if they have any projects that they specifically wish to have investigated. Some institutions provide lists of possible

DOI: 10.1201/9781003208471-6

projects or you can look on the various Zoo Association web pages for priority areas of research too.

Your research question will help to formulate your aims and objectives. You need to remember your aim is the overarching question that will help to shape the title of your project and links specifically to your research question. The objectives are the smaller steps that you need to take to reach your aim that usually start with "action verbs." For example, let's go with our original tiger enrichment project mentioned earlier. The overall aim would be to investigate how scent enrichment affects the behaviour of two Sumatran tigers housed at *(name of your chosen)* Zoo. The objectives could be: (1) Identify baseline behaviours performed by the tigers; (2) Examine the behaviour performance of the tigers with added scent enrichment; and (3) Evaluate any variations in behaviour according to the enrichment provision. As a rule of thumb for an aim like this, you should be looking for approximately three objectives.

5.1.2 Selecting methods

The appropriate method to collect your data will dictate whether you see success or struggle. Different methods are selected according to the aims and objectives of the research and are therefore appropriate for different types of data collection (Rose & Riley, 2021). When planning your methods, you will need to include two different types of rules, the sampling rules, and the recording rules. Sampling rules specify **which subjects to watch and when**. The recording rules identify **how** the behaviour is recorded. You will need to include which of each of the rules will apply to your data collection in your methods. Table 5.1 summarises the different recording techniques that can be applied to behavioural data collection in the zoo and for a deeper understanding of these, we recommend Bateson and Martin's (2021) *Measuring behaviour: an introductory guide*.

When you have decided which sampling and recording rules you will use, you will need to also specify the amount of time you are collecting the data for. Examples of these can be seen below:

> Continuous focal sampling was used to record behaviours outlined in the ethogram for 30 minutes. Individuals were selected using a predetermined random allocation that allowed for all individuals to be observed for a period of 20 hours before enrichment and 20 hours after.
>
> Instantaneous scan sampling of the group was used to record social behaviours every 30 seconds for 30 minutes. Data were collected for a total of 20 hours accounting for a maximum of 160 points of occurrence.

Pilot studies are a great way to test methods. For example, there may be observation restrictions, zoo or animal routine restrictions, or difficulty identifying individuals, and these may affect how feasible your selected data collection method is. We recommend putting aside some time to head to the zoo and collect some data as a test. Remember, the aim is to collect valid and reliable data, so it is important to get this right.

Being organised is key, structuring your data collection is vital to ensuring that you collect the right data at the right time, and you do not miss individuals for example. Designing a data collection timetable is the best way to manage this. Table 5.2 shows a data collection timetable that was used to ensure twelve individual bonobos *(Pan paniscus)* were recorded at each time slot twice for a behavioural project at Twycross Zoo, UK. A tip is to collect as much data as needed and plan for extra just in case.

5.1.3 Ethical approval

Any research taking place in zoos – whether it be on the visitors, staff, or animals – will need to be ethically assessed before it can be approved. The purpose is to ensure that any subjects of the research cannot be harmed physically or mentally by what you are going to investigate. We will focus here on animal behavioural research ethics. Different institutions will have different processes to review projects but generally, if you are observing zoo animal behaviour and not requesting any changes to the animals' environment or husbandry procedures then the project will unlikely come up against any problems.

If, however, you are hoping to add novel items of enrichment or make changes to the animals' enclosure design for example, you will need to consider what you would do if these alterations cause harm or distress to the animal concerned. Within your ethical approval, you would need to add caveats to your project such as any items provided to the

Table 5.1 Summary of behavioural data collection methods.

	Name	Description	Comments
Sampling rules	*Ad libitum*	Not systematic. Observer notes down visible behaviours when it seems appropriate.	Use for preliminary observations, producing ethograms, or recording rare behaviours.
	Focal	Observing one individual recording all incidences of its behaviour using a predetermined schedule of order.	Can sometimes lose data when individuals go out of sight.
	Scan	Whole groups of individuals are scanned and each behaviour of each individual at that incident is recorded.	Good to obtain lots of group level data across the same time frame/season. Can be used at the start of a focal sample and used in conjunction with focal to tell a broader picture.
	Behaviour	Occurrences of particular behaviours and the individuals involved recorded.	Use for recording rare specific behaviours such as fights or copulations. Can be used in conjunction with focal or scan.
Recording rules	Continuous	An exact record of the behaviours performed over a specified period of time as well as when those behaviours occurred, i.e., 6mins 10 s of resting then 2 mins 42 s of locomotion.	Enables a true picture of the pattern of behaviours as well as the exact time or frequencies of behaviours. Observer fatigue is possible due to continued observations so shorter sessions might be needed.
	Instantaneous	Behaviours are recorded **on** the instant that is predetermined, e.g., every 60 s, i.e., a buzzer goes off at 60 s intervals and on each beep, you record the behaviour being performed.	Not used for recording rare behaviours. The score obtained is expressed as a proportion of the sample point, e.g., in a 10 min observation period every 60 s, the behaviour occurred on 4 out of 10 beeps, and therefore scores 0.4 for that period.
	One-zero	A behaviour is recorded if within the predetermined time frame, e.g., every 60 s the behaviour has occurred during the proceeded sample, i.e., a buzzer goes off at 60 s intervals, you record the behaviour if it has been performed in the time preceding the buzzer.	The score obtained is expressed as a proportion of the sample points as above but within the beeps rather than on the beeps. Not a robust measure as can be seen to both over and underestimate data. Only to be used when no other method seems suitable.

animals should be removed or husbandry practices reverted back to normal if extreme issues arise due to your project (e.g., unwanted behavioural changes indicative of welfare compromise). You will need to discuss the ethical requirements with the institution directly and/or the university if you are a student undertaking a research project. In addition, if you wish to publish any of your research in academic journals, an ethical statement will be required to show that no animals were harmed as a result of your project. For more information see the ARRIVE guidelines (Percie du Sert et al., 2020).

Table 5.2 An example of a data collection timetable used for planning the data collection process. One individual, "Lina," is highlighted to show more clearly how to organise this.

9am	9.30am	10am	10.30am	11am	11.30am	1pm	1.30pm	2pm	2.30pm	3pm	3:30pm
Lina	Louisoko	Maringa	Likemba	Malaika	Cheka	Ndeko	Lopori	Keke	Rubani	Lucuma	Diatou
Diatou	Lina	Louisoko	Maringa	Likemba	Malaika	Cheka	Ndeko	Lopori	Keke	Rubani	Lucuma
Lucuma	Diatou	Lina	Louisoko	Maringa	Likemba	Malaika	Cheka	Ndeko	Lopori	Keke	Rubani
Rubani	Lucuma	Diatou	Lina	Louisoko	Maringa	Likemba	Malaika	Cheka	Ndeko	Lopori	Keke
Keke	Rubani	Lucuma	Diatou	Lina	Louisoko	Maringa	Likemba	Malaika	Cheka	Ndeko	Lopori
Lopori	Keke	Rubani	Lucuma	Diatou	Lina	Louisoko	Maringa	Likemba	Malaika	Cheka	Ndeko
Ndeko	Lopori	Keke	Rubani	Lucuma	Diatou	Lina	Louisoko	Maringa	Likemba	Malaika	Cheka
Cheka	Ndeko	Lopori	Keke	Rubani	Lucuma	Diatou	Lina	Louisoko	Maringa	Likemba	Malaika
Malaika	Cheka	Ndeko	Lopori	Keke	Rubani	Lucuma	Diatou	Lina	Louisoko	Maringa	Likemba
Likemba	Malaika	Cheka	Ndeko	Lopori	Keke	Rubani	Lucuma	Diatou	Lina	Louisoko	Maringa
Maringa	Likemba	Malaika	Cheka	Ndeko	Lopori	Keke	Rubani	Lucuma	Diatou	Lina	Louisoko
Louisoko	Maringa	Likemba	Malaika	Cheka	Ndeko	Lopori	Keke	Rubani	Lucuma	Diatou	Lina
Lina	Louisoko	Maringa	Likemba	Malaika	Cheka	Ndeko	Lopori	Keke	Rubani	Lucuma	Diatou
Diatou	Lina	Louisoko	Maringa	Likemba	Malaika	Cheka	Ndeko	Lopori	Keke	Rubani	Lucuma
Lucuma	Diatou	Lina	Louisoko	Maringa	Likemba	Malaika	Cheka	Ndeko	Lopori	Keke	Rubani
Rubani	Lucuma	Diatou	Lina	Louisoko	Maringa	Likemba	Malaika	Cheka	Ndeko	Lopori	Keke
Keke	Rubani	Lucuma	Diatou	Lina	Louisoko	Maringa	Likemba	Malaika	Cheka	Ndeko	Lo ri
Lopori	Keke	Rubani	Lucuma	Diatou	Lina	Louisoko	Maringa	Likemba	Malaika	Cheka	Ndeko
Ndeko	Lopori	Keke	Rubani	Lucuma	Diatou	Lina	Louisoko	Maringa	Likemba	Malaika	Cheka
Cheka	Ndeko	Lopori	Keke	Rubani	Lucuma	Diatou	Lina	Louisoko	Maringa	Likemba	Malaika
Wlaika	Cheka	Ndeko	Lopori	Keke	Rubani	Lucuma	Diatou	Lina	Louisoko	Maringa	Likemba
Likemba	Wlaika	Cheka	Ndeko	Lopori	Keke	Rubani	Lucuma	Diatou	Lina	Louisoko	Maringa
Maringa	Likemba	Waika	Cheka	Ndeko	Lopori	Keke	Rubani	Lucuma	Diatou	Lina	Louisoko
Louisoko	Maringa	Likemba	Waika	Cheka	Ndeko	Lopori	Keke	Rubani	Lucuma	Diatou	Lina

5.2 COLLECTING BEHAVIOURAL DATA

Once the project's aims, objectives, and methods are all in place, the next stage is to start collecting data. It is important to remember that no matter what the research, the needs, and welfare of the animals must come first. Animal care staff will likely be interested in the work that is being conducted but ultimately, they are there to provide for the needs of the animals and this must take priority. Some institutions provide a lead keeper contact, someone that researchers can get in touch with if they spot any issues or want to check in with in the morning. This can be extremely useful as keepers have a wealth of knowledge about the animals and are usually more than willing to share it. Additionally, on the occasion when animals may not be able to be studied that day (e.g., veterinary visits, enclosure maintenance), keepers could communicate this to save the researcher unnecessary visits to the zoo.

Published research on the implications of human-animal interactions has shown varying behavioural reactions of animals to zoo visitors and staff (Sherwen & Hemsworth, 2019). It is therefore vital that on arrival at the animal enclosure, consideration that even the presence of a researcher at the side of the enclosure may impact upon the animals' behaviour. Researchers need to be extremely careful that what they record is a result of the conditions currently being experienced rather than because of a prolonged observer. To mitigate this, a short period of habituation should be incorporated at the beginning of each day where possible. Alternatively, the use of a carefully positioned video camera or live stream might be feasible, which will reduce the issues associated with human presence.

With the development in technology, there are now sufficient technological tools that can help with behavioural data collection. Some are free to download and use (e.g., Behavioural Observation Research Interactive Software, BORIS), while others will have associated costs (e.g., Noldus Observer XT). "ZooMonitor" has been specifically designed by zoo researchers for zoo researchers and is free to download and use for Zoo Association member zoos such as BIAZA, EAZA, AZA, etc. This software can be downloaded onto a mobile phone or tablet and helps with collating data and exporting it whenever required. The software is monitored and updated regularly, so will go further to answering your aims and objectives.

As with any animal research, things can (and will) go wrong – the zoo setting is no exception. Even though the animals are more than likely to be within their enclosures, there are incidents that occur. Examples include surprise births, deaths, escapes, enclosure malfunction, social or hierarchical disruptions, and group alterations (e.g., incoming or outgoing individuals). Another example, in light of the recent pandemic, was the worldwide closures of zoos and aquariums for months at a time. The physical collection of behavioural data by external researchers was near impossible. However, researchers created novel ways of continuing which included using web camera footage (with approved use from the institutions) or keeper-collected data (Williams, Carter, Rendle, & Ward, 2021). With most problems comes a solution, it just might take a little while to find. Optimistically we like to think these problems create great opportunities for zoo researchers to build on their problem-solving skills and resilience as well as their research.

5.3 ANALYSING BEHAVIOURAL DATA

Most research in behavioural ecology is accompanied by methodological issues such as biases, and zoo research is no exception. Despite careful planning in experimental design and data collection, there are sometimes issues in zoo research that cannot be avoided. This does not prevent you from carrying out the research, but there will be considerations that will need to account for such issues, and this can be done during the data analysis stage. Data analysis is essential to every research project, it helps us understand these data and ultimately provides answers to the research questions in hand. There are many steps to data analysis such as data cleansing and data manipulation and we touch on these briefly (Section 5.3.2), but the main aim of this section is to provide guidance to performing statistical analysis on behavioural data.

5.3.1 Statistical tests

The most powerful way to test a scientific question is through the use of statistics. Statistical tests are designed to test statistical hypotheses based on observed data, and there are many available, one suitable for all types of data. So which tests are suitable for zoo data? Zoo animal behavioural data have typical characteristics. For example, they often have a non-normally distributed count or continuous time variable, which is common when recording the presence or absence of a behaviour. However, it really depends on the type of research question and type of data.

It is important to have a solid understanding of your collection methods. Figure 5.1 is a flowchart to help you determine which statistical test is most suitable for your study, and this can be very useful to simplify the potentially complex process. It can be broken down into three main steps:

1) The first key question of this flowchart: **Is your response variable quantitative (e.g., a count or percentage), or is it categorical (e.g., the different species in a zone within an enclosure)?** Popular behavioural methods (e.g., ethograms) use quantitative response variables such as the number of times (count) or the percentage of time that a behaviour was performed.
2) The next thing to think about is the research question itself: **Are you interested in the relationship between variables** (e.g., the effect of the number of zoo visitors on the percentage of time spent pacing)? Correlation and regression are statistical methods that investigate relationships between variables, e.g., the effect of X on Y. **Or does your research question compare groups**, such as sociality between male and female chimpanzees, then tests such as the t-test

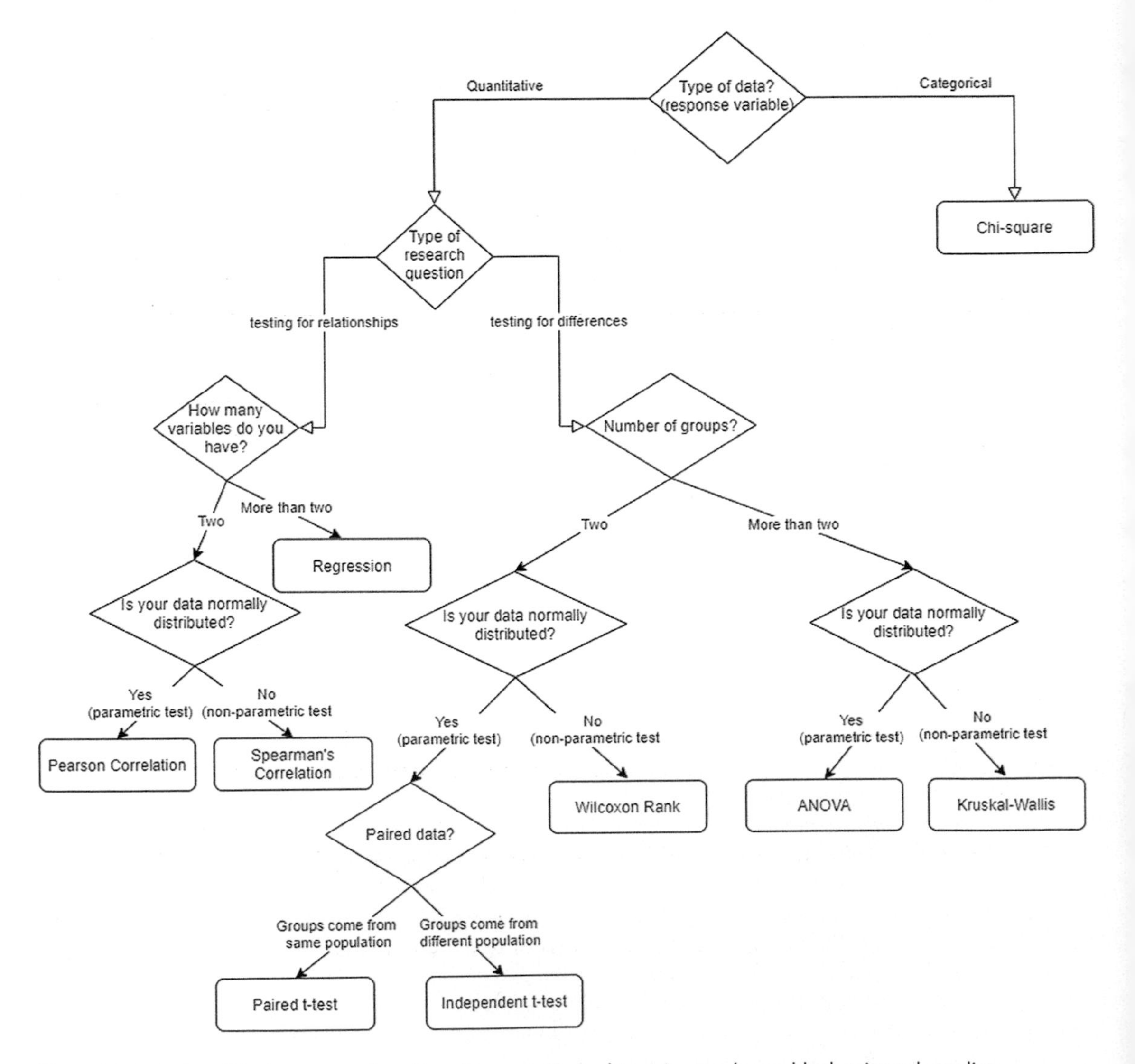

Figure 5.1 A simplified process for choosing a statistical test in zoo-based behavioural studies.

(see Figure 5.1 for paired and independent t-test) or ANOVA (or their non-parametric equivalents) test for a significant difference between the means of groups. But before you can choose one of these tests, you must check the distribution of your data.
3) The final deciding question: **Is your data normally distributed (parametric)?** Histograms are a useful way to visualise the distribution of data, and normality tests (such as Shapiro Wilk) will test if your data are significantly different from a normal distribution (e.g., a bell curve) or not. Non-normally distributed data can be analysed with non-parametric statistical tests that do not assume the data to be normally distributed.

If you follow those three steps for each research question, you should get an understanding of which test you need to use (Figure 5.1). For example, if you are interested in the relationship between two quantitative variables you can measure this using Pearson's correlation, or for non-normally distributed data a rank correlation coefficient such as Spearman's rank, which is less sensitive to non-normal distributions. Regression modelling is another popular statistical technique in ecology to analyse complicated relationships. However, if you are comparing groups and your data is not normally distributed, you can use Wilcoxon Rank for testing the difference between the means of two groups, and Kruskal Wallis for more than two groups.

A common statistical violation in zoo data is a lack of data independence. Independent data are when data points are not correlated i.e., the data points do not influence other data points. As we sample from the same enclosure and population, we need to consider the independence of data as most statistical tests assume independent data points. The importance of this really depends on the research question; however, data independence can be improved by increasing sampling intervals (e.g., a sampling period of one day is more than enough to obtain independent data points), sampling multiple populations from other zoos, or sampling multiple individuals within the same group. For further details on data independence and zoo animal research, please see Plowman (2008) for an explanation of BIAZA's Zoo Research Guidelines. Chapter 5 (Section 5.1) of these guidelines provides information specific to issues with data independence (Dow, Engel, & Mitchell, 2006).

5.3.2 R

R is a versatile and open source (free-to-use) software popular amongst biologists, which is capable of a variety of statistical analyses and building stunning visualisations. R is also a programming language where the analysis is performed by writing code. Therefore, learning how to code in R is a skill that takes some time to develop, but as statistics are essential to evaluating behavioural research, it makes sense to invest the time to learn the most effective ways to perform these analyses. The remainder of this section covers the steps to analyse behavioural biology data in R at a higher overview level, with the assumption that the reader knows R to at least a beginner level.

Imagine data collection has finished, and you no longer have to stand in the pouring rain and write down which giraffe *(Giraffa camelopardalis)* is standing by which. You now have a dataset and can begin data analysis, which can be broken down into five steps:

1. Import data
2. Clean data
3. Transform data
4. Statistical tests
5. Visualisation

We have included code snippets from each stage of the analysis from an example that uses a dataset of behavioural responses of rhinos *(Diceros bicornis)* following the introduction of a new animal.

Step one is to import your data. Your data may be stored in an Excel spreadsheet or CSV file on your computer or in the cloud. Depending on your data's file format you may use packages such as *xlsx* to help you do this.

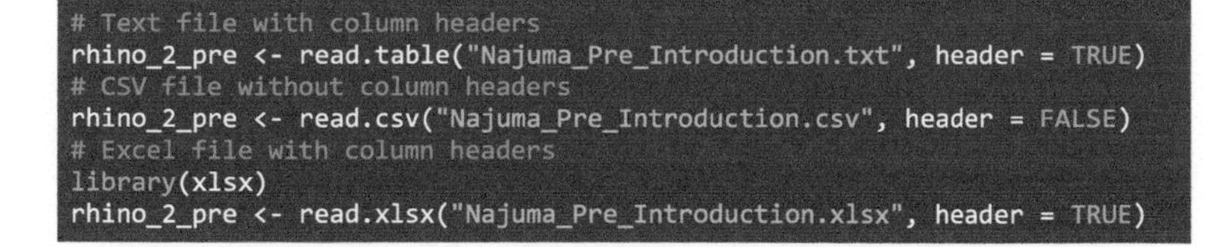

```
# Text file with column headers
rhino_2_pre <- read.table("Najuma_Pre_Introduction.txt", header = TRUE)
# CSV file without column headers
rhino_2_pre <- read.csv("Najuma_Pre_Introduction.csv", header = FALSE)
# Excel file with column headers
library(xlsx)
rhino_2_pre <- read.xlsx("Najuma_Pre_Introduction.xlsx", header = TRUE)
```

Once you have a data frame in R, you need to check for data issues such as outliers, typos, etc. Even if you are confident, it is important to clean your data, as this exercise helps you get familiar with your data and its structure. If you find a typographical error, correct it, and if you find an outlier, consider removing it.

```
# View a summary
summary(rhino_2_pre)
# Check the structures of your variables
str(rhino_2_pre)
# Look out for spelling mistakes by checking the unique values
names(rhino_2_pre)
unique(rhino_2_pre$Locomotion)
```

There are thousands of packages on CRAN (official R package library: https://cran.r-project.org/index.html), and it can be quite overwhelming to choose the right packages to use. In behavioural biology, a very useful suite of packages is the *tidyverse*. Within the *tidyverse* include the packages *dplyr* for data manipulation and *ggplot2* for data visualisation. In the example, the rhino introduction data is stored in two data frames, before and after the introduction of a new animal. We want to merge these data into one data frame and then transform it into a *long* format (one column for the observed variable, *behaviour* in the example below), as required by *ggplot2* and most statistical functions in R.

Now that your data are wrangled (tidied and transformed), you are ready for visualisation and testing. It can be helpful to create some visualisations first to help determine how to test your data, for example, a scatter plot to understand the relationships between two variables, or a histogram to check the normality of your data. Another package in the *tidyverse* is *ggplot2* and it is great for data visualisation. Visualisations may also be the last thing you do as you spend (literally) hours trying to get the perfect looking graph for your report.

Let's take the rhino data as an example. For simplicity, our main research question might be: *Has Rhino 2's locomotion time changed since another rhino was introduced?* As explained above, we need to check the features of these data to find an appropriate test for this question. (1) We know these data are quantitative because they are measured in time (seconds). (2) As expected, there is a left-skewed distribution towards zero seconds (i.e., not normally distributed). (3) We have two groups that we are

```
# Label the data sets with their treatment
rhino_2_pre$treatment <- "pre"
rhino_2_post$treatment <- "post"

# Merge and convert to long format
rhino_2_pre_post <- rhino_2_pre %>%
  merge(rhino_2_post, all = TRUE) %>%
  pivot_longer(!c("date", "treatment"), names_to = "behaviour",
               values_to = "time_sec")

> rhino_2_pre_post
# A tibble: 165 x 4
   date       treatment behaviour          time_sec
   <date>     <chr>     <chr>                 <dbl>
 1 16/09/2020 pre       Locomotion              121
 2 16/09/2020 pre       Stereotypy             2167
 3 16/09/2020 pre       Other                    30
 4 16/09/2020 pre       Environ_Interac         194
 5 16/09/2020 pre       Communication            52
 6 16/09/2020 pre       Feeding                 376
 7 16/09/2020 pre       Rest_Sleep                0
 8 16/09/2020 pre       Courtship.Mating        275
 9 16/09/2020 pre       Browse.present            0
10 16/09/2020 pre       Not.visible             385
# ... with 155 more rows
```

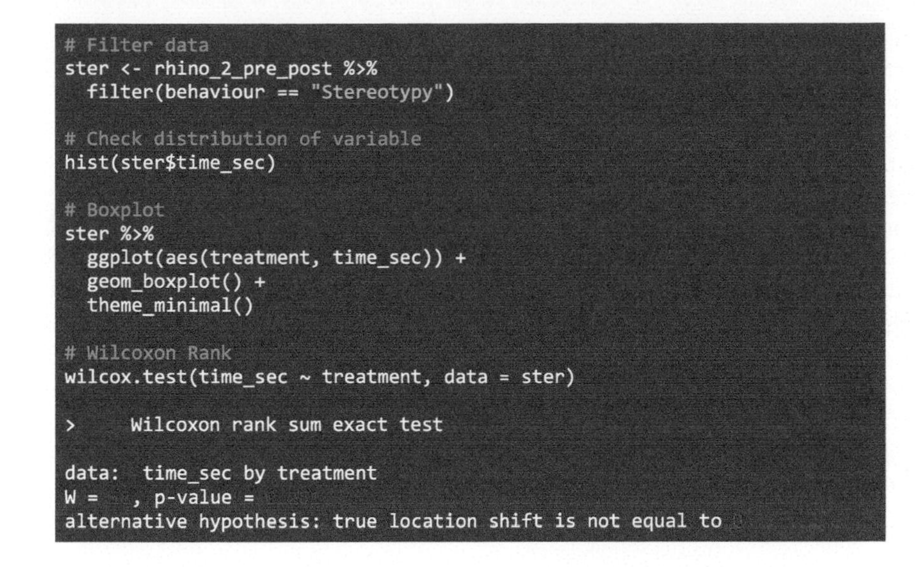

```
# Filter data
ster <- rhino_2_pre_post %>%
  filter(behaviour == "Stereotypy")

# Check distribution of variable
hist(ster$time_sec)

# Boxplot
ster %>%
  ggplot(aes(treatment, time_sec)) +
  geom_boxplot() +
  theme_minimal()

# Wilcoxon Rank
wilcox.test(time_sec ~ treatment, data = ster)

>     Wilcoxon rank sum exact test

data:  time_sec by treatment
W =    , p-value =
alternative hypothesis: true location shift is not equal to
```

comparing, pre-introduction and post-introduction. (4) Our samples are independent (enough) as there is only one per day. Therefore, by referring to the diagram (Figure 5.1), this brings us to a Wilcoxon Rank (Figure 5.2). As the p-value is less than 0.05, we can conclude that there is no difference between the treatment groups, i.e., the rhino introduction did not significantly change the amount of time Rhino 2 spent stereotyping. Remember, it would be unwise to repeat this test for each behaviour (Locomotion, Feeding etc.), as this increases the probability of a Type I error – falsely rejecting the null hypothesis.

For more in-depth guidance on analysis in R, we recommend R for Data Science (Wickham & Grolemund, 2016).

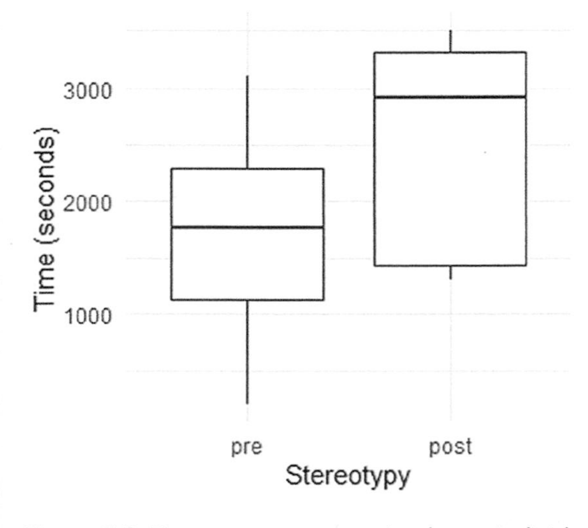

Figure 5.2 Time spent stereotyping by an individual rhino pre- compared to post-rhino introduction.

5.4 CONCLUSION

In this chapter, we have briefly described how to collect and analyse behavioural data from the zoo. We have covered a wide range of methods and touched on some examples. There are lots of methods out there, and one of the challenges in research is choosing appropriate techniques to enable you to answer the research question. A successful zoo-based research project is achieved with good planning and a thorough understanding of your study, focusing on the following key points.

- Decide on your research question(s), considering the target zoo and existing research to support your ideas.
- Think about what data you will need to answer these research questions.
- Plan how you will collect these data and then trial your data collection method(s) in a pilot study.
- Familiarise yourself with these data and then choose statistical testing that is most appropriate to the successful (and valid) answering of your research question(s).

REFERENCES

Bateson, P. & Martin, P. (2021). *Measuring behaviour an introductory guide* (4th edition). Cambridge University Press, Cambridge, UK.

Dow, S. Engel, J., & Mitchell, H. (2006). Autocorrelation, temporal independence, and sampling regime. In Plowman, A.B. (Ed.), *Zoo research guidelines: Statistics for typical zoo datasets*. BIAZA, London, UK.

Percie du Sert, N., Hurst, V., Ahluwalia, A., Alam, S., Avey, M.T., Baker, M., Browne, W.J., Clark, A., Cuthill, I.C., Dirnagl, U., Emerson, M., Garner, P., Holgate, S.T., Howells, D.W., Karp, N.A., Lazic, S.E., Lidster, K., MacCallum, C.J., Macleod, M., Pearl, E.J., Petersen, O.H., Rawle, F., Reynolds, P., Rooney, K., Sena, E.S., Silberberg, S.D., Steckler, T., & Würbel, H. (2020). The ARRIVE guidelines 2.0: Updated guidelines for reporting animal research. *PLoS Biology, 18*(7), e3000410.

Plowman, A.B. (2008). BIAZA statistics guidelines: Toward a common application of statistical tests for zoo research. *Zoo Biology, 27*(3), 226–233.

Rose, P.E., & Riley, L.M. (2021). Conducting behavioural research in the zoo: A guide to ten important methods, concepts and theories. *Journal of Zoological and Botanical Gardens, 2*(3), 421–444.

Sherwen, S.L., & Hemsworth, P.H. (2019). The visitor effect on zoo animals: Implications and opportunities for zoo animal welfare. *Animals, 9*(6), 366.

Wickham, H., & Grolemund, G. (2016). *R for data science*. O'Reilly Media, Sebastopol, USA.

Williams, E., Carter, A. Rendle, J., & Ward, S.J. (2021). Impacts of COVID-19 on animals in zoos: A longitudinal multi-species analysis. *Journal of Zoological and Botanical Gardens, 2*(2), 130–145.

PART II

Selected taxonomic accounts

DOI: 10.1201/9781003208471-7

6

The behavioural biology of primates

LISA M. RILEY
University of Winchester, Winchester, UK

6.1 INTRODUCTION TO PRIMATE BEHAVIOURAL BIOLOGY

The order Primates show remarkable diversity. Following the evolution of the first primates 65 million years ago, around 500 species, part of ~60 genus, and ~14 families, are now recognised (Mittermeier, Rylands, & Wilson, 2013). Table 6.1 shows the primate evolutionary tree from suborder to genus. Some primates exhibit typical anatomy and behavioural ecology, but it is worth noting that there are cases of extreme specialism, for example, the golden snub-nosed monkey (*Rhinopithecus roxellana*). As a Colobine (florivorous), these monkeys have a compartmentalised stomach to support bacterial digestion of lichen and leaves, but they also have dense fur and, as their name suggests, a much-reduced nose, possible adaptations to the unique snowy climate they endure. Yet even "typical" primates, fruit-eating, tropical forest-dwelling species, evolved to fulfil a specialised ecological niche – the location, feeding opportunity and "job" an animal does in the wider ecosystem. For example, red-faced black spider monkeys (*Ateles paniscus*) may aid forest regeneration via seed dispersal of fruiting species like *Brosimum lactescens*, a main constituent of the spider monkey diet. Selection pressures in the wild have shaped every extant primate, like all undomesticated animals on the planet. To understand the interaction between genetics and the environment, primate evolution must be considered.

Proto-primates, or *Plesiadapiformes*, were the first primate-like mammals to evolve. In the early Paleocene Epoch, the beginning of the Cenozoic Era 65–55 million years ago, small insectivorous, arboreal, placental mammals, the tree shews (order *Scandentia*), had evolved. Global changes in climate had driven the evolution of broad-leaf forest and this created an arboreal niche, driving primate evolution (the Arboreal Hypothesis). A vast proliferation of angiosperms (flowering plants) later provided an array of edible seeds, flowers,

DOI: 10.1201/9781003208471-8

Table 6.1 Primate evolutionary tree, after Perelman et al. (2011) and Mittermeier et al. (2013).

Suborder: Haplorrhini
Infraorder: Simiiformes
Parvorder: Catarrhini (Old World Monkeys, Gibbons, Great Apes)

Family	Subfamily	Genus
Hominidae (Great Apes)	Homininae	*Pan* (Chimpanzee)
		Gorilla (Gorilla)
	Ponginae	*Pongo* (Orangutan)
Hylobatidae (Lesser Apes)		*Hylobytes* (Lar gibbon)
		Hoolock (Hoolock gibbon)
		Nomascus (Crested gibbon)
		Symphalangus (Siamang)
Cercopithecidae (Old World Monkeys)	Cercopithecinae	*Papio* (Baboon)
		Theropithecus (Gelada)
		Mandrillus (Mandrill and drill)
		Lophocebus (Crested mangabey)
		Cercocebus (Mangabey)
		Macaca (Macaques)
		Erythrocebus (Patas monkey)
		Chlorocebus (Vervet)
		Allenopithecus (Allan's swamp monkey)
		Miopithecus (Talapoin monkey)
		Allochrocebus (Terrestrial guenon)
		Cercopithecus (Arboreal guenon)
	Colobinae	*Pygathrix* (Douc langur)
		Nasalis (Proboscis monkey)
		Rhinopithecus (Snub-nosed monkey)
		Semnopithecus (Lutangs, langurs, leaf monkey)
		Trachypithecus (Lutangs, langurs, leaf monkey)
		Presbytis (Surilis)
		Piliocolobus (Red colobus)
		Colobus (Colobus)

Parvorder: Platyrrhini (New World Monkeys)

Family	Subfamily	Genus
Callitrichidae		*Mico* (Marmoset)
		Cebuella (Pygmy marmoset)
		Callithrix (Marmoset)
		Callimico (Goeldi's monkey)
		Leontopithecus (Lion tamarin)
		Saguinus (Tamarin)
Aotidae		*Aotus* (Night monkey)
Cebidae		*Saimiri* (Squirrel monkey)
		Cebus (Capuchin monkey)
Atelidae		*Lagothrix* (Woolly monkey)
		Brachyteles (Muriquis)
		Ateles (Spider monkey)
		Alouatta (Howler monkey)

(*Continued*)

Table 6.1 (Continued) Primate evolutionary tree, after Perelman et al. (2011) and Mittermeier et al. (2013).

Family	Subfamily	Genus
Pitheciidae		*Cacajao* (Uakari)
		Chiropotes (Bearded saki)
		Pithecia (Saki monkey)
		Callicebus (Titi monkey)
Infraorder: Tarsiiformes		
Family	**Subfamily**	**Genus**
Tarsiidae		*Tarsius* (Tarsier)
Suborder: Strepsirrhini		
Infraorder: Lemuriformes		
Family	**Subfamily**	**Genus**
Lemuridae		*Lemur* (Lemur)
		Hapalemur (Bamboo lemur)
		Eulemur (True lemur)
		Varecia (Ruffed lemur)
Cheirogaleidae		*Cheirogaleus* (Dwarf lemur)
		Microcebus (Mouse lemur)
		Mirza (Giant mouse lemur)
Lepilemuridae		*Lepilemur* (Sportive lemur)
Indriidae		*Propithecus* (Sifaka)
		Avahi (Woolly lemur)
		Indri (Indri)
Infraorder: Chiromyiformes		
Family	**Subfamily**	**Genus**
Daubentoniidae		*Daubentonia* (Aye-aye)
Infrorder: Lorisiformes		
Family	**Subfamily**	**Genus**
Lorisidae		*Loris* (Slender loris)
		Nycticebus (Slow loris)
		Perodicticus (Potto)
		Arctocebus (Angwantibo)
Galagidae		*Galago* (Galago or Bushbaby)
		Otolemur (Greater galago)

and subsequently fruits creating further ecological niches causing tree shrews to evolve into colugos (order *Dermoptera*) and proto-primates (the Angiosperm-Primate Coevolution Hypothesis). By the Eocene Epoch (the late Cenozoic Era, 55–33 million years ago) true primates, early prosimians, had evolved and differed from their *Plesiadapiforme* cousins by having grasping hands and feet for greater climbing efficiency and slightly enhanced stereoscopic vision thanks to an increasingly forward-facing eye position, an adaptation most likely for judging distance in their arboreal habitat. With little competition prosimians underwent adaptive radiation as they exploited increasingly more microniches (fine-grade differentiation in feeding opportunities), developing ever larger brains and smaller snouts, and diversifying their diet to include the abundance of plant material surrounding them. During the very end of the Eocene Epoch monkeys, the first Anthropoid primates, evolved. For millennia, the global climate was warm and humid, with widespread temperate and tropical forests; perfect habitats to support the emerging primate radiations.

Primates have thus evolved to become a highly successful group of mammals that now occupy an impressive range of habitats and climate ranges such is their adaptive fitness. From high altitude Ethiopian grasslands where gelada baboon (*Theropithicus gelada*) graze, and the primary rainforests of South America where Bolivian squirrel monkey (*Saimiri boliviensis*) live, to transitional thorny desert forests in Madagascar where ring-tailed lemurs (*Lemur catta*) are found, and the Bornean coastal dipterocarp and mangrove forests in which proboscis monkeys (*Nasalis larvatus*) reside, the range of habitats, together with the range of genus and species, aptly shows the collective fitness of the order Primates.

In modern times, primates continue to evolve, though selection pressures have diversified. Human-led habitat change, global warming, pollution, and direct conflict between human and non-human primates have often devastating effects on primate populations – so quick and so great are these changes, primates have little time or opportunity to evolve. Mason et al. (2013) liken the selection pressures of captivity to human-induced rapid environmental change in the wild. Management strategies and husbandry practices, even with the best of intentions, cause phenotypic changes in captive-bred individuals within just one generation, diminishing wild-type fitness. If we are to better meet the needs of primates in captivity, we need to fully appreciate what a primate *is* and what changes in primate anatomy and behaviour are adaptive (have improved primate fitness).

6.2 PRIMATE ECOLOGY AND NATURAL HISTORY RELEVANT TO THE ZOO

There is no single physiological or behavioural trait present, or present to the same degree, in all primates. Wild selection pressures have created a set of primate-defining characteristics and the combination of these traits leads to classification within the order Primates. The evolution of each trait is inextricably linked to the ecological processes and habitat of the primate and traits often interact and share the same causal ecological process. Six such traits exist: (1) Big brains; (2) Social complexity; (3) Extended life histories; (4) Foraging complexity; (5) Visual dominance; (6) Hands and feet (summarised in Figure 6.1).

6.2.1 Big brains

Primates exhibit allometric brain expansion; they have larger-than-expected brains relative to their body size. Particularly, the neocortex, where complex thought, language and emotional processing occurs, has undergone considerable expansion in primates compared with most other animals. While all extant primates show this encephalisation, differentiation among primates exists; prosimians (lemurs, lorises and galagoes) have more typical mammalian-sized brains, while monkeys, lesser apes, and great apes show increasing brain expansion respectively. As well as the development of sophisticated foraging strategies, big brains are associated with sophisticated social interaction, tool use, culture, and greater conscious awareness of self and others. Primates are among the most sentient of animals, and when kept in captivity their cognitive, emotional, and social needs must be prioritised. To understand possible ways of achieving this, theories of brain expansion must be discussed, and therefore behavioural ecology must be considered. In their review, Dunbar and Shultz (2017) discuss technical ("instrumental") and social hypotheses of brain expansion. Primates developed large brains to allow food innovation, extractive foraging, tool-mediated foraging, to exploit patchily distributed food, and to develop complex social behaviours like Machiavellian tactical deception (Byrne & Whiten, 1994). Thus, ecology has both directly (via food availability) and indirectly (via sociality) affected brain expansion, leading to adaptive success.

6.2.2 Social complexity

Primates have complex social organisation (e.g., Swedell, 2012). Few primates are solitary (e.g., galagoes, mouse lemurs and orangutans), and monogamous pair-bonded groups are rare (though Hylobatidae and some members of Callitrichidae like titi monkeys are notable exceptions). Typically, one-male multifemale or multimale-multifemale groups, where polygyny and polygynandry are common, are formed. Most species are female philopatric (females remain in the natal group, males emigrate), few are male philopatric (*Pan* spp. both well-known exceptions), while for monogamous species both sexes emigrate. Primates therefore manage multiple relationships throughout their lifetime. While impressive, diverse social structure is not unique to primates – the sophisticated

Figure 6.1 Primate characteristics. Top: Crested macaque (*Macaca nigra*) demonstrating extractive foraging. Note the large brain case, dextrous hands and feet with nails and forward-facing eyes. The arms and legs of the macaque are proportionate and the upright position show this is a quadrupedal species. Bottom left: Ring-tailed lemur showing the muzzle-like nose profile typical of Strepsirrhines (prosimians), a feature retained despite early migration of the eyes to the front of the head. Bottom right: Suspensory posture of lar gibbon (*Hylobytes lar*): note the large brain case of this lesser ape and the elongated arms that are hypermobile with precision hand grip to aid brachiation.

(indicative of higher-order cognitive function) interactions between primates are unique.

Primates are political. As highly sentient beings they think about social gain and strategise profitable interactions (de Waal, 2007). Those in a social group must cooperate with each other forming alliances (friendships) reinforced by allogrooming, and in some species co-parenting (e.g., callitrichids), alloparenting (e.g., vervet monkey, *Chlorocebus aethiops*) and food sharing (e.g., male Taï Forest chimpanzees, *Pan troglodytes verus*) with non-kin. But primates also compete with group members and evidence tactical deception – deceiving dominant individuals to gain resource access or mating opportunity (Byrne & Whiten, 1994), scapegoating, and confiscating infants to avoid aggression (e.g., *Papio* spp., *Macaca* spp., *Gorilla* spp., and *Pan* spp.). Working out who to pledge allegiance to at times of social instability can spell social success or social ostracism, and ultimately life or death, and so primates think tactically to optimise political opportunity both now and in the future.

Providing social opportunity in captivity is essential for primate wellbeing. The focus should not solely be on group structure but on types of social interaction and designing enclosures that provide the opportunity for complex social cognition.

6.2.3 Extended life histories

Primates have extended life histories compared to other animals that are allometrically scaled with body size and brain size. The bigger the primate and the bigger their brain, the longer their life history. Life history refers to age at weaning, sexual maturity, and first birth, characteristics of the oestrus cycle, gestation length, birth interval, and life span. Protractive maturation in primates is likely due to brain expansion. Brains are energetically expensive, and it takes time to partition the required nutrients both *in utero* (hence longer gestations) and in infancy (hence delayed sexual maturity of the infant and extended interbirth interval for the mother), and time is needed to develop sophisticated cognitive characteristics associated with a big brain. Matsuzawa et al. (2008) discuss the Master-Apprenticeship hypothesis of tool-mediated oil palm nut-cracking behaviour in Bossou chimpanzees (*P. t. verus*) where an infant repeatedly observes their mother demonstrating such tool use and may, after three to five years acquire the skill proficiently via social learning. Extended life histories are adaptive – learning tool-mediated nut cracking allows utilisation of an important protein and fat source that would otherwise be inaccessible, hence in an evolutionary arms race, tool-using chimpanzees would theoretically be more likely to survive. Yet in terms of conservation, the generation length of primates is well beyond typical mammalian population recovery times, 25 years for chimpanzees, for example (Humle, Maisels, Oates, Plumptre, & Williamson, 2016). In captivity, we must value these extended life histories and consider them as a life-long learning opportunity.

6.2.4 Foraging complexity

Primates are typically omnivores and consequently the gastrointestinal (GI) tract in most species is generalised (like any monogastric mammal). Fruit is the mainstay for many primates, though many other items are also eaten. Some dietary specialists have evolved; the graminivorous (grass-eating) gelada baboon, gumnivorous callitrichids (*Saguinus* spp., *Callithrix* spp.), leaf-eating specialists (superfamily *Colobinae*) and those who are partly insectivorous (galagoes, *Galago* spp., and *Otolemur*; dwarf lemurs, *Cheirogaleus* spp.; sportive lemurs, *Lepilemur* spp.; aye-aye, *Daubentonia* spp.). Only the tarsier (*Tarsius* spp.) is carnivorous, eating insects, snakes, lizards, birds, and bats. By diversifying their diets, primate species can be sympatric (overlap in habitat) and use an array of microniches (e.g., from the terrestrial substrate to the emergent canopy, from tree trunk to terminal branches). GI tracts and dentition do show some adaptations, for example in the gum-eating marmosets the incisors are elongated and chisel-like to effectively gnaw tree bark, and the caecum is extended to aid absorption of exudates. This diet has also caused behavioural changes – gum eaters maintain territories with core tap-trees where gum is regularly harvested. Monogamy helps maintain territory and this means male callitrichids are present to share parental duties, leading uniquely in primates to the birth of twins. Thus, diet is a determinant of dentition, GI tract anatomy, feeding behaviour, ranging habits, social organisation, and reproductive output.

Due to their large brains and associated cognitive abilities, primates profit from a wide variety of food resources not directly accessible. Primates specialise at extractive foraging, overcoming a plant's defences to access food embedded in a matrix of inedible material – such as *Neesia* fruits that Sumatran orangutans (*Pongo abelii*) eat using a stick tool to pry the fruit open and rub off irritating hairs, thus accessing the edible seeds (van Schaik et al., 2003). Food innovation and complex foraging is seen throughout the order Primates but particularly in Anthropoids including baboons, macaques, and apes. Wild primates spend 40–60% of their day foraging, showing how important this behaviour is, and how important the cognition and habitat characteristics driving foraging are. In captivity food is prepared and delivered to primates with typically minimal opportunity for extractive foraging; as primates demonstrate contra-freeloading phenomenon (performing food-processing actions using food already processed), foraging appears to be a highly motivated behavioural need.

6.2.5 Visual dominance

Most mammals rely on olfactory communication, but primates show visual dominance. The eye location in primates is forward-facing not lateral as in other mammals and this coincides with a reduction in muzzle size (and one/two less molars). Olfaction remains important for all primates, but visual dominance brings several advantages. The migration of the eyes to the front of the head creates stereoscopic vision for judging distance, and primates have pronounced visual acuity, allowing precise prey capture and objects manipulation. All Old World monkeys and apes have trichromatic vision (long, medium, and short wavelengths of light detected by three cone types, meaning red, green, and blue-yellow colours are visible). Trichromacy is intermittent in New World Monkeys and Strepsirrhines, who, are otherwise dichromats. The purpose of trichromacy may be to distinguish ripe fruit or leaf buds, to detect predators or aid sexual selection. Hence, primate visual communication signals are multiple and trichromats have evolved striking markings, like the African guenons (*Cercopithecidae* Family) – a group of typically arboreal, frugivorous monkeys ranging from the aptly named moustached guenon (*Cercopithecus cephus*) and putty-nosed guenon (*C. nictitans*) to the striking white brow ridge, chest and hip strips of the Diana monkey (*C. diana*) to the some-what wizard-like quality of the red and black cap and long white beard of the De Brazza's monkey (*C. neglectus*). De Brazza's monkeys live in pairs while typically guenons form one-male multi-female groups. The bright coloration and distinctive markings of these animals have important roles to play in individual and species recognition, helping guenons avoid hybridisation, and may have a role to play in male-to-male competition and sexual selection as we know skin coloration in primates, like red-faced bald uakaris (*Cacajao calvus*) does. When one lives in large groups, in forests where food sources allow multiple similar species to co-exist, visual communication signals, and recognising kin and friends is vitally important for survival. In captivity, the selection pressures that led to this suite of visual changes need to be maintained and therefore exact habitat type, social group structure, and community ecology are important husbandry considerations.

6.2.6 Hands and feet

Primates have five digits on each hand and foot, that, along with wrists/ankles are highly mobile compared with other mammals. With opposable thumbs and big toes, and flat nails rather than claws, primates have precision grips and incredible manual dexterity which supports extractive foraging and an arboreal lifestyle. Greater mobility in the arm provided by the clavicle and positioning of the arms to the side of the body further helps with arboreal locomotion and extractive foraging. This coincides with the upright posture of primates – the foramen magnum (hole in the skull through which the spinal cord passes) shows migration from the back to the base of the skull, hence primates can sit upright freeing their hands for foraging. These adaptations have also influenced primate locomotion. Prosimians are vertical clingers and leapers propelling themselves from tree to tree using their elongated legs while the back remains vertical. Most other primates, including the Anthropoid monkeys and apes, are typically quadrupeds (either arboreal or terrestrial) where legs are used for propulsion and arms for steering. For balance the legs and arms are of similar proportion, providing speed and agility. Suspensory locomotion is seen in some species including lesser apes (gibbons and siamangs), spider monkeys and muriquis (*Brachyteles* spp.). In dense forest canopies where fruit is well-dispersed this energetically expensive mode of locomotion is quick and saves time, with the sugary fruits able to replace lost energy. Suspensory primates have elongated arms, hypermobile shoulder and wrist joints with long slender fingers to cup neatly round horizontal branches when moving at speed. In captivity, enclosure design must encourage appropriate modes of locomotion but also the extent of locomotion – if, in the wild, the primate leaps, climbs or swings great distances then for general health and fitness this needs to be possible in captivity not in terms of distance travelled but in terms of energy expenditure and motivation.

Any species of primate can have any combination of these traits and to understand how these traits combine for effective functionality, Table 6.2 summarises behavioural ecology for representative species commonly kept in captivity globally.

Table 6.2 Behavioural ecology (life history variables, habitat, and niche) and associated husbandry considerations of representative species or subspecies of the order Primates. (Information on life history, habitat and niche sourced from Rowe et al. 1996; IUCN Red List (www.iucnredlist.org)). Body mass: male, female. Group size: average or typical range.

Classification	Behavioural ecology	Husbandry consideration
Genus: *Pan* Subspecies: Western chimpanzee (*P. t. verus*) Body mass: 40–60 kg, 32–47 kg Life span: 53 years Oestrus cycle: 36 days Female sexual maturity: 135 months Gestation: 240 days IUCN Red List Status: Endangered	Range country: West Africa, Senegal to Ghana Habitat type: Tropical rainforest, woodland, and savannah Diet type: Omnivorous (up to 75% fruit) Social structure: Fission-Fusion Group size: 35 Philopatry: Females emigrate Mating System: Promiscuous Behavioural adaptations: Large brain. Cooperative hunting and meat-sharing, tool use, culture, self-awareness, tactical deception, reconciliation behaviour. Allomothering. Diurnal. Arboreal (feeding), terrestrial (travelling)	Provide a complex, diverse, and large enclosure allowing terrestrial patrolling and arboreal feeding Create enclosure zones each with abundant nesting opportunities and feeding platforms for individuals to disperse and affiliate when so motivated Provide feeding enrichment – puzzle feeders, simulated hunting and tool use, problem-solving opportunity – to stimulate cognition and feeding behaviour Encourage culture – provide problem-solving, cooperation, and competition tasks
Genus: *Hylobates* Species: Lar gibbon (*H. lar*) Body mass: 4.9–7.6 kg, 4.4–6.8 kg Life span: 44 years Oestrus cycle: 27 days Female sexual maturity: 108 months Gestation: 204 days IUCN Red List Status: Endangered	Range country: North Sumatra, Peninsular Malaysia, Myanmar, Thailand, Southern China Habitat type: Evergreen, semi-evergreen, and mixed evergreen-deciduous forest in emergent and mid-canopy Diet type: Frugivorous Social structure: Pair-bonded family group Group size: two to five Philopatry: Both emigrate Mating System: Monogynous Behavioural adaptations: Brachiation and bipedal walking on branches. Diurnal and arboreal. Territorial, with territorial duetting. Suspensory feeding. Large brain	Multiple high-mid-canopy height branch complexes (dense collections of branches) with horizontal branching of variable circumference to promote brachiation Sufficient space to allow sustained brachiation Swings between branch complexes Aerial feeding platforms suspended under a single branch to encourage suspensory feeding (entire body freely hanging under one/two arms) Allow territory to be established in visual isolation from other pairs No ground feeding

Genus: *Macaca* Species: Crested macaque (*M. nigra*) Body mass: 6–10 kg, 3.5–5.5 kg Life span: 20 years Oestrus cycle:36 days Female sexual maturity: 49 months Gestation: 174 days IUCN Red List Status: Critically endangered	Range country: Sulawesi, islands of Manado Tua and Takise Habitat type: slightly elevated rainforest, dense vegetation Diet type: Frugivorous (70% of diet) Social structure: Multimale-multifemale or one-male-multifemale Group size: five to 25 Philopatry: Males emigrate Mating System: Promiscuous Behavioural adaptations: Female ano-genital swelling. Quadrupedal, diurnal, terrestrial (60%) and arboreal (40%). Form large social groups. Large day ranges. Genus-level evidence of tool use and innovation.	Furniture should include ground and understory quadrupedal walking opportunity connected by boardwalks or paths in vegetation – ground cover is important for privacy and to encourage ranging. Long enclosures with dispersed aerial food platforms would encourage wild-type feeding and climbing to reach food. Diverse substrates include grass, shrubs, rocks, and dirt paths to encourage quadrupedal running and climbing Some problem-solving opportunities (e.g., food out of reach or hidden) should be supplied
Genus: *Colobus* Species: Black and white colobus (*C. guereza*) Body mass: 13.5 kg, 8.5 kg Life span: 22 years Oestrus cycle: 24 days Female sexual maturity: 4 years Gestation: 158 days IUCN Red List Status: Least concern, but decreasing	Range country: Across central Africa Habitat type: Primary and secondary deciduous and evergreen forest, wooded grassland. Rest in emergent canopy Diet type: Florivorous. Low variation of plant species Social structure: Unimale-multifemale (rarely multimale-multifemale) Group size: three to 15 Philopatry: Males emigrate Mating System: Polygynous Behavioural adaptations: Arboreal, diurnal, territorial. Social groups are aggregations with interaction sparse Gut specialisms (sacculated stomach) allow mature leaf eating. Vestigial thumbs. Quadruped, leaping	Provide complex branching clusters (variable circumference, variable angle of suspension) at mid-high enclosure heights with forks for resting and large gaps between complexes to encourage leaping Food (high fibre, leaves) should be well-dispersed and branch complexes should be large to allow group members to sit apart from each during feeding Food should be presented at height with leaves on twigs so they can be picked. Food in abundance, can be lower quality, and at times limited in type

(*Continued*)

Table 6.2 (Continued) Behavioural ecology (life history variables, habitat, and niche) and associated husbandry considerations of representative species or subspecies of the order Primates. (Information on life history, habitat and niche sourced from Rowe et al. 1996; IUCN Red List (www.iucnredlist.org)). Body mass: male, female. Group size: average or typical range.

Classification	Behavioural ecology	Husbandry consideration
Genus: *Saimiri* Species: Black-crowned Central American squirrel monkey (*S. oerstedii*) Body mass: 0.6–1.25 kg, 0.6–0.9 kg Life span: 15–20 years Oestrus cycle: Seasonal (synchronised female receptivity) Female sexual maturity: 2.5 years Gestation: 145 days IUCN Red List Status: Endangered	Range country: Costa Rica, Panama Habitat type: Seasonally inundated forests, river edge forest, floodplain, and secondary forests Diet type: Insectivorous-frugivorous. Small fruits from mid to low canopy. Can spend 70% of day hunting insects. Pick insects off leaves Social structure: multimale-multifemale, low aggression-egalitarian outside of breeding season Group size: 20–75 Philopatry: Females emigrate Mating System: Polygynandry Behavioural adaptations: Male 20% weight gain two months prior to breeding season. Quadrupedal, diurnal, mostly arboreal prefer branches 1–2 cm in diameter. Highly cohesive group	Need dense vegetation covered enclosure, particularly at understory and lower canopy level. Small branches suspended at multiple angles Live insect food supplemented with fruit. Insects should be free ranging in the enclosure Feeding must increase 2-months prior to breeding season for males to bulk-up Large enclosure to support large group sizes
Genus: *Daubentonia* Species: Aye-aye (*D. madagascariensis*) Body mass: 2.5–2.8 kg Life span: 20 years Oestrus cycle: 21–65 days (non-seasonal) Female sexual maturity: 2 years Gestation: 172 days IUCN Red List Status: Endangered	Range country: Madagascar, entire coastal perimeter of island Habitat type: Highly fragmented. Primary rain forest, deciduous forest, secondary growth, dry scrub forest, and mangrove swamps Diet type: Seeds of ramy (*Canarium* spp.). Insect larvae Social structure: Solitary Group size: one to two Philopatry: Both sexes emigrate Mating System: Data deficient Behavioural adaptations: Percussive foraging, with adapted middle finger with claw for larvae extraction (taps finger along wood, detects vibrations, uses incisors to gnaw bark and middle finger to extract larvae). Nocturnal. Mostly arboreal, will travel on ground. Huge (200 ha) home range. Nest in vine tangles or tree forks during the day. Continually growing incisors. Little social interaction	Large enclosures, solitary living Naturalistic feeding opportunity – insect larvae filled branch feeders (drill hole in branch, fill with larvae, seal with wood shavings and branch exudate or edible glue). Arboreal feeding opportunity Nest boxes, vines, branch forks for resting Reversed lighting Nuts and seeds, shells intact, for nutrition and good dental health Complex branching with vines and undergrowth Grass and wood bark substrate to facilitate terrestrial travel Scent marking and territory – cleaning should not eliminate all scent marks

6.3 ENCLOSURE CONSIDERATIONS BASED ON BEHAVIOURAL EVIDENCE

From their most basic anatomical characteristics to their locomotor and behavioural traits, the external environment determines what primates look like, how they move, what and how they eat, their social complexity, and generally how they utilise their physical and social environment. When maintaining primates in captivity, we must therefore plan enclosures with the specie's specific ecological niche and associated adaptations at the forefront of our minds and be sensitive to the biological needs of the species. Table 6.2 shows husbandry ideas for a range of primates in relation to their behavioural biology.

Proposed here is a philosophical shift in our approach to primate husbandry – the Autonomy Approach (AA). This strives to give control over ranging, social interaction, and feeding back to individual primates (within obvious captive limits), rather than the expression of these behaviours being dictated by the limitations of traditional enclosures and husbandry strategy. For example, food is prepared and delivered directly to primates at particular times under traditional husbandry procedures, rather than the primate determining when they would like to forage and being able to act upon that motivational need. The AA maximises the expression of behavioural adaptations, designing husbandry strategies and enclosures based on adaptive fitness and with the ultimate aim of promoting autonomy. This approach further aims to prioritise primate social needs and cognition, and by stimulating ranging behaviour will promote better physical health. The AA is based on a "podular" enclosure design (Figure 6.2) with multiple enclosure "pods" joined by aerial and/or terrestrial tunnels (dependent upon the locomotory mode of the primate). Indoor and outdoor pods share equal importance (in size and complexity/resources) and are interconnected. As primates are often not housed in range countries, indoor pods offer an opportunity for primates to be maintained in range-like temperatures and humidity while outdoor pods offer fresh air, and vitamin D exposure, hence both indoor and outdoor pods combine to promote good skin and body condition.

A podular enclosure design where primates have physical access to all areas but cannot directly see all areas simultaneously, stimulates exploration and provides reason to travel. If areas can be separated, podular enclosures provide a flexible management tool if an individual needs to be isolated for health or reproductive reasons and offers futureproofing as infants mature or social tensions rise. Pods allow patrolling in territorial animals but also provide personal space and privacy. Groups can disperse and associate when individuals want but individuals can seek privacy when needed – primates need social opportunity to promote good welfare but, the ability to distance oneself and be private when needed is also imperative. Personal space in the wild can be easily achieved and is regularly sort but in captivity providing personal space is traditionally challenging. Social stress is natural, and it is important that captive primates experience social stress (not distress) to develop coping strategies, preparing individuals for future social stressors. Hence, individuals need enclosures where they control how close, or distant, they are to others. Under the AA approach, the podular enclosure design would mimic foraging activity in the wild as all areas need to be explored throughout the day to determine if food is available. This stimulates cognitive function. Key to the AA is promotion of independent feeding and natural feeding opportunity. Crucially, individuals should be able to choose when they eat and within limits what they eat. Independent feeding opportunity should be available in both outdoor and indoor pods planted with species-appropriate food items including arboreal and terrestrial herb gardens, fruiting trees, edible climbers, access to insect colonies, and in general encourage nature (insects, amphibians, birds). In addition, an array of feeding devices should be used to ensure species-appropriate feeding. In these ways, podular enclosure designs promote autonomy. They give control over multiple aspects of life and when to act back to the primate and encourage behaviours relevant to behavioural ecology and evolution. Exact furnishings and enrichment devices must be designed specifically with the behavioural ecology of the resident primates in mind.

Podular enclosure design would also work well for mixed-species exhibits, which are increasingly popular as zoos strive to provide more engaging and educational opportunities. Zoo visitors like naturalistic enclosures – dioramas with an abundance of nature. When sympatric species are housed together in captivity, in addition to considering behavioural ecology of each species,

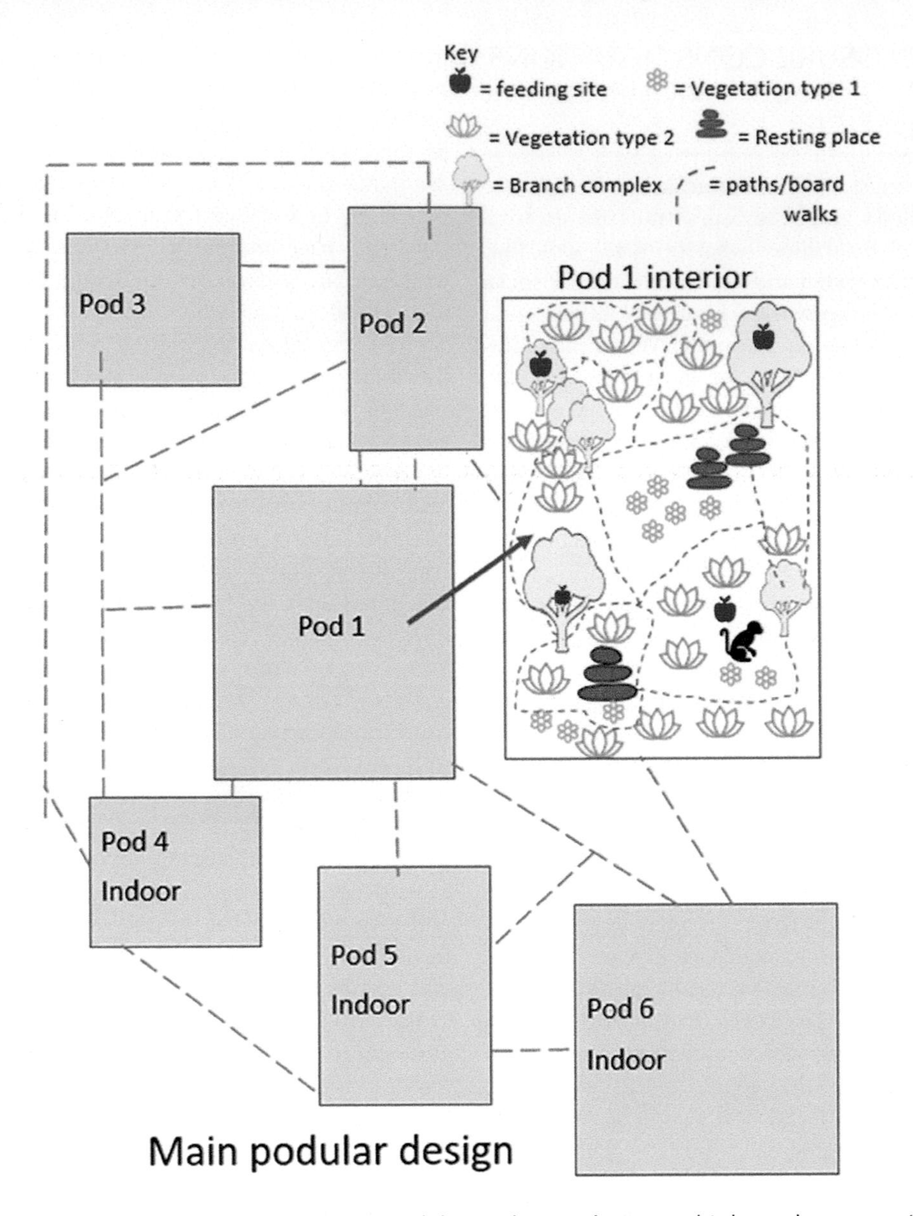

Figure 6.2 "Podular" enclosure design. In a podular enclosure design multiple enclosures or "pods" are linked via terrestrial or arboreal tunnels. This facilitates ranging and social interaction. Under the proposed Autonomy Approach each pod would contain independent feeding opportunity (food present, primate determines when to eat and must forage for the food) and be designed with species adaptations in mind. Pod 1 design would be suitable for a terrestrial/semi-terrestrial quadrupedal primate-like many baboon or macaque species.

community ecology also needs consideration and provision made in each pod. In the wild, sympatric species mitigate direct competition by evidencing niche separation. For example, in Budongo Forest, Northern Uganda, galago (one unidentified dwarf galago), potto (*Perodicticus pottoibeanus*), red tail monkey (*C. ascanius schmidti*), blue monkey (*C. mitis stuhlmanni*), black and white colobus (*Colobus guereza occidentalis*), olive baboon (*P. anubis*) and eastern chimpanzee (*P. t. schweinfurthii*) are sympatric but achieve niche separation. Galago and potto are nocturnal insectivores while all other species are diurnal. Black and white colobus are specialist folivores foraging on mature leaves indigestible to the other species. Chimpanzee, red-tail monkey, and blue money utilise many of the same arboreal forest layers, though chimpanzees also travel terrestrially, and all are omnivorous-frugivorous. While chimpanzees can directly dominate a food resource due to their large body mass, their dietary strategy is food guarding while fruit ripens. Redtail monkey and blue monkey harvest under-ripe fruit and store it in their cheek pouches and can exploit fruit on terminal branches more easily than chimpanzees due to their smaller body sizes.

In captivity, a successful mixed-species podular exhibit needs space to afford opportunity for spatial and temporal distancing and the creation of wild-like microniches to support species-specific feeding, locomotion, rest, and social interaction, and provide species-appropriate challenge/learning.

6.4 BEHAVIOURAL ECOLOGY AND PRIMATE WELFARE

Welfare concerns the biological function, naturalness, and feelings of animals (Fraser, Weary, Pajor, & Milligan, 1997). To achieve good welfare the animal must encounter natural challenge suitable for their adaptations. Thus, aligning husbandry with behavioural ecology would be an excellent platform on which to build better welfare for captive animals.

Primates are active thinkers, when they get the chance to interact, engage, or explore, they do so. As an order they are inquisitive and like to problem solve, innovate, and express their highly developed emotional capacities. Given the long life and cognitive capabilities so characteristic of primates, special consideration of their welfare is required as the suffering they can ensure is profound and can last many years. It is not sufficient to assume that a primate is "typical" and requires a version of a standardised enclosure, usually with a fixed wooden climbing frame, rope or firehose climbing structures, short grass substrate, and resting areas. This default enclosure design may provide basic opportunity but not necessarily the biologically relevant opportunities primates need to thrive. In captivity, evolution does not simply become irrelevant, primates remain motivated to perform behavioural needs. When their enclosure prohibits expression of behavioural needs, primates develop abnormal and maladaptive behaviours. Thus, husbandry must allow for the development and performance of highly motivated behaviours throughout the lifetime of the primate, allowing primates to experience positively valanced emotions and a life worth living (Mellor, 2016). Ultimately the animal will perceive its own state of welfare (Ohl & Putman, 2018) but being in an enclosure that affords choice and control with options to express highly motivated behavioural needs and act in ways a primate is adapted for, give the best chance in captivity of a primate perceiving it is in a good welfare state. Figure 6.3 outlines how this could be achieved when designing an enclosure for gelada.

6.5 SPECIES-SPECIFIC ENRICHMENT FOR PRIMATES

The provision of environmental enrichment (EE) is an essential part of primate care. In general, EE based on behavioural biology will stimulate exploration and provide learning opportunities, promote affiliative social interaction, playfulness, and potentially joy. For primates therefore, EE must provide extractive foraging opportunities, where food innovation, problem-solving, and tool use are encouraged for relevant species. Examples include artificial termite feeders for chimpanzees, artificial gum feeders for callitrichids and artificial percussive foraging opportunities for aye-ayes. Nesting opportunities should be provided in the form of substrates (e.g., blankets, wood wool, shredded paper) and nesting platforms (e.g., wide, flat shelves, large metal nesting dishes) for nest-building great apes, and next boxes or climber/vines for the smaller bodied prosimians and New World monkey species who nest in the wild. Primate behavioural biology suggests EE such as simulated hunting of mammals, birds, amphibians, and reptiles in great apes, Old World monkeys, and some New World primates, or hunting of insects in prosimians, and New World monkeys, should be provided. Hunting allows competition (e.g., threat display) and cooperation (e.g., food sharing) and help primate autonomy, regulation of emotional expression, and promotion of physical exercise.

Behavioural biology can be used to modify generic climbing frames so branch type, size, angle of suspension, height, and degree of complexity encourage wild-type modes and amounts of locomotion and foraging, and species-specific feeding postures. The behavioural biology literature suggests the most important EE for primates is independent feeding. This can be achieved by growing fruit trees (hawthorn, willow, fig, wild cherry),

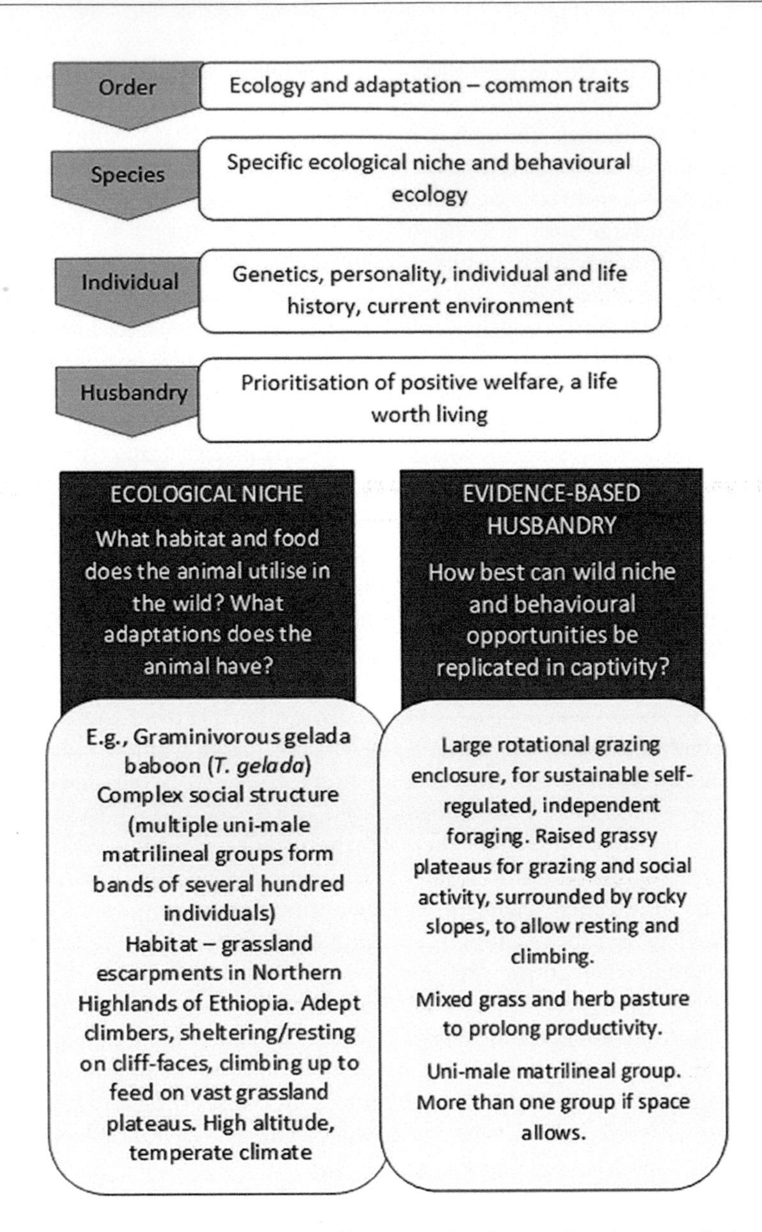

Figure 6.3 Behavioural biology and primate welfare. Levels of consideration needed to achieve behavioural biology-informed evidence-based husbandry are outlined, using gelada husbandry as an example.

shrubs/vines (red/black currant, grapes), and plants (raspberries, loganberries) inside enclosures with protective mesh covers to stop primates from over-foraging. Independent feeding may reduce regurgitation and reingestion, over-grooming of self and others, and ease general frustration and boredom.

Table 6.2 outlines other possible enrichment for named primate species while the further reading suggested in Box 6.2 contains a wealth of primate-specific EE ideas.

6.6 A NOTE ON PRIMATE CONSERVATION

Primate reintroductions typically have poor success rates and consequently conservation biologists may consider it inappropriate to include zoo-reared primates in reintroductions, preferring to translocate wild-born animals. Yet zoos are custodians of primate evolution, breeding individuals to safeguarding against genetic diversity

BOX 6.1: Where next for behavioural biology-informed primate husbandry?

There is an abundance of literature on primate behavioural adaptations, but the application of such information to inform husbandry practice requires further consideration. Below are suggested areas of further research:

1) How feasible is the autonomy-based approached advocated here? Giving control of ranging, feeding and social behaviour back to captive primates is vital for their welfare but how practical and achievable is this approach in the real world? Will it require a top-down or bottom-up culture shift to modern zoo keeping? Is there support for this from zoo staff and the zoo-going public?
2) What knowledge gaps exist in primate behavioural biology relevant to the zoo? Some horizon scanning research is likely needed, involving consultation with zoo practitioners. Primates in general are extremely well studied and such future scoping my reveal gaps in accessibility to knowledge in addition to actual gaps in our understanding.
3) How can we determine what aspects of behavioural ecology have adaptive value in the zoo, given zoo populations show phenotypic difference to wild individuals? In the zoo primates must tolerate close human-contact and other management practices, is there a balance to be sort between replicating wild opportunity and incorporating tolerance of management practices? And what are the consequences and ethics of this?
4) How can researchers work best to facilitate the use of behavioural ecology-based husbandry? How can collaboration between zoo practitioners and zoo researchers be fostered to embed behavioural biology-based husbandry into general zoo practice?

BOX 6.2: Primate ecology and husbandry resources

For more information on primate adaptations, primate species characteristics (range, habitat, life history), and primate husbandry, search the resources below:

- Galán-Acedo, C., Arroyo-Rodríguez, V., Andresen, E., & Arasa-Gisbert, R. (2019). Ecological traits of the world's primates. *Scientific Data, 6*(1), 1–5. Discusses a new database on primate ecological traits.
- Fleagle, J.G. (2013). *Primate adaptation and evolution* (3rd Ed). San Diego, USA: Academic Press. An excellent text on anatomical adaptations and evolution of primates.
- All the World's Primates website, hosted by Primate Conservation, Inc., includes useful resources on primate behaviour. This website accompanies the book "All the World's Primates" that replaces Rowe, Goodall, and Mittermeier (1996) https://alltheworldsprimates.org/Home.aspx
- University of Wisconsin-Madison, Wisconsin National Primate Research Centre, Primate Info Net website provides species factsheets discussing taxonomy, habitat, diet, social group structure, and behaviour of wild primates. https://primate.wisc.edu/primate-info-net/
- Animal Welfare Institute Refinement Database provides a reference list for laboratory animal (including macaque and marmoset) welfare and enrichment https://awionline.org/content/refinement-database
- NC3Rs The Macaque Website discusses husbandry, management, and welfare assessment of macaques. https://www.nc3rs.org.uk/macaques/captive-management/
- Wolfensohn, S. & Honess, P. (2005). *Handbook of primate husbandry and welfare.* Oxford, UK: Blackwell Publishing. A comprehensive guide to primate care.

loss. Captive breeding programmes maintain outbreeding but focus on conservation of genetics. This species-level approach overlooks behavioural adaptations expressed at the species and individual level. Conservation of behaviour is important as with each subsequent zoo-born generation, primates lose behavioural skills relevant to wild survival. Mason et al. (2013) suggest, within one generation, there are phenotypic changes in mammalian offspring compared to the parental phenotype. First-generation (F1) individuals begin the long process of adaptation to captivity, where selection pressures are artificial. This is where consideration of behavioural biology can have a positive impact beyond the welfare of the individual and towards conservation of behaviour, and for many primate species, conservation of cultural traditions (Riley, 2018).

Cultural traditions are behavioural skills acquired via social learning and transmitted across generations, that are not genetically determined nor dependent on the ecological provision. In the wild, chimpanzees show the most expansive culture (other than our own), often involving tool use. Chimpanzees with equal resource access develop differential behavioural strategies to acquire those resources. After initial innovation by an individual, the behavioural skill is observed, emulated (a social learning process where goals and subgoals of a task are learnt from a demonstrator and reproduced) and acquired by otherwise naïve individuals, thus establishing a cultural tradition. Zoo husbandry must support behaviour conservation, preserving behavioural plasticity with biologically relevant (aligned with behavioural biology) opportunity. Captive primates do not lose the capacity to learn skills, they seek opportunity to develop skills that have fitness value. By creating a captive environment that offers more natural opportunity, to feed, move, be social and learn, more captive primates will be suitable for reintroduction. Thus, aligning husbandry with behavioural biology can benefit welfare and conservation.

6.7 CONCLUSION

This chapter argues that husbandry practices relating to primate care should be informed by behavioural biology. The Autonomy Approach outlined focuses on promoting independence by providing opportunity for captive primates to control when they feed, travel, and interact socially as they would in the wild. This chapter has shown that primates have a set of shared characteristics that evolved due to climate and habitat changes, dominated by brain expansion and that influence social interaction and group structure, habitat utilisation, and feeding behaviour. These traits should, as advocated here, form the basis of primate husbandry with species-specific adaptations further determining enclosure design and wider husbandry. Presenting biologically relevant problems and allowing a primate to independently solve those problems provides appropriate challenge and supports positive welfare states (Meehan & Mench, 2007). This is especially important for primates given their cognitive capacity to suffer. To inspire the reader, Box 6.1 considers the next steps of implementing behavioural biology-based husbandry in primates, while Box 6.2 leads the reader to further resources including books and websites. Given the plight of wild-living species to survive increasing anthropogenic pressures, captive primates truly represent hope for the future but their welfare and relevance to conservation must be every zoo's priority if this hope is to be realised. This chapter presents ideas that will make help make hope a reality.

REFERENCES

Byrne, R., & Whiten, A. (1994). *Machiavellian intelligence*. Oxford, UK: Oxford University Press.

De Waal, F.B. (2007). *Chimpanzee politics: Power and sex among apes*. Baltimore, USA: John Hopkins University Press.

Dunbar, R.I.M., & Shultz, S. (2017). Why are there so many explanations for primate brain evolution? *Philosophical Transactions of the Royal Society B: Biological Sciences, 372*(1727), 20160244.

Fraser, D., Weary, D.M., Pajor, E.A., & Milligan, B.N. (1997). A scientific conception of animal welfare that reflects ethical concerns. *Animal Welfare, 6*, 187–205.

Humle, T., Maisels, F., Oates, J.F., Plumptre, A., & Williamson, E.A. (2016). *Pan troglodytes* (errata version published in 2018). The IUCN Red List of Threatened Species 2016: e.T15933A129038584. https://dx.doi.org/10.2305/IUCN.UK.2016-2.RLTS.T15933A17964454.en. Accessed on 31 January 2022.

Mason, G., Burn, C.C., Dallaire, J.A., Kroshko, J., Kinkaid, H.M., & Jeschke, J.M. (2013). Plastic animals in cages: Behavioural flexibility and responses to captivity. *Animal Behaviour, 85*(5), 1113–1126.

Matsuzawa, T., Biro, D., Humle, T., Inoue-Nakamura, N., Tonooka, R., & Yamakoshi, G. (2008). Emergence of culture in wild chimpanzees: Education by master-apprenticeship. In Matsuzawa T. (Eds.), *Primate origins of human cognition and behavior*. Tokyo, Japan: Springer.

Meehan, C.L., & Mench, J.A. (2007). The challenge of challenge: can problem solving opportunities enhance animal welfare? *Applied Animal Behaviour Science, 102*(3–4), 246–261.

Mellor, D.J. (2016). Updating animal welfare thinking: Moving beyond the "Five Freedoms" towards "a Life Worth Living." *Animals, 6*(3), 21.

Mittermeier, R.A., Rylands, A.B., & Wilson D.E. (2013). *Handbook of the mammals of the world: Volume 3 primates*. Barcelona, Spain: Lynx Edicions.

Ohl, F., & Putman, R. (2018). *The biology and management of animal welfare*. Dunbeath, Scotland: Whittles Publishing Limited.

Perelman, P., Johnson, W.E., Roos, C., Seuánez, H.N., Horvath, J.E., Moreira, M.A., Kessing, B., Pontius, J., Roelke, M., Rumpler, Y., & Schneider, M.P.C. (2011). A molecular phylogeny of living primates. *PLoS Genetics, 7*(3), e1001342.

Riley, L.M. (2018). Conserving behaviour with cognitive enrichment: A new frontier for zoo conservation biology. In Berger, M., & Corbett, S. (Eds.), *Zoo animals: Husbandry, welfare and public interactions*. (pp. 199–264). New York, USA: Nova Science Publishers Inc.

Rowe, N., Goodall, J., & Mittermeier, R. (1996). *The pictorial guide to the living primates*. New York, USA: Pogonias Press.

Swedell, L. (2012). Primate sociality and social systems. *Nature Education Knowledge, 3*(10), 84.

Van Schaik, C.P., Ancrenaz, M., Borgen, G., Galdikas, B., Knott, C.D., Singleton, I., Suzuki, A., Utami, S.S., & Merrill, M. (2003). Orangutan cultures and the evolution of material culture. *Science, 299*, 102–105.

7

The behavioural biology of ungulates and elephants

IAN HICKEY
Chester Zoo, Chester, UK

PAUL ROSE
University of Exeter, Exeter, UK
WWT, Slimbridge Wetland Centre, Slimbridge, Gloucestershire, UK

LEWIS ROWDEN
Zoological Society of London, London, UK

7.1 INTRODUCTION TO UNGULATE AND ELEPHANT BEHAVIOURAL ECOLOGY

The biological and ecological similarities of elephants and ungulates (odd- and even-toed hoofed mammals) mean that many of the challenges experienced with captive husbandry are also shared between these two basic "groups." Although aspects of elephant biology, such as body size, can mean maintenance of optimal captive husbandry is challenging, the many similar biological and behavioural characteristics between ungulates and elephants warrant a shared consideration of their captive care. Factors such as social structure, reproductive strategy, feeding ecology, and cognitive ability are just some aspects of behavioural ecology that are featured in this chapter. The combined taxonomic diversity of ungulates and elephants is vast. Species exhibit variation in ecological niche specialisation, habitat selection, and morphology that can be compared. As of 2022, the International Union

DOI: 10.1201/9781003208471-9

Table 7.1 Summary of taxonomic orders and families for ungulate and elephant groups, including representative species that are maintained in captive populations.

Order	Family	Representative species
Proboscidea	Elephantidae	African savannah (*Loxodonta africana*) and Asian (*Elephas maximus*) elephants
Perissodactyla	Equidae	*Equus* species, e.g., zebra, wild asses, and wild horses
	Tapiridae	*Tapirus* species, e.g., South American tapir (*Tapirus terrestris*), Malayan tapir (*Tapirus indicus*)
	Rhinocerotidae	African (*Ceratotherium and Diceros* sp.) and Asian (*Rhinoceros* spp.) rhinoceros
Artiodactyla	Camelidae	Bactrian (*Camelus bactrianus*) and dromedary (*C. dromedarius*) camels. South American camelids, e.g., guanaco (*Lama guanicoe*) and vicuna (*L. vicugna*)
	Suidae	Pig species and relatives, e.g., babirusa (*Babyrousa babyrussa*), red river hog (*Potamochoerus porcus*), Visayan warty pig (*Sus cebifrons*)
	Tayassuidae	Peccary species, e.g., collared peccary (*Dicotyles tajacu*)
	Hippopotamidae	Common (*Hippopotamus amphibious*) and pygmy hippopotamus (*Choeropsis liberiensis*)
	Tragulidae	Chevrotain and mousedeer, e.g., lesser mousedeer (*Tragulus kanchil*)
	Giraffidae	Giraffe, okapi (*Okapia johnstoni*)
	Antilocapridae	Pronghorn antelope (*Antilocapra americana*)
	Moschidae	Musk deer (*Moschus* spp.)
	Cervidae	Deer species, e.g., southern pudu (*Puda puda*), reindeer (*Rangifer tarandus*), Visayan spotted deer (*Rusa alfredi*), European elk (*Alces alces*)
	Bovidae	Antelope, e.g., impala (*Aepyceros melampus*), dik-dik (*Madoqua* spp.), bongo (*Tragelaphus eurycerus*). Cattle, e.g., American bison (*Bison bison*), lowland anoa (*Bubalus depressicornis*), gaur (*Bos gaurus*) Goats, e.g., Alpine ibex (*Capra ibex*) Sheep, e.g., barbary sheep (*Ammotragus lervia*) Takin (*Budorcas taxicolor*)

for Conservation of Nature (IUCN) recognises three extant species of elephant. Ungulate classification is more debatable, with estimates ranging from 250 to 450 described species. Table 7.1 summarises this breath in taxonomy and highlights some of the species groups in taxonomic orders that are found in zoos worldwide. Figure 7.1 highlights the extreme diversity of form and function, but also "family resemblances" across the different species of ungulate.

Such taxonomic diversity means this chapter does not aim to provide detail on all species; instead, a comprehensive overview including relevant examples is used to demonstrate aspects of behavioural biology important to modern zoo husbandry. A species-specific example of holistic considerations of behavioural biology for captive husbandry is demonstrated in a case study on giraffe (*Giraffa camelopardalis*) ecology and zoo welfare – a commonly housed ungulate with specialised, but often overlooked, care needs. It is vital to adopt an evidence-based approach to the husbandry and management of all these species, not only to meet the many conservation goals of ex-situ populations but also to ensure optimum animal welfare. Knowledge of the behavioural ecology of ungulates and elephants is the foundation of the evidence-based approach to viable captive animal management.

Figure 7.1 Artiodactyla (even-toed ungulates) galleries at the Natural History Museum at Tring, UK, illustrating the speciose nature of this taxonomic group.

7.2 UNGULATE AND ELEPHANT ECOLOGY, AND NATURAL HISTORY RELEVANT TO THE ZOO

When considering the captive needs of ungulates and elephants, there is a vast array of relevant natural history information available for evaluation. Topics such as natural home range sizes, patterns of activity, feeding strategy and reproductive behaviours all interlink and should be considered in a holistic manner when reviewing literature on wild ecology of these species. Despite significant variation in body sizes amongst ungulates and elephants, the natural home range size of a species typically exceeds that which is available to them when housed in captivity. Being the largest of extant land mammals, elephants understandably have been the focus of a great deal of attention regarding space provision in zoos. Indeed, it is accepted that no captive environment can provide the comparable space available to wild elephants (Rees, 2021). Elephants are known to cover vast areas of land in their need to feed regularly throughout the day and this should be one of the primary considerations when creating their captive environment and husbandry plans.

Similar comparison of wild habitat sizes and spatial occupancy is relevant for captive ungulates too, with even some of the smaller species like the Java mouse deer (*Tragulus javanicus*) having reported mean home range sizes of 5.9 hectares (Matsubayashi, Bosi, & Kohshima, 2003) – exceeding the typical space provided in zoos. There is also expected to be variation in the home range sizes of ungulate and elephants, with migration patterns observed in many species that follow resource availability across many hundreds of kilometres and habitat types. As it is well documented that these species travel great distances throughout their home ranges, available space should be one of the primary considerations for enclosure design. Space (both indoors and outdoors) provided to captive ungulates and elephants should be maximised as much as possible to allow for the greatest opportunity for movement and travelling.

For the majority of ungulate and elephant species, movement ecology throughout a home range relates most keenly to the need to access food resources. All species of elephant and most ungulate species are classed as obligate herbivores. It is primarily the Suidae (pigs) that are classed as omnivores, regularly consuming non-plant material and possessing a monogastric digestive system (Pérez-Barbería & Gordon, 1999). The varied feeding strategies exhibited across other ungulates and elephants relate to morphological differences in the digestive systems, specifically dentition and stomach structure. This in turn manifests in the selection of vegetative material as part of each species' core diet. Broadly, ungulates and elephants are assigned to one of three categories of feeding strategy (Hofmann, 1989) based on foraging ecology and digestive morphology (often along a sliding gradient):

1. Concentrate selectors (completely browsing species).
2. Intermediate feeders (a mixture of browsing and grazing).
3. Grass or roughage consumers (completely grazing species).

Concentrate selectors, also known as browsers, select diets that contain >75% dicotyledonous foliage and tree or shrub stems/foliage; with examples being European elk, red duiker (*Cephalophus natalensis*) and black rhino (*Diceros bicornis*). On the opposite end of the scale, grass consumers (grazers) are species whose diet contains less than 25% of dicotyledonous browse material, instead consuming monocotyledonous grasses which are high in silica and lower in protein compared to browse foliage. Species representatives from this foraging category include American bison, white rhinoceros (*Ceratotherium simum*) and Grevy's zebra (*Equus grevyi*). Between these two categories fall the intermediate feeders whose diet consists of approximately even proportions of browse and grass foliage, often with seasonal variation, and variation related to habitat availability. Elephants fall within this category, alongside many ungulates including European bison (*Bison bonasus*), red deer (*Cervus elaphus*), Nubian ibex (*Capra nubiana*), banteng (*Bos javanicus*) and greater one-horned rhinoceros (*Rhinoceros unicornis*). This variation in feeding niche ecology means that habitat selection is key in order to ensure that there is appropriate food resource available, and this may change seasonally and sees the browsers and grazers tending to have more specialised habitat selection than the more flexible intermediate feeders. Despite the variation in specific types of plant material consumed, all browsers, grazers and intermediate feeders need to spend a significant proportion of their daily activity budgets in the wild consuming plant matter in order to meet nutritional needs and ensure healthy operation of the digestive system. Theorised to be linked to this need for very regular food consumption, elephants and most ungulates tend to be active throughout much of the 24hour period. There are often peaks of activity around crepuscular periods of the day, but it is not uncommon to observe active behaviour across a full 24-hour cycle.

Both elephants and ungulates exhibit variation in social behaviours and structures and a species-specific social structure is a highly relevant aspect of behavioural ecology when considering captive needs. Most species employ a polygynous mating system, whereby one male is reproductively active with multiple females, and this relates to how conspecific social groupings are often maintained. For example, some breeding bighorn sheep (*Ovis canadensis*) rams will be socially dominant and maintain a harem of ewes. Meanwhile, subordinate rams will maintain less fixed social relationships to this main breeding group, being more peripheral. Variation around this general principle does exist and there is often temporal separation of sexes, with females and offspring maintaining more permanent social bonds whereas male animals are more reproductively and socially transient. For example, and as seen in many ungulate species too, elephant social groups are female dominant, with males integrating into a herd only during the mating season. Both African and Asian elephant bulls spend the majority of their time alone or with male-only groups (Hartley, Wood, & Yon, 2019), mainly coming together with family herds during breeding. The reproductive strategy employed generally features male-male competition for reproductive access, with social aggression observed in many species, and more complex male-led dominance displays, such as lekking as seen in red lechwe (*Kobus leche*), also apparent.

As well as intraspecific social interaction, ungulates and elephants engage in interspecific interactions as part of their natural ecology. These interactions occur across many taxonomic groups and geographic settings, with hypothesised benefits for mitigating predation risk and maximising foraging efficiency. For example, behavioural ecology research shows that interspecific groups of ungulates on the African savannah show distinct patterns of vigilance and groupings relative to predation risk when feeding in a shared area (Creel, Schuette, & Christianson, 2014). More indirect relationships also exist between species groups, for example changes in elephant presence and feeding have been shown to predict the community of foraging ungulates in Africa – both as a result of physical feeding activity and general elephant behaviour (de Boer, Van Oort, Grover, & Peel, 2015)

Perhaps associated with this social complexity, it is important to consider the communication and cognition aspects of natural ecology in these taxa. The communication intricacies of elephants have been extensively studied and the intelligence of the species is also widely recognised. Examples of cognitive abilities in elephants are well documented with Figure 7.2 showing how an individual elephant was able to manipulate surrounding objects within its captive environment to reach an otherwise unreachable food resource. Although cognitive abilities have been less studied in most ungulate

Figure 7.2 Example of tools used by a zoo-housed Asian elephant.

species, there are recent examples of tool use being documented in species including the Visayan warty pig (Root-Bernstein, Narayan, Cornier, & Bourgeois, 2019). Discoveries like these suggest that cognitive abilities to achieve an end goal may be present in other species within the ungulate taxa and therefore intelligence should be considered when reviewing species ecology.

BOX 7.1: The giraffe – a case study for the considerations of best-practice care of captive ungulates

Many hoofed mammals with a long history of captive housing, still pose challenges to successful in-zoo care because of their specialised ecological and evolutionary adaptations. Familiar and recognisable does not mean easy-to-look-after. One of the most familiar of zoo-housed ungulates, if not of all zoo-housed species, is the giraffe – a browsing specialist from sub-Saharan Africa, the world's tallest mammal and largest of all ruminants. Giraffe have fascinated humans for millennia; depicted in cave art from 8000 years ago and were first displayed live in Europe by Julius Caesar in 46 BC. Post-Roman Empire, examples of living giraffes arriving from Africa to Europe were rare. A single animal was presented to the Medici family in Florence in 1487 and 340 years later three giraffes were gifted to the respective monarchs of the United Kingdom, France, and Austria to much public interest. Sadly, none of these animals lived very long. Giraffe were not kept for any length of time in European zoos until the newly formed Zoological Society of London obtained four animals for its Regent's Park Zoo in 1836, from Sudan. These animals bred profusely, with the first ever zoo-born giraffe arriving in 1839 and a further 16 calves following up to 1867. So good was London Zoo at giraffe reproduction, that these original four animals were able to populate Regent's Park with live giraffe until 1881. Giraffe are indeed popular with visitors and therefore worthy of investment; after the death of the last of these Sudanese pioneers, the Zoological Society was without giraffe until 1895 when it obtained an import of a southern giraffe for the princely sum of £500 (nearly £68,000 by today's standards). Since this expensive important, Regent's Park has never been without a herd of giraffe. However, in spite of their popularity and the fascination they hold over zoo visitors and managers, and the determination expended by zoos on keeping them, the giraffe is not an easy animal to cater for. In fact, they are probably the best example of a ubiquitous zoo-housed species whose captive husbandry does not match its natural ecology or evolutionary history, and can end up be detrimental to the health and wellbeing of the individuals it is designed to care for (Hofmann & Matern, 1988).

The giraffe is an obligate browser (O'Kane, Duffy, Page, & Macdonald, 2013) and concentrate selector (Clauss, Lechner-Doll, Flach, Wisser, & Hatt, 2002) – it must consume leaves from trees and shrubs to gain optimal nutrition and keep the microbe populations of its chambered stomach (used for cellulose digestion and production of volatile fatty acids, VFA, that are metabolised for energy) healthy and functioning. Unusually for a herbivore, giraffe select foliage based on nutritional quality, rather than availability, consuming a wide range of different species of tree and shrub and demonstrating seasonal variability in vegetation browsed. As a tropical ungulate, the giraffe is not cold tolerant (Potter & Clauss, 2005). It has limited capacity for thermoregulation during prolonged periods of cold weather. Giraffe have limited subcutaneous fat stores and, as an adaptation to foraging during the hottest parts of the day, lose heat rapidly to avoid hyperthermia. Consequently, zoo giraffe in temperate climates must be kept warm. Many issues of "sudden death" (formally termed peractute mortality) are attributed to a chronic period of poor diet, leading to ruminal acidosis and inefficient VFA production, and a resulting energy deficit whereby the animal is expending more on homeostatic demand than it is assimilating from food. Adding further environmental pressure (i.e., a requirement to thermoregulate in cold conditions) on top of this nutritional challenge pushes the giraffe to the limits of any metabolic coping strategies it may possess, and eventually the animal collapses. Mortality of collapsed giraffe is high and preventative strategies (warmth, adequate diet, ad-lib forage, limited social stress, opportunities for rumination) are much better management approaches than attempts at curative treatment. Discrepancy between the wild and the zoo is illustrated by Figure 7.3, and examples of giraffe husbandry challenge (with potential solutions) are outlined in Table 7.2.

Table 7.2 Examples of typical challenges associated with the husbandry of giraffe, a commonly housed but specialist ungulate species.

Challenges with captive giraffe husbandry:	Solutions to these husbandry challenges?
Peracute mortality ("sudden death")	Avoid cold shock, prolonged exposure to low temperature (below 10°C), provide quality forage and measured amount of browser-specific concentrate.
Serous fat atrophy (evidence of chronic physiological stress)	Health check animals (e.g., body condition score) to ensure that physiologic and energetic demands are met by diet. Avoid cold temperatures, poor quality forage, and non-specific concentrate as per the above.
Worn teeth	Do not feed grass hay. Ensure animals are provided with browse and palatable, good quality lucerne/alfalfa hay.
Reduced digestive surface area of rumen	Provide meaningful quantities of browse to stimulate the movement of the ruminal wall. Do not feed grass hay. Measure concentrate pellet according to individual animal needs.
Cold intolerance	Heat the giraffe's indoor housing to maintain a high ambient temperature (minimum 18°C). Reduce time outside when temperatures fall below 10°C. Keep giraffe inside when the external temperature is below 5°C.
Energy deficits	Ensure giraffe can meet physiologic demand from diet offered. Do not add extra environmental pressure due to cold exposure. Avoid excess stress due to poor social group structure or constant visitor presence.
Weight loss/apparent wasting	Check past records of nutrition. Feeding of grass hay or access starch may have caused damage or blockage of rumen. Feed more browse, measure pellet, and provide warmth. Encourage rumination in an attempt to move any blockage through the rumen.
Abnormal behaviour patterns and altered time budgets	Promote rumination by providing increasing amounts of browse and *ad-lib* access to quality lucerne forage. Provide social enrichment and maintain stable social groupings. Allow giraffe to feed in different parts of the enclosure. Promote rumination, and hence saliva production, to buffer any acidotic changes to rumen, as this can cause excess licking or oral stereotypic behaviour.

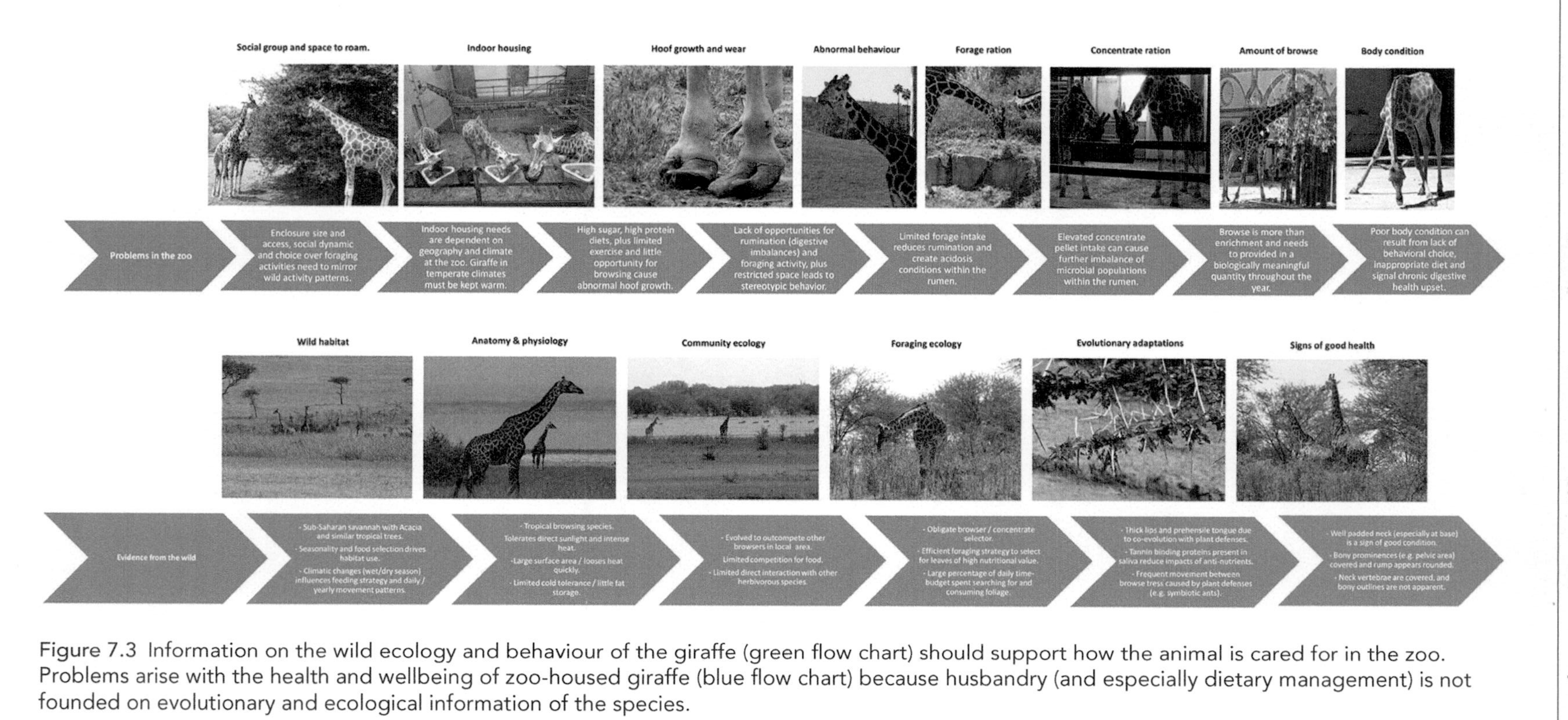

Figure 7.3 Information on the wild ecology and behaviour of the giraffe (green flow chart) should support how the animal is cared for in the zoo. Problems arise with the health and wellbeing of zoo-housed giraffe (blue flow chart) because husbandry (and especially dietary management) is not founded on evolutionary and ecological information of the species.

7.3 ENCLOSURE CONSIDERATIONS FOR UNGULATES AND ELEPHANTS BASED ON BEHAVIOURAL EVIDENCE

As with all zoo-housed species, it is essential that key aspects of elephant/ungulate ecology be considered in enclosure design to enhance animal welfare. Although, as previously mentioned, it is not usually possible to provide directly equivalent space provision as would be available in the wild, the quality of space can be maximised by considering behavioural needs. Elephants and ungulates in zoos can be housed in a range of enclosures that promote different aspects of behavioural repertoires to provide rich and diverse daily time-activity budgets (Figure 7.4).

Wild elephants for example move primarily in search of food and available resources, and this can be replicated in the captive setting through the provision of numerous and varied feeding stations throughout an enclosure. This encourages individuals to move together as a wild herd would do, while the unpredictable timing of food supply will encourage more movement throughout the available space. Similarly, the type of environment made available in captivity should reflect natural habitat selection for the species. For example, species such as the banteng that would naturally inhabit a mixture of open grassland space as well as more dense forest and scrub, should have a comparative gradient in structural complexity provided to them in captivity. This choice is essentially important for the maintenance of optimum welfare and will facilitate the expression of a more comprehensive range of natural behaviours. By providing appropriate complexity in terms of structure and available resources, the comparative reduction in available space will be mitigated against. For both ungulates and elephants, the importance of allowing and encouraging free movement between varied enclosure resources cannot be underestimated. This is particularly true across both diurnal and nocturnal conditions, to reflect the varied activity patterns of many species of ungulate and elephant. In housing systems where resources can be provided safely overnight, it is preferable to allow ungulates and elephants free access between all indoor and outdoor spaces (Powel & Vitale, 2016). If free access cannot be provided, the indoor enclosure should provide the most comprehensive choice of resources as logistically possible, to allow the performance of natural behavioural repertoires over a 24hour period.

Considering the nutritional and digestive specifics of many species of ungulates and elephants, it is vital that husbandry practice incorporates best-practice knowledge of natural feeding ecology. Browsing animals should have continual access to appropriate leguminous forage, whereas grazers should have equally consistent access to grass-based forages, e.g., hay. The delivery of these food resources is also a key consideration, with the goal to provide opportunities for extended periods of foraging without compromising the nutrient balance needed to be maintained in zoo settings. A common method of food presentation in the modern zoo, particularly for elephants, is the use of high-level hay nets that can be either manually or automatically lowered and raised at various times throughout the 24-hour period. A good enclosure should provide a number of these in several locations, able to be deployed at various intervals regardless of human caregiver presence. Feeding enrichment is one of the best ways to motivate and keep captive elephants occupied. To help increase time spent feeding at each net, quality hay can be mixed with straw of a lower nutritional value. Elephants will naturally seek out the more highly palatable forage and will consequently spend longer at each feeding point.

It is also important to allow as much space around each hay net as possible, to allow individuals to feed at the same time. An additional benefit of raised nets for elephants is the opportunity for calves to feed with the herd, as residual hay will naturally fall to the floor as adults feed. Beyond these commercially produced forages, the provision of naturally occurring browse (tree material, often with leafy material) is a vital component of the management of concentrate selectors in zoos. This is particularly true of obligate browsers such as giraffe, okapi, black rhino, and elk, with available browse giving the opportunity for appropriate behavioural and physical expression as well as correct nutrition.

The design of all aspects of the physical environment in a zoo enclosure should consider the behavioural biology of the species present. For elephants and ungulates, adequate and appropriate substrate is a crucial element of enclosure furnishing as these species are prone to foot health issues

Figure 7.4 Examples of elephant and ungulate enclosures that provide for important aspects of behavioural ecology. Top: A wetland enclosure for sitatunga (*Tragelaphus spekii*), swamp-dwelling species adapted to a wetland environment, that allows foraging and wading in water. Centre top: An enclosure for Asian elephants showing a range of topographies and areas of exploration. Centre bottom: bharal (*Pseudois nayaur*) enclosure showing wide expanses of differing terrains, topography, and grazing opportunities. Bottom: Klipspringer (*Oreotragus oreotragus*) enclosure that provides multiple areas for jumping and climbing on steep rocks, enabling this antelope to utilise its unique hoof anatomical adaptations to life on rocky outcrops.

(e.g., overgrown hooves, poor weight bearing, and cracks and fissures in their feet). Sand substrate is widely regarded as the best medium for a captive environment for many ungulates, and elephants in particular. For these and other species such as rhinos, deep sand will provide both support and cushioning when walking and assist with the wearing of the foot pads and nails as they walk throughout the day, reducing the need for regular hands-on foot care. Sand should be of a fine grade and free draining to minimise water retention and allow easy manipulation by both animals and caregivers alike. Sand provides a comfortable substrate for sleeping on and allows individuals to maintain vital social bonds, especially when there is appropriate space available to sleep in close proximity to conspecifics. For elephants, studies highlighting the importance of sleep have shown that sleeping in close proximity to conspecifics improves rest (Williams, Bremner-Harrison, Harvey, Evison, & Yon, 2015).

Deep sand can be manipulated to create mounds to increase comfort and make rising from sleep easier, especially for older animals. Sand mounds are also a valuable environmental component for young calves that will use them to express play behaviour, and caregivers should manipulate and change these mounds regularly to keep the environment novel. A range of substrates in one enclosure space will allow for animals to engage in species-relevant natural behaviours. For example, access to soil and mud encourages self-maintenance behaviours such as bathing/wallowing for many species including rhino and common hippopotamus, which fulfils a behavioural need as well as providing health benefits, e.g., protection against ectoparasites. Consider also ungulates that naturally inhabit wetland habitats such as the lechwe, and design enclosures to include water bodies to allow for the expression of natural behaviours (aquatic foraging) and maintenance of hoof health.

When designing enclosures and husbandry plans, the social aspects of ungulate and elephant ecology must also be considered. Group demographics should reflect natural settings, with appropriate group sizes and sex ratios for each species across the year. Facilities should allow for the natural separation of individuals and subgroups, but also the facilitation of efficient social integration at times when this is required. When this is not appropriately managed, undesired behaviours such as aggression may become hugely detrimental to the welfare of individuals and breeding success of entire groups. For example, male elephants will engage in aggressive competition behaviours if housed together in the presence of females (particularly at times of reproductive activity) and so separate housing away from the main herd (indoor and outdoor) should be provided, while also considering the ability to house subadult male elephants in a bachelor group upon attainment of sexual maturity (Schmidt & Kappelhoff, 2019). However, for most individuals of a captive elephant or ungulate group, it is important to minimise social separation wherever possible. Historically, it was typical for caregivers to separate animals and intervene during important events such as births, but the importance of allowing the group to experience these moments together cannot be underestimated. The majority of elephant births occur at night (Rees, 2021) and it is important that all members of the herd have the opportunity to be together for this key event. Adult females will assist by keeping young bulls at an appropriate distance but leaving female calves closer, allowing them to learn vital reproductive behaviours for the future. This “babysitting” behaviour is not restricted to birthing events, however. Adult females and even subadult females will regularly exhibit allo-mothering behaviours, sharing childcare responsibilities with the mother and even sucking non-related calves. This unique aspect of elephant behavioural ecology should considered when designing facilities to ensure that, wherever possible, calves are not separated from any member of their herd during day-to-day husbandry practices.

Behaviour training has long been a fundamental part of captive elephant husbandry, and is now also becoming increasingly used in the management of many ungulate species. Historically, these animals have been managed through “free contact” management methods, which often required the animal to be physically separated from other members of the group, even restricting the movements of mothers and calves. With zoos now more widely adopting the “protected contact” method, where caregivers manage the animals behind a protective barrier, there is a greater requirement to factor in behavioural training and management areas into captive facilities. At times where individual separation cannot be avoided, allow calves to circulate between the training area and the rest of the herd. This can be achieved by closing animal doors part way, restricting movement for adults but allowing free movement for calves. Training participation should always be through choice, and there should be alternative options for individuals such as the provision of hay nets while other members of the herd are taking part in training.

Taking the fact that both ungulates and elephants engage in naturally occurring interspecific interactions in-situ, it makes sense to incorporate mixed-species exhibition into modern zoo design. Although the practice would benefit from more extensive research to provide an informative evidence-base, it would be fair to assume that there is some level of enrichment benefit to the species housed together as it allows for the expression of some behaviours that would not be possible without the presence of a different species. Enclosure design in these cases is key to ensure successful mixed-species practice, providing enough space but also the correct resources. Where there may be competing demands for physical structures, environments, and substrates across species, it is essential that a sufficient range of these opportunities are available for all species and individuals to use. This may include "creep" spaces, where access is provided only to certain species within the exhibit so that specific needs can be catered to and a choice for conspecific only interaction is available.

7.4 UNGULATE AND ELEPHANT BEHAVIOURAL BIOLOGY AND WELFARE

To examine how ungulate welfare is linked to core understanding of animal ecology and biology, Figure 7.5 again uses the giraffe as an example of underlying behavioural and physiological principles for essential to the attainment of good welfare.

Figure 7.5 explores key wellbeing considerations that underpin effective giraffe husbandry around the principle of WELFARE, with each letter of the word WELFARE describing a specific area of husbandry challenge and basic need.

Warmth – giraffes houses must be heated and well insulated (depending on zoo location) and giraffe kept indoors when the weather is cold and there is wind chill. Giraffe struggle to maintain stable body temperatures in cold climates and have little subcutaneous fat to aid thermoregulation. In temperate climates, animals need to be kept indoors when temperatures fall into single figures to prevent homeostatic disruption, which can be fatal.

Enrichment – wild giraffe forage over wide areas, live in complex social groups and are active across daylight, crepuscular, and nocturnal hours. Daily husbandry must include opportunities for relevant nutritional, occupational, physical, sensory, and social enrichment across the full 24hr cycle.

Figure 7.5 Giraffe health and wellbeing in the zoo can be considered against the acronym WELFARE. Figures on the top row, marked with the sad face emoji identify indicators of poor giraffe welfare. Aspects of poor health and wellbeing occur when giraffe WELFARE is not considered. Figures on the bottom row show aspects of good giraffe health and wellbeing, when husbandry and housing fulfil all WELFARE considerations.

Leaves – giraffe evolved specifically to eat tree leaves, from height. Cut tree branches ("browse") is not enrichment for captive giraffe, but an essential component of daily diets. All captive giraffe must have access to a meaningful amount of browse every day of the year irrespective of season.

Feeding – the use of bespoke browser concentrate pellets is a necessary evil for captive giraffe as research shows that zoo-housed animals cannot maintain energy needs on a forage ration alone. Browser pellet should be feed on an individual basis and a weighed amount provided to determine each animal's daily intake. Changes to body condition or hoof overgrowth need to be considered alongside of the pelleted ration offered and exercise levels of the animals.

Alfalfa – zoo-housed giraffe must only be fed a legume-based hay, e.g., alfalfa/lucerne (*Medicago sativa*). Grass hays wear down teeth and result in stratified rumen contents, leading to ruminal blockage. Legume hays are the best currently available forage for giraffe but are no complete replacement to browse. Forage must be available *ad libitum* for all animals in a herd.

Rumination – an appetitive behaviour of a strong, underlying motivational drive and a key indicator of positive giraffe welfare. Giraffe must be encouraged to ruminate in captivity by ingesting large volumes of forage and browse, known amount of concentrate and no high sugar foods (e.g., fruits, vegetables, cereals, and bread). Rumination satiates an animal, calming it, and enabling assimilation of nutrients and energy.

Exercise – giraffe enclosures should be large enough to enable the herd to come together as one for social activities such as rumination or browsing. Enclosures must provide space for individuals to separate away from the herd and to associate preferentially. Wild giraffe walk between feeding trees, and relationships between trees and their defences against browsing, keeps giraffe on the move. Enclosures should allow a giraffe to walk across different substrates to naturally wear down hooves.

Examples of poor giraffe wellbeing caused by husbandry that is not based around WELFARE include those in the top row of images in Figure 7.5. From left; Frustration behaviour caused by a desire to reach a highly valued resource (in this case tree leaves) that the animal can see but cannot gain access to. A lack of opportunities to browse and collect leafy material (processing with the lips, tongue, and teeth) leads to oral stereotypic behaviours, such as tongue playing, vacuum chewing, or mouthing of enclosure structures). Note the empty "enrichment" device in the left of the photo. Poor body and skin condition and the animal eating from the floor, indicative of poor diet and housing, and a lack of browsing opportunities. Pacing and repetitive walking. Individual giraffe will pace along a fence line if anxious to enter indeed stalling or if they have been locked out on show with no choice to enter housing. Enclosures need to provide meaningful space to move around and free opportunities to enter and leave indoor housing. Overgrown hooves caused by a combination of poor diet, laminitic changes to hoof structure, lack of exercise, or poor enclosure substrate that does not wear hooves evenly. Examples of good giraffe wellbeing where a zoo is following WELFARE include those images in the bottom row of Figure 7.3. From left; Excellent body condition, especially of the giraffe's neck that is wide and fleshy at the base. Expansive indoor housing, multiple accessible valued resources, and opportunities for browse to be provided as routine husbandry and not only as enrichment. Choice over social associates and high levels of rumination. Giraffes assume a characteristic posture for rumination, usually standing with the head and neck forward at around 45°. Ad-lib access to high quality alfalfa forage. A giraffe pulls forage from a hay rack that can be elevated to the height of the giraffe, simulating a natural stance and position for browsing.

A key feature to ensure positive welfare of captive ungulates and elephants is the facilitation of appropriate social groupings. Elephants in particular thrive when part of multi-generational family units and should be housed with related individuals wherever possible. There are however instances where this group structure may not be possible and, whether holding a multi-generational family herd, male-only group or group of unrelated individuals, managers need to consider the long-term impact of any husbandry changes or enclosure modifications on each individual animal throughout their life. For species such as elephants with considerably longer lifespans than many other zoo-housed animals, this is arguably of even greater importance. An animal's needs will change at various stages throughout its life, and individual welfare plans should be implemented to record this. Factors that may inform an individual's welfare plan can include age, health, reproductive status, past experiences, and social bonds. Behavioural welfare assessment tools have been developed to monitor the welfare state and needs of individual elephants over time (Yon et al., 2019) and these can be used in conjunction with individual welfare plans to allow managers to tailor husbandry practices for each individual.

Similarly, zoo ungulates may benefit from the application of Qualitative Behavioural Assessment, QBA (a whole animal approach to inferring the emotional components of welfare) that can be useful for welfare assessments in other zoo-housed taxa (Rose & Riley, 2019), and because QBA has its foundation in the world of domestic ungulate welfare (Wemelsfelder & Lawrence, 2001) many of the descriptors of behavioural expression (emotions and body postures or changes to orientation that signal mood and feelings) could be an excellent starting point for the definition of lists of descriptors applicable to zoo-housed ungulates.

7.5 SPECIES-SPECIFIC ENRICHMENT FOR UNGULATES AND ELEPHANTS

It is important to engage the full range of an animal's physical capabilities and provide opportunity to express a comprehensive range of natural behaviour, both for elephant and ungulate species. Figure 7.6 shows examples of different types of feeding methods at various levels within an enclosure, encouraging a diverse range of behavioural complexity that can be used with captive elephants. Although these examples, summarised underneath the figure itself, are visualised for elephants many of them are applicable to ungulates and their behaviours, and so should be considered accordingly. Surveys of ungulate husbandry practices (Rose & Roffe, 2013; Rose & Rowden, 2020; Rowden & Rose, 2016) show variation in enrichment schedules across species. For some ungulates, enrichment may be part of routine but for

Figure 7.6 Illustration of elephant feeding methods promoted by environmental enrichment. Examples are: 1. Hay net, 2: Browse, 3: Bark pile, 4: Feeder wall, 5: Buried food, 6: Scatter feeder, 7: Ground-level tunnel, 8: Movable furniture. (Credit: Hannah Thompson, based on an original sketch by Ian Hickey).

others, it may be provided infrequently or never, and often simply living in a social setting is the singular source of behavioural enrichment accessible. Although the importance of appropriate social groupings cannot be understated, additional enrichment inputs to aid behavioural management should be considered more frequently for more species.

Both the type and number of feeding opportunities are important to challenge animals mentally and physically. Different methods of food provision encourage animals to use a range of muscles and work cognitive abilities to problem solve. It should be noted, however, that while encouraging movement throughout the day is important, many species of ungulates and elephants will also stay in one place and feed together when resources are abundant. When planning food distribution from day-to-day, this should be considered so that there are days where individuals remain together around foraging patches. Wherever logistically possible, adding one or more large piles of browse into an enclosure encourages social cohesion and group feeding, allowing the herd to feed from one abundant location rather than moving from feeding point to feeding point.

Figure 7.6 describes the following methods of elephant feeding methods to promote excellent health and wellbeing in captive animals.

1. **Hay net** – Ideally hung from electrical hoist so too allow operation from at a distance from animals to discourage anticipatory behaviour. Set at heights to allow individuals of different sizes to reach. Extends feeding time, requires physical exertion and allows multiple individuals to feed together, including calves, as hay drops to the floor.
2. **Browse** – Can be set in large piles to encourage group feeding or hung throughout the enclosure to encourage travelling and foraging. Variation in browse species day-to-day provides novelty and different species can also require animals to feed in different ways.
3. **Bark pile** – Ideally created with the use of a mechanical vehicle (e.g., tractor/digger) to make it as large as possible. Feed pellets and chopped vegetables can be interspersed throughout a large pile of bark or wood chippings, encouraging extended periods of time foraging and allowing group feeding.
4. **Feeder wall** – Timed feeders can be placed out of view behind holes in a wall of the enclosure or building. Feeders can be off-the-shelf products (widely produced for farm and equine husbandry) or can be built if ready-made products are not available. These can be pre-programmed to be deployed at varying times throughout the day and night, encouraging animals to forage.
5. **Buried food** – A deep sand substrate allows whole food items such as fruit or vegetables to be buried, which elephant seeks out and digs up. This encourages forging behaviour and requires the animal to work for reward.
6. **Scatter feeder** – Automatic feeders that scatter pellets and other small items of food over a large area of the enclosure to encourage ground-level foraging. These can be pre-programmed to be deployed at varying times throughout the day and night. Consider placing above bark pile to lengthen foraging time.
7. **Ground-level tunnel** – To encourage an additional range of motion and muscle use, create long ground-based tunnels that animals can only access by kneeling.
8. **Movable furniture** (e.g., wobble tree) – Mimics the natural environment allowing elephants to push against trees to get a reward. Think about putting food items such as pellets or vegetables in a hollow near the top or hanging browse that the animals can only reach by pushing the post.

7.6 USING BEHAVIOURAL BIOLOGY TO ADVANCE UNGULATE AND ELEPHANT CARE

Ungulates are commonly housed in zoological facilities across the globe, and elephants are one of the flagship species that are persistently popular with the visiting public. Despite this ubiquity of presence and popularity, there are persistent gaps in our knowledge to ensure best-practice animal care. There have been significant advancements in the behavioural basis for elephant management in zoos in recent years, with evident improvement in welfare status following mitigations and interventions (e.g., significant changes to accepted practice) informed by behavioural biology (Finch, Sach, Fitzpatrick, Masters, & Rowden, 2020).

To achieve comparable advancement in the management of ungulate species, it will be necessary to increase research outputs focussing on species that are comparatively understudied. By having the same level of focus as dedicated to elephants, distinguishing the important nuances of behavioural biology for individual ungulate species (recognising that one ungulate's needs are not the same as all others) will become commonplace. Moving away from ill-fitting domestic models, which have commonly been used as a baseline for zoo ungulate husbandry, will occur more swiftly if research activity increases on ex-situ animals. It would be especially useful to advance knowledge in areas such as the:

- validation of available (or development of novel) welfare assessment tools applicable to zoo-housed ungulates.
- investigation of activity patterns in mixed-species exhibits, specifically in relation to behaviour management practices (e.g., to reduce aggression and provide ecologically relevant exhibit features).
- evaluation of changes to elephant management practices to ensure that i) uptake of appropriate husbandry occurs across all facilities and ii) new methods and approaches continue to be relevant to the animals.

BOX 7.2: Further information to inform the care of ungulates and elephants

Husbandry guidelines for a range of species (of ungulate and elephants) have been produced by zoo membership organisations. For example, optimal care of elephants and rhinoceros species can be found at https://www.eaza.net/conservation/programmes/#BPGandBIAZA's elephant welfare working group can be foundathttps://biaza.org.uk/elephant-welfare-group.

An excellent online resource for "all things" ungulate natural history and ecology is the Ultimate Ungulate website http://www.ultimateungulate.com.

7.7 CONCLUSION

The taxonomic diversity of ungulates and elephants is extreme, with varied morphological traits and key aspects of behavioural biology that allow these taxa to exist in many geographic ranges and habitats. As more information on wild behaviour emerges, it is increasingly clear that this variation includes assorted dietary strategies, complex social organisation, and often overlooked means of communication and skilled cognitive abilities. It is vital to consider all of these intricate aspects of species biology when planning for optimal care of both ungulates and elephants, recognising that each species will have unique requirements that must be catered for to achieve best-practice husbandry. Moving away from the management approach that provides the same environment to all captive ungulates (regardless of specific biological niche and species behaviour), towards an evidence-based approach and scientific evaluation of captive care will ensure that lessons learnt on these taxa can be shared widely, benefitting individual animals, and enhancing the conservation value of populations on a long-term basis.

REFERENCES

Clauss, M., Lechner-Doll, M., Flach, E.J., Wisser, J., & Hatt, J.-M. (2002). Digestive tract pathology of captive giraffe. A unifying hypothesis. *Proceedings of the European Association of Zoo and Wildlife Veterinarians, 4*, 99–107.

Creel, S., Schuette, P., & Christianson, D. (2014). Effects of predation risk on group size, vigilance, and foraging behavior in an African ungulate community. *Behavioural Ecology, 25*(4), 773–784.

de Boer, W.F., Van Oort, J.W., Grover, M., & Peel, M.J. (2015). Elephant-mediated habitat modifications and changes in herbivore species assemblages in Sabi Sand, South Africa. *European Journal of Wildlife Research, 61* (4), 491–503.

Finch, K., Sach, F., Fitzpatrick, M., Masters, N., & Rowden, L.J. (2020). Longitudinal improvements in zoo-housed elephant welfare: A case study at ZSL Whipsnade Zoo. *Animals, 10*(11), 2029.

Hartley, M., Wood, A., & Yon, L. (2019). Facilitating the social behaviour of bull elephants in zoos. *International Zoo Yearbook, 53*(1), 62–77.

Hofmann, R., & Matern, B. (1988). Changes in gastrointestinal morphology related to nutrition in giraffes *Giraffa camelopardalis*: A comparison of wild and zoo specimens. *International Zoo Yearbook, 27*(1), 168–176.

Hofmann, R.R. (1989). Evolutionary steps of ecophysiological adaptation and diversification of ruminants: A comparative view of their digestive system. *Oecologia, 78*(4), 443–457.

Matsubayashi, H., Bosi, E., & Kohshima, S. (2003) Activity and habitat use of lesser mouse-deer (*Tragulus javanicus*). *Journal of Mammalogy, 84*(1), 234–242.

O'Kane, C.A.J., Duffy, K.J., Page, B.R., & Macdonald, D.W. (2013). Effects of resource limitation on habitat usage by the browser guild in Hluhluwe-iMfolozi Park, South Africa. *Journal of Tropical Ecology, 29*, 39–47.

Pérez-Barbería, F.J., & Gordon, I.J. (1999) The functional relationship between feeding type and jaw and cranial morphology in ungulates. *Oecologia, 118*(2), 157–165.

Potter, J.S., & Clauss, M. (2005). Mortality of captive giraffe (*Giraffa camelopardalis*) associated with serous fat atrophy: A review of five cases at Auckland Zoo. *Journal of Zoo and Wildlife Medicine, 36*(2), 301–307.

Powel, D.M., & Vitale, C. (2016). Behavioral changes in female Asian elephants when given access to an outdoor yard overnight. *Zoo Biology, 35*(4), 298–303.

Rees, P.A. (2021). *Elephants under human care: The behaviour, ecology and welfare of elephants in captivity*. Academic Press, London, UK.

Root-Bernstein, M., Narayan, T., Cornier, L., & Bourgeois, A. (2019). Context-specific tool use by *Sus cebifrons*. *Mammalian Biology, 98*(1), 102–110.

Rose, P., & Riley, L. (2019). The use of qualitative behavioural assessment to zoo welfare measurement and animal husbandry change. *Journal of Zoo and Aquarium Research, 7*(4), 150–161.

Rose, P.E., & Roffe, S.M. (2013). A case study of Malayan tapir (Tapirus indicus) husbandry practice across 10 zoological collections. *Zoo Biology, 32*(3), 347–356.

Rose, P.E., & Rowden, L.J. (2020). Specialised for the swamp, catered for in captivity? a cross-institutional evaluation of captive husbandry for two species of lechwe. *Animals, 10*(10), 1874.

Rowden, L.J., & Rose, P.E. (2016). A global survey of banteng (*Bos javanicus*) housing and husbandry. *Zoo Biology, 35*(6), 546–555.

Schmidt, H., & Kappelhoff, J. (2019). Review of the management of the Asian elephant *Elephas maximus* EEP: Current challenges and future solutions. *International Zoo Yearbook, 53*(1), 31–44.

Wemelsfelder, F., & Lawrence, A.B. (2001). Qualitative assessment of animal behaviour as an on-farm welfare-monitoring tool. *Acta Agriculturae Scandinavica, Section A-Animal Science, 51*(S30), 21–25.

Williams, E., Bremner-Harrison, S., Harvey, N., Evison, E., & Yon, L. (2015). An investigation into resting behavior in Asian elephants in UK zoos. *Zoo Biology, 34*(5), 406–417.

Yon, L., Williams, E., Harvey, N.D., & Asher, L. (2019). Development of a behavioural welfare assessment tool for routine use with captive elephants. *PLoS One, 14*(2), e0210783.

8

The behavioural biology of carnivores

KERRY A. HUNT
University Centre Sparsholt, Winchester, UK

8.1 INTRODUCTION TO CARNIVORE BEHAVIOURAL BIOLOGY

This chapter will cover terrestrial mammals in the order Carnivora, including the families Felidae (cats), Canidae (dogs), Ursidae (bears), Procyonidae (raccoons), Mustelidae (weasels, otters), Mephitidae (skunks), Herpestidae (mongoose), Hyaenidae (hyena), Eupleridae (Malagasy carnivores), Ailuridae (red pandas), and Viverridae (civets). The semi-aquatic members of this order, including polar bears (*Ursus maritimus*) and pinnipeds are covered in Chapter 9. There are approximately 300 extant species in this order (Hassanin et al., 2021) and Table 8.1 provides example species to show the wide variation in diet and social structure across this order.

This order includes some of the most wide-ranging terrestrial species in the animal kingdom, this is due to many of these species being generalists when it comes to their diet and habitat requirements. The grey wolf (*Canis lupus*) was once the most widely distributed land mammal, with its range spanning across Canada, North America, Europe, and Asia. This decreased due to poaching predominantly due to human–animal conflict based on the perceived risk to people and livestock, however the IUCN considers wolf populations stable based on the latest (2018) Red List evaluation. Carnivores can be found on every major land mass and in all habitats, from artic regions through to rainforest and deserts.

Species are classified into the order Carnivora based on specific morphological features including:

- Elongated canine teeth, as well as carnassial teeth for shearing meat.
- Third upper molars are missing from adult members of this order.
- Clavicle is either very small or completely missing.
- Three of the carpal bones in the front limbs are fused into a single bone called the scapholunar.

DOI: 10.1201/9781003208471-10

Table 8.1 Selected ecological information on representative species of Carnivora useful for planning husbandry management.

Species	Feeding behaviour	Social behaviours	Other information
Cheetah *Acinonyx jubatus*	Feeding times are flexible dependent on activity/ distribution of larger predators in the areas. However, a preference for dawn and dusk for solitary animals, earlier for coalitions is noted. Most prey is medium-sized ungulates, e.g., impala (*Aepyceros melampus*). Hunting strategy is stalk and sprint short distances. Fasts after meals for two to five days.	Females tend to be more solitary, unless caring for cubs. Related males will stay together in coalitions of two to four individuals. Housing of females in social groups might lead to reproductive suppression. Cubs tend to stay with mother until 18 months old.	IUCN Red List: Vulnerable Length: 1.1–1.5 m Speed 80–130 km/h Weight 34–64 kgs.
Bush dog *Speothos venaticus*	Tends to hunt during the day, in small social groups with most common prey being large rodents, e.g., agouti (*Dasyprocta* spp.)	Groups sleep in close contact, whilst there is a breeding pair there is not a clear hierarchical structure to the group. Usually live in family groups of a breeding pair and offspring.	IUCN Red List: Near Threatened Length 57–70 cm Weight 5–7 kg Gestation period 68 days.
Sun bear *Helarctos malayanus*	Foraging takes up a large part of activity budget. Uses front paws to dig through rotten logs for invertebrates including termites and beetles. Fruit and vegetables are also consumed and will opportunistically feed on small vertebrates.	Generally, an elusive animal so limited information is known about social behaviours. Wild bears appear to be mostly solitary except for females with cubs. Rare occurrences of multiple animals coming together to feed from the same fruit tree.	IUCN Red list: Vulnerable. Length 1.2–1.5 m Weight 27–65 kg. Smallest species in the bear family.
Racoon *Procyon lotor*	As opportunistic omnivores, raccoon diets include invertebrates, berries. Vegetation and a range of vertebrate species. Have also been shown to raid crops. Those in urban areas also scavenge in human rubbish.	Social groupings in raccoons varies considerably from coalitions to strictly territorial. Resource availability impacts upon this and generally the older, larger animals dominated in any aggressive interactions.	IUCN Red List: Least Concern Length 40–70 cm not including tail, which can be another 40 cm long. Weight 4–9 kg.
Smooth-coated otter *Lutrogale perspicillata*	Diet is primarily fish supplemented with invertebrates (including crabs and prawns), eggs, and other small vertebrates. Studies have shown dietary flexibility and opportunistic feeding in this species.	Lives in family groups consisting of a breeding pair and their offspring. The breeding female is the dominant individual in the group.	IUCN Red List Vulnerable Length 70–80 cm. Weight 7–11 kgs.

Striped skunk *Mephitis mephitis*	Opportunistic omnivore. Strong claws enable effective digging for food. Skunks are not a fast-moving species so hunting behaviour is limited to slow moving species.	A solitary species in the wild although they do manage occasional social contacts. Female co-denning behaviour has been observed.	IUCN Red List: Least Concern Length 57–80 cm including the tail, which can be up to 38 cms. Weight up to 6.5 kgs..
Meerkat *Suricata suricatta*	Actively hunt several species of invertebrate, including scorpions (Scorpiones). They learn to handle venomous and toxic species. Most food items are underground, so meerkats are effective diggers.	Lives in large social groups of up to 50 animals. Sentry duties and use of creches are well documented in this species. Meerkats have a strict matriarchal society.	IUCN Red List: Least Concern Length 25–35 cm Weight 0.62 to 0.97 kg.
Striped hyena *Hyaena hyaena*	Whilst known for scavenging behaviours, striped hyenas are efficient hunters, taking large prey including sambar (*Rusa unicolor*). As a facultative carnivore they will also consume vegetation and will raid fruit crops.	Often considered solitary as hunts alone, however, associates in small family groups (two to four animals) back at their den. This usually consists of one female and with one to three males. Females solely responsible for care of offspring.	IUCN Red List; Near Threatened Length 85–130 cm Weight 22–55 kg.
Binturong *Arctictis binturong*	Diet is primarily fruits and vegetables however, they also consume birds, small mammals, and carrion.	Binturong males are mostly solitary, females live in small family groups with offspring or alone. Often kept in pairs in captivity with no issues. Limited knowledge on wild behaviour however territory is marked with a scent that to humans resembles popcorn.	IUCN Red List: Vulnerable Length 71–84 cm Weight 11–32 kg.
Red panda *Ailurus fulgens*	Red pandas are known for their diet of bamboo with a focus on the leaf tips and new growth shoots, which tend to be the more nutritious parts of the plant. Whilst this can make up to 95% of the animal's diet, they will also forage for roots, fruits, and invertebrates. Have been observed occasionally killing and consuming birds and small mammals.	Generally solitary except during the breeding season. Both sexes mark territory with fluid from glands on the base of the feet and the anus. Males show more territorial behaviour than females and tend to have large home ranges. However, EAZA Best Practice Guidelines promote keeping in breeding pairs all year round.	IUCN Red List: Endangered Length 50–65 cm without the tail, tail can be 30–50 cm long. Weight 3–6.2 kg.
Fossa *Cryptoprocta ferox*	An excellent hunter, they predate on lemur species as well as other small to medium-sized mammals like rodents, birds, and wild pigs. An ambush hunter, they catch their prey using their retractable claws.	Females are generally solitary unless caring for offspring. Males are often solitary but can also be found in coalitions of two or three males. Males share overlapping territories with each other and females. Whereas female territories only overlap with related females, e.g., daughters.	IUCN Red List: Vulnerable Length: 61–80 cm Weight: 7–12 kgs.

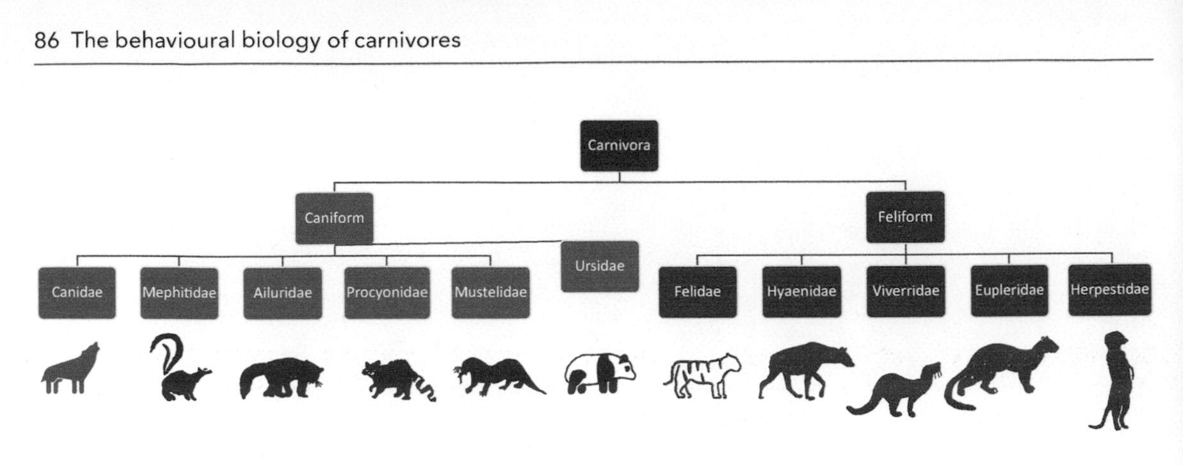

Figure 8.1 The division of the different families in the order Carnivora.

The order is then divided down into two major suborders; these are Caniformia and Feliformia. Feliform species are primary classified by double-chambered auditory bullae where the two bones are joined by a septum, however there are a few other features to distinguish Feliforms from Caniforms – this includes shorter rostrums, more specialised carnassial teeth, retractable or at least semi-retractable claws, and more digitigrade locomotion. Caniforms have a single-chambered or partially divided auditory bullae, and generally tend to be more omnivorous in their diet. Figure 8.1 shows the division of the families into these suborders.

8.2 THE APPLICATION OF CARNIVORE ECOLOGY AND NATURAL HISTORY TO ZOO MANAGEMENT

Research into carnivore ecology can be used to guide captive management, specifically the identification of biologically relevant resources, management of breeding pairs, and appropriate social groups, nutritional requirements, and feeding ecology.

8.2.1 Social interactions

As shown in Table 8.1 there are a wide range of social groupings in carnivores ranging from solitary species, such as leopards (*Panthera pardus*) through to spotted hyena (*Crocuta crocuta*) that can live in groups of over 100 individuals. Many of these species, including several of the more solitary species are often kept in pairs in captivity as well as in single sexed groups, however there are ethical concerns with some of these population management techniques, mostly around them limiting the expression of natural behaviours if animals are not being kept in the same groups seen in the wild. With several species naturally living in either family groups where the males (predominantly) undergo natal dispersal upon reaching breeding age, as seen in wolves, or residing in harems of one or two males and many females, as seen in lions (*Panthera leo*), often this leads to an excess of males. Batchelor groups are then maintained as an option for housing these excess males. While several species in the wild will form coalitions of related individuals, or bachelor groups, these tend to be unstable, with animals moving out once they reach sexual maturity. So, the maintenance of these groups long-term is an abnormal social grouping that can lead to increased aggression, so should be carefully monitored in captive environments.

The management of species in correct social groupings is important for encouraging breeding success. It is recommended for red pandas that breeding pairs are kept together continuously, whereas to encourage successful breeding in cheetahs it is recommended that the females are kept individually prior to breeding. This is due to mixed-species groups outside of breeding times being unnatural in the wild and subsequently rare in captivity, also social housing with other females can lead to ovarian suppression, reducing reproductive success.

Many species of Carnivora readily breed in captivity with Species360 (an international organisation that maintains an online database of captive wild animals) data from January 2022 showing that 561 meerkats were born in zoos that use its ZIMS (a records keeping programme) in the preceding

12 months. That said, research suggests as little as 20% of recommended captive breeding pairs result in successful offspring, with factors including mate choice, maternal age, and previous breeding experience impacting on reproductive success in painted dogs, *Lycaon pictus* (Yordy & Mossotti, 2016). Consideration of all these factors as well as how to introduce each new pairing are important, with many carnivore species using complex olfactory communication to provide intraspecific information of territory boundaries, individual identification, or reproductive status. With amount of scent marking being linked to reproductive success in several species, adequate opportunities for scent marking and sharing of scents should be provided prior to and during the establishment of new social groupings. Due to such variability in social grouping seen across this order, there is not one clear recommendation for social groupings, although some examples have been provided in Table 8.1. Species-specific research, feeding into best practice guidelines, is needed when developing suitable social groupings.

Most of the larger species of carnivore are kept in single-species exhibits, however there have been some successful mixed-species exhibits including the mixing of bears with wolves. Smaller species are more commonly kept in mixed-species enclosures and some species, like meerkats, have been included in walkthrough exhibits (for example, at Longleat Safari Park). Some Carnivora species, such as dhole (*Cuon alpinus*), are not recommended at all for mixed-species exhibits due to their ability to hunt very large prey items.

BOX 8.1: Mixed species exhibits and carnivores

Mixed species exhibits are increasingly utilised in zoological collections, with evidence highlighting benefits to the animals as well as the visiting public. However, the correct establishment of social groups, species choice, and enclosure design to facilitate this is important. As previously mentioned mostly the smaller carnivore species are used in mixed species so Asian short-clawed otters (*Aonyx cinereus*) and binturongs being housed with slivery langurs (*Trachypithecus cristatus*) at Colchester Zoo, cheetahs have been kept with rhinoceros and striped skunks and porcupines (*Erethizon dorsatum*) mixed at Los Angeles Zoo. Dorman and Bourne (2010) highlight that one of the risks to housing carnivores in mixed exhibits includes injury or death if incompatible individuals are housed together. It is worth noting that this is also the case in mixed species exhibits of other (non-carnivore) taxa and this can occur even when mixing individuals of the same species, so it is worth reviewing each situation individually.

Woburn Safari Park was the first zoological collection in the UK to house black bears (*Ursus americanus*) and wolves together. Ensuring a sufficiently large exhibit with suitable visual and physical barriers, along with escape areas for each species, is needed to enable successful mixing. Consideration of the individual animals as well as species implications is important with research indicating that hand-reared individuals are less suitable in mixed species exhibits as they often lack the required inter- and intraspecific social communication mechanisms. Several studies that have investigated interactions between canid and ursids in mixed species exhibits suggest that, more often than not, both species avoid direct interactions but when conflict did occur it was infrequent (with 86% of aggressive interactions being resolved through threat displays).

The benefits of mixed species exhibits include efficient use of space, often facilitating animals in larger exhibits than if they are housed in single-species exhibits, and environmental complexity and social interaction diversity can be increased compared to those seen in single-species housing. Beneficial activity levels are often higher in mixed species exhibits, which is particularly important for carnivores that can often suffer from obesity or muscle loss in captivity due to limited opportunities for exercise.

Whilst not that common, some small carnivore species (e.g., meerkats) can be kept in walkthrough exhibits. However, there is a lack of research on the impact of living in walkthrough exhibits for these species, as well as the overall suitability of such species for this type of enclosure.

8.2.2 Nutrition and diet

Some species (e.g., meerkats) as well as being prone to the obesity that many in this order suffer from when kept in captivity, have also been documented to have high cholesterol due to the fat content of whole prey diets. Dobbs, Liptovszky, and Moittie (2020) used blood samples to monitor the impacts of a dietary change on cholesterol levels, finding that, initially, all meerkats had cholesterol levels well above values found in wild individuals. High cholesterol in this species leads to hypercholesterolemia and meningeal cholesterol granulomas, so where possible, specific species serum cholesterol and other biochemical testing should form part of routine monitoring.

Regular monitoring of an individual's mass should be completed to monitor fluctuations and review nutritional impacts on health – weighing being the clearest method to perform this. However, for many of the larger carnivore species, weighing apparatus are not built into exhibits, so the use of other measures of estimating body mass need to be considered – such as body condition scoring. Body condition scoring sheets for some species (e.g., cheetah) are available in best practice guidelines. Many captive carnivores are obese, so regular monitoring of body condition can help identify if this occurs, ensuring dietary changes can be made to reduce the calorific intake. A decrease in body condition score can be indicative of a range of health issues, so regular monitoring ensures these are detected earlier to uphold better welfare. Many captive carnivores have reduced activity, so monitoring the impact of diet is important to maintain a healthy body weight and avoid joint issues.

Generally, carnivores have simple digestive systems, with obligate carnivores like tigers (*Panthera tigris*) having a monogastric stomach, small colon, and a limited caecum. This is adapted for high protein, limited fibre and glucose, and digestion. There are prominent members of this order that have a vegetarian diet, e.g., giant panda (*Ailuropoda melanoleuca*), who have evolved adaptations accordingly, however most of these adaptations do not involve the digestive system, which is still simple and short.

The wild diet of species in the order Carnivora varies substantially (see Table 8.1) and recommendations for feeding also varies considerably. However, a focus should be on replicating the nutritional content of the wild diet, even if the specific dietary items cannot be replicated. For those that usually predate on other species, a diet containing whole carcasses is recommended with research showing this can reduce diarrhoea, vomiting, and gastritis. In the wild these species would spend time processing their food, with time spent on removing fur, pulling at flesh, and gnawing through bones. Provision of whole carcasses can impact on activity budgets to better replicate these behaviours. Finch, Williams, and Holmes (2020) showed that a move to a whole carcass diet increased the total amount of time spent feeding, ultimately decreasing pacing behaviours, in Asiatic lions (*Panthera leo persica*). Recommendations towards less frequent larger meals for species like lions are common, with other behavioural impacts including a reduction in agonistic behaviour, as this is commonly seen on feed days regardless of if the animals are fed daily or not. An increase in carcass feeding has also been associated with more solid faecal scores in comparison to animals fed a commercial diet. An increase in whole carcass feeding generally leads to an increase in consumption on feed days, facilitating further the need for starve days around this.

8.2.3 Feeding frequency

In the wild large carnivores would not successfully hunt every day, so they are commonly adapted to a gorge-and-fast method of digestion. However, captive animals are often fed daily, sometimes even twice a day. Altman, Gross, and Lowry (2005) moved lions from a daily feeding schedule, with one starve day per week, to a gorge-and-fast schedule where animals were fed three times a week. This improved digestibility of fats, protein, and dry matter. A recommendation of starve days for many species now is common, however the benefits are controversial, with some suggesting fasting days decrease stereotypic behaviour and others linking starve days to an increase in pacing behaviour. Overall amount of food for the week should stay consistent and a larger meal provided prior to starve days. If moving to introduce starve day(s) it is recommended that behavioural observations are conducted to ensure that there is no increase in abnormal repetitive behaviours (ARBs) on fast days. Altman et al. (2005) also suggest that randomisation of any feeding/fasting schedule would be beneficial to decrease anticipatory pacing behaviour.

Many smaller carnivore species are adapted to multiple small meals a day, so once or even twice a day feeding may not be suitable for these species either. Research on domestic cats (*Felis catus*) suggests that multiple feeds are preferred, with up to 12 small meals per day selected when food is provided ad libitum, each with a calorie content of approximately 23 kcals. Meerkats usually feed throughout the day, with many collections offering several feeds, however the Association of Zoo and Aquariums (AZA) care manual for these species recommend feeding the majority of the diet in the morning, as animals will then return to the food throughout the day mimicking their wild behavioural repertoire. Food should be spread around the exhibit or, if placed in feeding stations, there should be one feeding station for every three meerkats to reduce aggression. Providing food in small portions throughout the day has been linked to increased aggressive interactions in this species so a large meal in the morning and a later afternoon feed may be the most beneficial for social interactions.

Research in coyotes (*Canis latrans*) concluded that terrestrial carnivores would benefit from a feeding regime that consists of a mixture of unpredictable and predictable elements with the unpredictable elements integrated into a more predictable framework. This is assumed to help replicate seasonal fluctuations within an otherwise stable environment (Gilbert-Norton, Leaver, & Shivik, 2009).

8.3 ENCLOSURE CONSIDERATIONS BASED ON BEHAVIOURAL EVIDENCE

Enclosure design and furnishing needs to consider the specific needs of the animals, keepers, and visitors to be functional and sustainable. Carnivores need the ability to remove themselves from visitor areas. Williams, Carter, Rendle, and Ward (2021) highlighted that Amur leopards (*Panthera pardus orientialis)* showed an increase in negative human–animal interactions and visitor avoidance when zoos reopened following the COVID-19 lockdowns, suggesting that leopards can be stressed by visitor presence. If animals are unable to avoid or remove themselves from this stressor then there is an increased chance of the development of ARBs. Visual barriers in the exhibit can help with this. Visual barriers are also beneficial for animals to avoid each other within their enclosure. This can be particularly advantageous for social species, like meerkats, as it allows subordinate individuals to show avoidance behaviours when needed. Including naturalistic elements like shrubbery, trees, hills, rocks, and termite mounds can allow animals to hide whilst also providing additional shelters within the exhibit. Visual barriers can also be beneficial to reduce aggressive behaviours, as well as competition for resources. The ability for animals to have natural dens or at least a choice of denning sites can help with reproductive success. This space being away from visitor areas has also been linked to improved reproductive success (Miller, Ivy, Vicino, & Schork, 2019). This is not surprising, with many species of carnivore in the wild hiding their young in response to potential predation. Dens for social species also need to be of sufficient size for all members of the group to huddle together, however a choice of dens must be provided to allow individuals to avoid conspecifics if required.

Many species of carnivore are arboreal or have a preference for an elevated viewing area, for example, meerkats for sentry duty or cheetahs sitting on platforms and the roof of vehicles in game reserves, so consideration of vertical as well as horizontal space is important. This also allows species to exhibit climbing and jumping behaviours that promote activity and increase the exhibit's complexity. Opportunities for vertical activity also increase the amount of useable space in the enclosure.

A variety of natural substrates, such as sand, soil, and bark within the exhibit can be beneficial for species to just add variation, as well as help with drainage. The use of concrete in exhibits has been linked to rheumatism in dholes so should be avoided. Smaller species like Asiatic golden cats (*Catopuma temminckii*) benefit from cubbing/sleeping boxes in various locations around the exhibit to hide in.

Consideration of areas for training, whether on or off show, should also be included in the design of the exhibit as these are beneficial for regular health and welfare checks and potentially to facilitate some veterinary diagnostics or treatments.

8.4 BEHAVIOURAL BIOLOGY AND CARNIVORE WELFARE

Carnivores, especially the big cats, have been well studied regarding their captive welfare and behaviour, with various topics including enrichment, keeper interactions, and visitor impacts being reviewed for a range of species. Enrichment is covered specifically in Section 8.5. Many of the welfare issues in captive carnivores can be linked to a lack of activity due to reduced requirements to maintain a territory or forage and hunt for food. As with many other taxa, a change in behavioural patterns is often the first sign of potential welfare concerns. Therefore, regular monitoring of behaviour is important.

Several studies have looked at the accuracy of keeper behavioural and/or welfare assessments, with most of these concluding that keeper assessment of their animal's behaviour and welfare is accurate. Both AZA and the World Association of Zoos and Aquaria (WAZA), encourage internal welfare assessments by suitably qualified members of staff. In one example Khadpekar, Whiteman, Durrant, Owen, and Prakash (2018) compared keeper assessment of sloth bear (*Melursus ursinus*) behaviour to their own formal behavioural observations, finding that keeper feedback was an effective tool to gather behavioural data.

Many of the species in this order are also used regularly for Keeper for the Day and other up-close experiences with members of the public. So, ensuring that they have regular positive keeper interactions around this will improve the keeper–animal bond and overall animal welfare in other interactions with people. Training individuals to participate in routine husbandry checks (when undertaken correctly) is a way to increase these positive human–animal interactions, as well as monitor overall health and welfare.

The European zoo association (EAZA) Felid Taxon Advisory Group has published specific guidelines on the training of cats for use in public demonstrations, however the considerations in this document could be applied to most other carnivore species. This includes:

- Use of positive reinforcement.
- Training must promote the use of natural behaviours.
- Physical touching of animals should only be to condition for veterinary treatments, e.g., habituation to touching of tail for blood draw.
- Animals should be provided with choice on participation in any training.

Choice is linked to increased positive welfare states in a range of species and training in carnivores has been shown to reduce the prevalence of some abnormal repetitive behaviours, with Shyne and Block (2010) showing reduced pacing in painted dogs following training sessions. There is also a suggestion that positive interactions via training can also reduce animals' negative responses towards visitors. This would be particularly important for species that are being used in close encounters with the public. Use of an online search engine for UK and European zoo websites using the search terms, "zoo encounters" and "keeper for the day," showed the following species being used in "Keeper for the Day" or other one-to-one or close encounter experiences:

- Lion
- Red panda (many of which include the option to go into the enclosure)
- Meerkat (again often going into the enclosure)
- Cheetah
- Hyena
- Asian short-clawed otter
- Tiger
- Wolf
- Leopard

Many of these experiences include protected contact training or feeding sessions with these animals. There is limited research into the impact of these sessions on the behaviour and welfare of these animals, with the few studies that have been completed generally having very small sample sizes and conflicting findings. Some research suggests that interactions with the public in the role of ambassador animals had no impact on the behaviour and heart rate of cheetahs, with other studies suggesting an increase in anticipatory behaviours in species used for public interactions. However, there was no suggestion of aggressive or antagonistic behaviours towards visitors. Careful monitoring of the behaviour of any animals used in animal encounters is required to ensure their suitability for this type of interaction and continue to promote positive human–animal interactions with unfamiliar individuals.

The behaviour and attitudes of the keeper impact on the behaviour of the animals in their care, with Carlstead (2009) showing that unexpected noises and movements from keepers while trying to initiate interactions with animals increased the amount of aggressive and apprehension behaviours shown by both maned wolves and cheetahs. This suggests that some habits may lead to more negative human–animal interaction. Ensuring keepers undertaking any training with animals have the relevant practical and theoretical understanding is essential to facilitating effective training. However, Carlstead (2009) noted that in many cases it was more experienced keepers that had developed these bad habits, so regular monitoring and integrated professional development would be beneficial to avoid this. Good keeper–animal interactions have been shown to increase reproductive success in big cats.

With improvements in nutrition, husbandry, and veterinary care many zoo animals are living longer, meaning they are spending more time in the geriatric stage of life. To optimise welfare in these stages then additional care requirements might be needed. Dental disease has been identified in jaguar (*Panthera onca*) and multiple bear species. This is often linked to diet provided as well as the methods of feeding.

Most of the big cats and grey wolves are prone to degenerative joint issues, which can affect overall mobility and subsequently impact on interaction with enrichment items. Other modifications to enclosure space many be needed as animals age, including reduced inclines and more bedding. Consideration should also be made during training sessions to ensure that tasks being requested are not outside the animal's abilities. For example, if an animal engages with training but does not lie down when requested it is likely due to joint stiffness or pain preventing completion of the behaviour. Ageing in most species is also associated with an overall decline in activity level and changes in the amount of interaction with conspecifics and zoo staff. Most bear species also show changes in grooming behaviours in old age (Krebs, Marrin, Phelps, Krol, & Watters, 2018). These changes however tend to be gradual as age progresses, any sudden changes in these behaviours could indicate a health issue and should be investigated. Research has also linked age with cognitive decline in a range of species, with associated amyloid protein plaques being found in the brains of bear and canid species. This could also impact on training programmes, particularly when training in new skills and in interactions with novel cognitive enrichment.

Assessments of emotional state and personality in big cats have been researched on several occasions, with a focus on the use of keeper assessments. Phillips, Tribe, Lisle, Galloway, and Hansen (2017) used 29 adjectives (such as excitable, playful, anxious, and stubborn) to describe big cat personality, asking cheetah and tiger keepers to assess their animals daily. Findings showed that keepers could effectively apply these descriptors to the animals in their care, however these did not seem to correlate with behavioural changes observed from the addition of environmental enrichment. Research shows differences in the expression of these personality traits between sexes, and correlated some of these traits to reproductive success, behavioural diversity, and circulating cortisol.

Many captive carnivores have been shown to suffer from stress in captivity, however, this can be reduced through suitable husbandry. Assessments of behaviour in carnivores then should employ a holistic approach covering health, behaviour, nutrition, impacts of zoo staff and visitors, enclosure use and the exhibit itself, and measures of mental state.

8.5 SPECIES-SPECIFIC ENRICHMENT

There are a huge range of options for providing enrichment for carnivore species (Figure 8.2). This is an area that has been well covered in the scientific literature due to the propensity for some of the larger members of the order to display ARBs that are easily identifiable by the public, e.g., pacing (Skibiel, Trevino, & Naugher, 2007). Clubb and Mason (2003) suggest that species with a larger ranging distance tend to show higher levels of stereotypic behaviour in captivity – with lions showing higher than expected levels of stereotypic behaviour and Arctic fox (*Alopex lagopus*) showing lower than expected levels of stereotypies. This study also showed that species that are wide-ranging in the wild were not moving around their exhibits as much, suggesting enclosures were not encouraging wild activity budgets at this point. Much of the research into enrichment for carnivore species is focused on encouraging natural behaviours, particularly those related to foraging and feeding behaviours. Captive carnivores can also suffer from obesity, so increasing activity level, amount of locomotion, and time taken to feed could also help malnutrition and in improving overall welfare.

Enrichment for carnivore species can include:

- Whole carcass feeding
- Automatic feeders

Figure 8.2 Several forms of enrichment are provided to captive carnivore species. Top left – a physical enrichment item provided to a captive tiger. Top right – social enrichment based on interactions and associations with a pride of lions. Bottom left – occupational enrichment to promote foraging opportunities for a captive bear. Bottom right – scratching and territorial marking opportunities for a captive tiger (photos: Z. Phillips & L. Burden).

- Ice blocks that contain frozen food or be made from blood, broth, or juice depending on the target species
- Plastic balls with holes in, so food can be hidden inside
- Live invertebrates, depending on the target species
- Fleeces or faeces from appropriate prey species
- Training
- Platforms to allow a change in height and viewing opportunities
- Digging opportunities within a range of substrates
- Scent enrichment
    - Differing scents have different effects on species can include perfumes, herbs, and spices. With catnip and nutmeg being shown to reduce the amount of inactive behaviour in black-footed cats, *Felis nigripes* (Wells & Egli, 2004), however catnip and cinnamon had no impact on the activity of oncilla, *Leopardus tigrinus* (Resende et al., 2011).
- Frozen fish
- Bones
- Buried food
- Management of a suitable social group if appropriate to the species
- Pole feeding (height varies depending on species, but most are 3 and 6 m high) to replicate the burst of energy needed to bring down prey
- Zip lines for simulating chasing of prey
- Blood trails
- Wood piles/rotten logs
- Run courses, also called lure coursing, is the chasing of a mechanical lure around a circuit, used to encourage sprinters such as cheetah to run in short bursts of high speed. Quirke, O'Riordan, and Davenport (2013) evaluated three of these courses, showing they varied in shape from a 70 m straight line to a 200 m^2 square.

These are just some examples, consideration of both species-specific behaviours and individual requirements should be made whenever looking to introduce novel enrichment. Wagman et al. (2018) has shown that when enrichment is provided also impacts on its effectiveness, with variable time enrichment having more of an impact on exploratory behaviour in bears than enrichment provided at fixed times.

Research on the impact of enrichment on activity levels varies in how much activity level is impacted. This can be affected by the type of enrichment, novelty, species being used with an individual animals' preference for specific enrichment items. However general trends suggest that the addition of suitable enrichment does positively impact on the activity levels of the animals as well as decrease incidences of ARBs across a range of carnivore species including black-footed cats, bush dogs, painted dogs. Skibiel et al. (2007) gave bones, frozen fish, and a range of spices to six different species of Felid, finding that all options increased the activity level in the big cats, and frozen fish and the species also decreased the amount of pacing behaviour seen.

Contrafreeloading behaviour is where animals prefer to work for their food, as opposed to being given it with no effort. Domestic dogs (*Canis lupus familiaris*) have been shown to exhibit contrafreeloading behaviour, as have many other species in the order Carnivora (brown bear, maned wolves, meerkats), one of the main exceptions being domestic cats who have been shown to prefer freely given food items over those that they need to work for (Delgado, Han, & Bain, 2021). Consideration of the behaviours, as well as the energetic requirements and muscles used/developed during foraging and hunting activities, should be factored into some of the enrichment devices offered to encourage contrafreeloading and promote health. Carnivores in captivity are generally less physically fit than their wild counterparts, with lower muscle mass, increased body mass, and higher levels of heart disease, as well as skeletal degradation. Enrichment is a way to increase fitness and welfare in these species, with Law and Kitchener (2019) showing that pole-fed tigers have arthrosis scores that are on average four times less than tigers who were not fed in this manner.

Enclosure design and development should consider the behavioural needs of the species including options to vary some of the features of the exhibit to increase variation. Suitable exhibit and enrichment design can lead to an increase in behavioural diversity which is a measure of positive welfare.

8.6 USE OF BEHAVIOURAL BIOLOGY TO ADVANCE CARE

This chapter has highlighted some of the specific requirements of species in the order Carnivora. Animals in this order have a wide variety of social, dietary, and behavioural requirements making it challenging to suggest specific management techniques that would advance care across the board.

These are some suggestions of current gaps in our evidence for good carnivore care where research would be beneficial moving forwards:

- Use of (suitable) carnivore species in walk-through exhibits.
- Generalist species, such as grey wolves, tend to cope better in zoos due to their overall behavioural flexibility, so more specialised species (e.g., aardwolf) would benefit from further research to enhance their care.
- With increasing numbers of excess males in captivity, management regimes to house these individuals, while encouraging the widest possible species-specific behaviour pattern, needs evaluation.
- Some of the Carnivora are living for significantly longer than they would in the wild, so application of suitable enrichment, training, and husbandry practices suitable for geriatric animals needs to be investigated.
- Impacts of animal encounters on behaviour, welfare, and preference for human–animal interactions over other forms of activity or resource.

Overall, some species of carnivore have been very well researched however issues with management and welfare are still evident, e.g., high rates of ARB in some long-ranging species. Continued research and application of the evidence base to current husbandry practices will ensure the best welfare and enhance conservation and educational messages with these species.

BOX 8.2: Further information to inform carnivore care

IUCN/SSC Specialist group Large Carnivore Initiative for Europe. Database of publications and updates on conservation project/reintroductions for wild carnivores. Large Carnivore Initiative for Europe > Home (lcie.org)

European Association of Zoos and Aquaria's Felid Taxon advisory groups guidelines on the use of Felid species in public demonstrations in zoos. Microsoft Word – 2017_Felid TAG demonstration guideline_final_approved (eaza.net)

Overview of the different carnivore species used in mixed-species exhibits. 2020_svabik_mixed_weasels_skunk.pdf (zoolex.org)

Resources to help with observing and learning more about carnivore behaviour:

Meerkat webcam from Newquay Zoo Webcams – Newquay Zoo

Black-footed ferret webcam Ferret Cams | Black-Footed Ferret Recovery Programme (blackfootedferret.org)

Edinburgh Zoo Sumatran tiger camera Live Tiger Cam | RZSS Edinburgh Zoo | Edinburgh Zoo

Taronga Zoo in Australia live stream their lion exhibit via their YouTube channel Lion Cam – YouTube

Painted dog TV live stream a hyena den from Kruger National Park Live: New Pridelands Hyena Den – YouTube

For more information on carnivores being used in visitor education and outreach programmes, see this example paper: Whitehouse-Tedd et al. (2021). Assessing the visitor and animal outcomes of a zoo encounter and guided tour programme with ambassador cheetahs. *Anthrozoös*, 10.1080/08927936.2021.1986263

8.7 CONCLUSION

The Carnivora is a large and diverse order of commonly kept and popular species in zoos. Their large range of social, behavioural, and dietary requirements makes it challenging to form generalised suggestions for husbandry and management. That said, all species would benefit from the provision of species-specific environmental enrichment and increases to the complexity of their enclosures. Further consideration of exhibit design and quality space, and the suitability of mixed species enclosures (where realistic) should be a priority. Overall, research outputs from species specific behavioural ecology study should be applied to all aspects of Carnivora management plans including enclosures, training, population management, and nutrition.

REFERENCES

Altman, J., Gross, K., & Lowry, S. (2005) Nutritional and behavioral effects of gorge and fast feeding in captive lions, *Journal of Applied Animal Welfare Science*, *8*(1), 47–57.

Carlstead, K. (2009). A comparative approach to the study of keeper-animal relationships in the zoo. *Zoo Biology*, *28*(6), 589–608.

Clubb, R., & Mason, G. (2003). Captivity effects on wide-ranging carnivores. *Nature*, *425*, 473–474.

Delgado, M.M., Han, B.S.G., & Bain, M.J. (2021). Domestic cats (*Felis catus*) prefer freely available food over food that requires effort. *Animal Cognition*. https://doi.org/10.1007/s10071-021-01530-3

Dobbs, P., Liptovszky, M., & Moittie, S. (2020). Dietary management of hypercholesterolemia in a bachelor group of zoo-housed Slender-tailed meerkats (Suricata suricatta). *Journal of Zoo and Aquarium Research*, *8*(4), 294–296.

Dorman, N., & Bourne, D.C. (2010). Canids and ursids in mixed species exhibits. *International Zoo Yearbook*, *44*(1), 75–86.

Finch, K., Williams, L., & Holmes, L. (2020). Using longitudinal data to evaluate the behavioural impact of a switch to carcass feeding on an Asiatic lion (Panthera leo persica). *Journal of Zoo and Aquarium Research*, *8*(4), 283–287.

Gilbert-Norton, L.B.., Leaver, L.A., & Shivik, J.A. (2009). The effect of randomly altering the time and location of feeding on the behaviour of captive coyotes (Canis latrans). *Applied Animal Behaviour Science, 120*, 179–185.

Hassanin, A., Veron, G., Ropiquet, A., Jansen van Vuuren, B., Lecu, A., Goodman, S.M., Haider, J., & Nguyen, T.T. (2021). Evolutionary history of Carnivora (Mammalia, Laurasiatheria) inferred from mitochondrial genomes. *PloS One, 16*(3), e0249387.

Khadpekar, Y., Whiteman, J.P., Durrant, B.S., Owen, M.A., & Prakash, S. (2018). Approaches to studying behavior in captive sloth bears through animal keeper feedback. *Zoo Biology, 37*(6), 408–415

Krebs, B.L., Marrin, D., Phelps, A., Krol, L., & Watters, J.V. (2018). Managing aged animals in zoos to promote positive welfare: A review and future directions. *Animals, 8*(7), 116.

Law, G., & Kitchener, A.C. (2019). Twenty years of the tiger feeding pole: Review and recommendations. *International Zoo Yearbook, 54*(1), 174–190.

Miller, L.J., Ivy, J.A., Vicino, G.A., & Schork, I.G. (2019). Impacts of natural history and exhibit factors on carnivore welfare, *Journal of Applied Animal Welfare Science, 22*(2), 188–196.

Phillips, C.J.C., Tribe, A., Lisle, A., Galloway, T.K., & Hansen, K. (2017). Keepers' rating of emotions in captive big cats, and their use in determining responses to different types of enrichment, *Journal of Veterinary Behavior, 20*, 22–30.

Quirke, T., O'Riordan, R., & Davenport, J. (2013) A comparative study of the speeds attained by captive cheetahs during the enrichment practice of the "cheetah run". *Zoo Biology, 32*(5), 490–6.

Resende, L.S., Gomes, K.C.P., Andriolo, A., Genaro, G., Remy, G.L., & Ramos Júnior, V.A. (2011). Influence of cinnamon and catnip on the stereotypical pacing of oncilla cats (*Leopardus tigrinus*) in captivity. *Journal of Applied Animal Welfare Science, 14*, 247–254.

Shyne, A., & Block, M. (2010). The effects of husbandry training on stereotypic pacing in captive African wild dogs (Lycaon pictus). *Journal of Applied Animal Welfare Science, 13*, 56–65.

Skibiel, A.L., Trevino, H.S., & Naugher, K. (2007). Comparison of several types of enrichment for captive felids. *Zoo Biology, 26*, 371–381.

Wagman, J.D., Lukas, K.E., Dennis, P.M., Willis, M.A., Carroscia, J., Gindlesperger, C., & Schook, M.W. (2018). A work-for-food enrichment program increases exploration and decreases stereotypies in four species of bears. *Zoo Biology, 37*(1), 3–15.

Wells, D.L., & Egli, J.M. (2004). The influence of olfactory enrichment on the behavior of black-footed cats, *Felis nigripes. Applied Animal Behaviour Science, 85*, 107–119.

Williams, E., Carter, A., Rendle, J., & Ward, S.J. (2021). Impacts of COVID-19 on animals in zoos: A longitudinal multi-species analysis. *Journal of Zoological & Botanical Gardens, 2*, 130–145.

Yordy, J., & Mossotti, R.H. (2016). Kinship, maternal effects, and management: Juvenile mortality and survival in captive African painted dogs, *Lycaon pictus. Zoo Biology, 35*, 367–377.

9

The behavioural biology of marine mammals

LOUISE BELL
University Centre Myerscough, Preston, UK

MICHAEL WEISS
University of Exeter, Exeter, UK

9.1 INTRODUCTION TO MARINE MAMMALS

Marine mammals are aquatic mammals that primarily reside within the ocean and are obligately tied to a marine ecosystem. Found globally, they include more than 130 species across four different groups, including some of the world's largest species (ITIS, 2021). Marine mammals are grouped into cetaceans (dolphins, porpoises, and whales), pinnipeds (seals, sea lions, and walrus, *Odobenus rosmarus*), sirenians (manatees and dugong), and the paraphyletic marine fissipeds (polar bear, *Ursus maritimus* and sea otter, *Enhydra lutris*). Highly specialised adaptations allow these species to thrive in extreme temperatures, depths, pressures, and darkness, fulfilling a range of different ecological roles, ranging from filter feeders (baleen whales), to herbivores (manatees), and top predators (killer whale, *Orcinus orca*). Between 1960 and 2021, a total of 16,324 individual marine mammals were recorded throughout the global zoo and aquarium community, as represented by the Zoological Information Management System (ZIMS) database. Eared seals

DOI: 10.1201/9781003208471-11

and sea lions (Otariidae) account for 51% of all these animals; earless seals and true seals (Phocidae) account for 16%; and polar bears account for 15%. The California sea lion (*Zalophus californianus*) is the most common marine mammal held in zoological collections (27% of all marine mammals kept).

9.2 MARINE MAMMAL ECOLOGY AND NATURAL HISTORY RELEVANT TO THE ZOO

With great ecological diversity across marine mammals, this chapter focuses primarily on the three taxonomic groups of pinnipeds, cetaceans, and marine fissipeds. Some of the most well-known of these marine mammals include sea lions and fur seals, harbour seals *(Phoca vitulina)*, bottlenose dolphins (*Tursiops truncates*), killer whales, polar bears, and sea otters. The natural history of these species is provided in Table 9.1, highlighting their life history and key ecological behaviour.

Pinnipeds are exclusively, and highly effective, aquatic predators. For example, a 100 kg seal can consume at least 5–7 kg of food daily (Nowak, 2003). Most pinnipeds are generalist foragers, predating fish, squid, and crustaceans but some southern species are more specialised, including the South American sea lion (*Otaria byronia*) that predates on fur seals (Arctocephalinae), and the New Zealand sea lion (*Phocarctos hookeri*) that predates on penguins (Sphenisciformes). All species of pinniped can respond to fluctuating prey availability, with Phocids able to fast for long periods of time. All pinniped species require some form of terrestrial habitat to haul out on for resting, birthing, and moulting for days or weeks at a time. Consequently, these species do not range far from land (Reeves & Stewart, 2003). Of the pinnipeds, the Otariidae (eared seals and sea lions) and Phocidae (earless seals and true seals) are the most commonly represented marine mammals in zoological collections.

Extremely geographically diverse, the order Cetacea consists of the baleen whales (Mysticeti), and the toothed whales (Odontoceti); which includes the family Delphinidae (oceanic dolphins). Odontoceti have greater taxonomic diversity than the Mysticeti, consisting of ten families, 34 genera, and more than 70 extant species. Ecology and natural history vary greatly, with the delphinids (dolphins) being the largest and most diverse family with more than 35 recognised species (Reeves & Stewart, 2003). The life histories of Cetacea are varied but primarily correlate with feeding strategy (single prey versus filter feeders) and body size (larger species having slower life history). All cetaceans are highly social to some degree, however only 10 genera are maintained in captivity, with two delphinids discussed in this chapter: the bottlenose dolphin (*Tursiops truncatus*) and killer whale.

Marine fissipeds (polar bears and sea otters) also have an inherent reliance upon the marine environment, however sea otters occupy much smaller geographic ranges compared to polar bears. Both polar bears and sea otters have toes which are separated from each other, unlike pinnipeds, which have digits joined within webbed flippers. Both species are efficient predators with movement patterns maximising prey availability and breeding opportunities (Hamilton, Kovacs, Ims, Aars, & Lyderson, 2017). As an extremely efficient ambush predator, the polar bear has exceptional auditory, olfactory, and visual acuities for hunting, and can detect prey up to 32 km away. They also use these senses to detect the presence, and reproductive status, of other bears within a vast home range (Owen et al., 2015). Polar bears typically prey upon ringed seals (*Pusa hispida*) and bearded seals (*Erignathus barbatus*), however their overall diet can include more than 80 different species spanning 21 plants, 18 birds and six fish species (Derocher, 2012). Polar bears exhibit physiological variation in response to the movement of the Arctic sea ice, with substantial seasonality in individual body mass and condition. Sea otters eat 20–33% of their body weight daily, with their diet consisting primarily of molluscs, crustaceans, echinoderms, and occasionally fish (Hunter, 2011).

9.3 ENCLOSURE CONDITIONS FOR MARINE MAMMALS

Due to their large size, ranging, and social behaviour, housing marine mammals can be challenging and expensive, requiring complex systems and expertise. Enclosure design has progressed from historical conditions based primarily on practical logistics and close visitor encounters (Figure 9.1), to modern naturalistic housing with complex management systems. Species-specific space requirements and husbandry practices differ across zoological collections, with extreme variations in enclosure design.

Table 9.1 Natural history and ecology of selected marine mammal species.

Family	Species & IUCN red list status (as of February 2022)	Distribution & home range		Key ecological behaviour	Social behaviour	Breeding behaviour			Life history [years] & size (length [cm]/ weight [kg])
		Small	Large			Sexual maturity [y]	Breeding group	Unique ecology	
Otariidae (eared seals, sea lions)	California sea lion Least Concern	✓ rarely >16 km from land	X	Groups of 15–20 swim, social play key for juveniles	Social and congregate in large groups with polygynous harems	F – 6–8 M – 9	✓	Sexually dimorphic bulls maintain breeding territories for 27 d with a M:F ratio of 1:16 individuals	30 years F – 150–200 / 50–110 M – 200–250 / 200–400
	Galapagos sea lion (*Zalophus wollebaeki*) Endangered	✓	X	15-hour foraging trips				75% of lactating females hunt nocturnally	
	South American sea lion Least Concern	✓	X	Not migratory but M may spend significant time at sea after breeding (some 1,600 km away)		F – 4 M – 6	✓	Males travel far	16 years F – 180–200 / 140 M – 216–256 / 200–350
	New Zealand sea lion Endangered	✓ Can move up to 2 km inland	X	Generalist forager but also predates penguins		F – 3–5 M – 6	Pairs	Sexually dimorphic, bulls have dark long mane. Small territories (2 m circle) of 1,000 on sandy beaches. Juveniles males may travel.	~23 years F – 160–200 / 61–104 M – 200–250 / 300
Phocidae (earless seals, true seals)	Harbour seal Least Concern	✓	X	Limited movements for foraging and breeding. Resides along shorelines, not associated with ice	Solitary but aggregates for breeding with polygynous bulls. Social cohesion only between pup and dam	F – 3–5 M – 6	✓	Juveniles can disperse ~300 km	~34 years F – 120–150 / 50–150 M – 150–200 / 70–170

(Continued)

Table 9.1 (Continued) Natural history and ecology of selected marine mammal species.

Family	Species & IUCN red list status (as of February 2022)	Distribution & home range		Key ecological behaviour	Social behaviour	Breeding behaviour			Life history [years] & size (length [cm]/ weight [kg])
		Small	Large			Sexual maturity [y]	Breeding group	Unique ecology	
	Australian/South African fur seal (*Arctocephalus pusillus*) Least Concern	✓ Range <200 km offshore	X	Feeds 70% fish & 20% cephalopods	Generally solitary but females form groups of 7–9 individuals at breeding times	F – 3–7 M – 4–5	✓	Bulls weigh x4 more than females. Females can have one pup per year until 23 y. Bulls cannot win territory until ~11 y	18 years F – 136–176 / 36–122 M – 184–234 / 134–363
	Northern fur seal (*Callorhinus ursinus*) Vulnerable	X	✓ more than 10,000 km	No regular haul out sites so range widely		F – 3–7 M – 5–6	✓	Bulls are unable to maintain territory until ~10–12 y, but breed until 15–20 y. Pups weaned at 1–2 y	30 years F – 142/ 43–50 M – 213 / 181–272
	South American fur seal (*Arctocephalus australis*) Least Concern	✓	X	Generalist forager but also predates fur seals	Aggregate	F – 3 (full size at 10) M – 7	Pairs	Bulls breeding territory of 50 m²	~30 years F – 143 / 30–60 M – 189 / 150–200
Odontoceti (Toothed whales)	Common bottlenose dolphin Least Concern	✓	✓	Range varies by species with resident coastal and inshore populations but others migrating long distances. Daily movement ranges lager than any terrestrial mammals	Social structure patterns: fluid social (few strong relationships between individuals); community-structured fission-fusion societies (some strong relationships & highly dynamic); multilevel social (closed, stable units associating with one another); Super pods (can be ~1,000 individuals in open-ocean populations)	F – 5–14 M – 9–13	✓	M disperse at maturity and forms all M "alliances"	F – 26 years M – 19 years F & M – 175–400 / 150–200

	Harbour porpoise (*Phocoena phocoena*) Vulnerable	X	✓	Frequent coastal waters, estuaries & large river mouths	Pairs or small groups but can often be found in large social groups of ~100	F & M – 3	Pairs or groups 5–10	No sexual dimorphism. Reproduces almost continually.	6–10 years (rarely >13) F & M – 150–160 / 45–60
	Killer whale Data Deficient	X	✓	Most extensive distribution of any non-human mammal. Is highly mobile and can have home ranges of several 1,000 km²	Highly social with matrilineal systems	F – 13–19 M – 15	Groups consist of matrilineal kin with centrality of the mother-calf bond (see Box 9.1)	2–6 y interbirth intervals	F – ~90 years M – 50–60 years F – ~850 / 5,500 M – 980 / 9,000
Ursidae	Polar bear Vulnerable	✓ Female bears travel less with daily ranges of 0.5–40 km/d and ~5,472 km/year	✓ M bears travel 14–18 km/d with a ~150–300 km annual range which can be 50,000 km² on stable ice and 250,000 km² on drift ice	19 subpopulations based on movements. Movement driven by prey type with swimming speeds of ~6.5 km/h for up to 65 km across open water	Solitary	F – 5–6 M 10–11	X F seek remote areas due to infanticide & cannibalism risk from M	Excessive sexual dimorphism with M up to ~800 kg & F up to ~450 kg. Pregnant F – 49% body mass fat composition, late spring when seal pups peak F dig snow dens 8–60 km from coast and inland, giving birth to ~2 cubs	F – ~32 years M – 25–30 years (with mortality 1–4%) Cub mortality ~25–65% in 1 st y and ~1% cub dispersal F – 200–250 / 150–300 M – 200–250 / 300–800
Mustelidae	Sea otter Endangered	✓ ~1 km from shore	X	Nocturnal and diurnal feeding and dives 25–40 m for 60–120 s. No subcutaneous fat layer to keep warm so uses air trapped in thick fur, with a very high metabolic rate	Solitary	F – 4 (earliest breeding 4–6) M – 5–6 (earliest breeding 6–8)	X	Solitary but can be found aggregating in groups ~2,000 with high-quality squid patches. Breed year-round with birth peaks differing by location. Pup is born at sea and carried by dam until 2 m.	F – 23 years M – 15 years Pup mortality varying by location – 17–53% F & M – 100–120 body & 25–37 tail / F – 15–32 & M – 22–45

F = female, M = male (Adapted from Kooyman & Trillmich, 1986; Nowak, 2003; Reeves & Stewart, 2003; Hunter, 2011; Matthews, Luque, Petersen, Andrews, & Ferguson, 2011; Derocher, 2012; Hamilton et al., 2017; Hempstead & Larson, 2019; Rendell, Cantor, Gero, Whitehead, & Mann, 2019; Weiss, Ellis, & Croft, 2021)

Figure 9.1 "Cuddles," a captive killer whale in a public talk at Dudley Zoo, UK, in 1971 (Photo: Dudley Zoo & Castle, 2021).

9.3.1 Environmental complexity

Taxa and species-specific housing requirements vary greatly, with large, long-lived, and far-ranging species, like the killer whale, difficult to house compared to localised low-ranging species, like the southern fur seals or sea lions. Despite all species requiring some form of access to water and land, spatial trade-offs often occur due to limited space availability and the varying land-to-water ratio requirements across species. Establishing the appropriate behavioural needs, and subsequent environmental complexity, for each species is therefore of critical importance. For example, tank and pool design are highly influential in meeting the full behaviour repertoires of marine mammals in captivity. In particular, differences in pool depth are considered behaviourally relevant for bottlenose dolphins; with dolphins displaying higher incidences of social swimming and play in moderate pool depths of 4–8 m, rather than in shallow or deeper areas (Lauderdale et al., 2021). In contrast, bottlenose dolphins housed in "open" sea pen facilities, appear to spend less time floating and swimming in circular patterns than those housed in a "closed" traditional pool facility (Ugaz, Valdez, Romano, & Galindo, 2013). This suggests that spatial design and complexity, and water access are important components of captive marine mammal welfare as they allow for the instigation of complex behaviours.

9.3.2 Housing and pool design

Daily activities, including foraging, playing, socialising, and travelling, consume a large proportion of the time of wild marine mammals. The design of housing and pools in zoos is critical for instigating such highly motivated behaviours in captive animals. For example, tank and pool conditions (e.g., construction materials, width, and depth) are highly influential in encouraging the performance of swimming and play. Species that forage at greater depths (e.g., bottlenose dolphins) generally need deeper pools to encourage diving and social swimming (Lauderdale, Shorter, et al., 2021), compared to shallow-living species, like sea otters. Access to wild resources can be particularly influential, with the use of naturalistic sea pens for cetaceans, and natural seal and sea lion pools, permitting access to wild substrates that create greater foraging opportunities and environmental complexity. However, such enclosures do not always permit water quality maintenance (including seasonal temperature variation and salinity levels) and can compromise the welfare of individual animals.

In reality, tank or pool design is often planned around the use of complex filtration systems that allow for water quality to be discreetly monitored and maintained to ideal conditions, promoting behavioural diversity while reducing potential health issues. For example, saline-induced ocular problems are common and affect half the population of captive seals and sea lions. These issues appear to increase with age, suggesting the maintenance of water quality is key across all life stages, with artificial environments being easier to control (Higgins & Hendrickson, 2013). Despite the importance of filtration systems, they also occupy a substantial amount of space and can create loud noises and vibrations inconsistent with these species' unique modes of communication. These undesirable noises and vibrations can also be affected by the design, layout, and construction materials used for the tank or pool which can cause sound amplification and reverberation. To prevent such issues all tanks and pools should be designed to provide additional depth and space, utilising irregular enclosure shapes and surfaces (Couquiaud, 2005). The need for specialised infrastructure, including quarantine facilities and access for veterinary and emergency evacuation procedures, can further limit the available water area available for animals, and should be carefully planned in addition to the housing requirements of the animals to ensure adequate space is available at all times.

9.3.3 Environmental conditions

Environmental conditions have great ecological importance for marine mammals, with thermal, chemical, and light considerations acting as key factors for the instigation of natural behavioural and physiological activities. The instigation of highly motivated behavioural states, including breeding, hibernation, and migration, can be triggered by such environmental conditions and thus the strict maintenance and control of these conditions is required in captive settings. As behaviour performance and efficiency are controlled by water temperature, mimicking the natural conditions of a species is especially important, however this can be difficult for species which naturally occur in different thermal zones throughout the year. In such cases, or where species are not being housed within their normal range, the artificial maintenance of relevant thermal zones within an enclosure is important to mimic such wild conditions. For example, bottlenose dolphin husbandry guidelines state that the minimum ambient water temperature of indoor facilities for adults should be no less than 12°C, which is only 1°C higher than their lowest identified critical temperature, increasing to 14°C for dolphins with calves (Rose, Snusz, Brown, & Parsons, 2017). Similarly, air temperatures for polar bears housed indoors should be no greater than 0°C for at least three winter months, and 12°C maximum for the rest of the year to mimic wild temperature variation (Rose et al., 2017).

Alongside of complex filtration and temperature control systems, wider and deeper pools, haul-out rocks, and different substrate types, can all provide greater thermoregulatory opportunities by promoting natural basking and cooling behaviours. For example, captive walrus enclosures should have periods of low temperatures during the year, and dry resting substrates should be cooled to freezing temperatures for the winter period to replicate the conditions of wild haul out areas (Rose et al., 2017). Similarly, lighting is also an important consideration, particularly for many polar species. For example, it is recommended in Rose et al. (2017) that all polar marine species be kept in at least 18 hours of uninterrupted darkness in any 24 hour period during the three winter months. This artificial graduation between seasons can replicate natural polar conditions and encourage wild-type behaviour. Maintenance of suitable environmental conditions across the 24 hour period is key to the comfort, health, and behavioural welfare of marine mammals; this includes water access schedules, relevant indoor air temperatures, the provision of shade opportunities etc. and should be continually monitored and adjusted when necessary.

9.3.4 Space use and proximity

Most marine mammals travel large distances and occupy multiple habitats, ranging from coastal and estuarine habitats, to pelagic and polar habitats. Replicating such complex and heterogenous environments is extremely difficult, if not impossible, but it is necessary to maximise behavioural opportunities. Replication can be achieved through the provision of appropriate water access, haul out areas, conspecific interaction opportunities, and indoor housing for thermoregulation. However, the unique husbandry and daily management requirements for marine mammals can restrict the complexity and layout of enclosures. Practical logistics (such as keeper access and water maintenance) complicate space use and availability of enclosure areas, preventing the performance of certain natural behaviours. For example, the need for animal and keeper retreat areas, specific animal training areas, and separation walls and gates, housing for life support systems, are all essential for daily management but can restrict opportunities for the display of natural behaviours. As space in the zoo is at a premium, this often results in a compromise between enclosure utility, and safety for both animals and keepers. Similarly, suitable viewing areas for both keepers and the public, often including large public seating areas for educational displays, are also required, which can consume a large proportion of useable space.

The design of marine mammal enclosures must consider the natural history of the species concerned, throughout all of its life stages, and should also consider both the present and future captive group composition, using the species' natural social behaviour as a guide. Regardless of the species, all enclosures should be designed with both access and utility in mind, including construction properties that allow for interchanging space use and complexity, utilising moveable gates to offer differing opportunities for individual animals at ecologically relevant times, such as breeding or pupping seasons. These systems are often used with species that can be managed in free contact style systems, such as dolphins and sea lions, where keepers and animals are in the same useable space. Such systems are not as feasible for species requiring permanent stable structures in accordance with their category of risk, such as polar bears.

Even when "expansive" spaces are provided, they must be considered in parallel with other ecological and social requirements of each species. For example, the provision of a deep pool may not replicate wild movement, foraging, and social play behaviours if the social group composition is not correct. Defining what "suitable" is can be extremely challenging, as the needs of many marine mammals are poorly understood, often relying on a limited number of wild and captive studies for each species. Immersive exhibits (Figures 9.2 and 9.3) can be suitably ecologically rich, including large open areas that promote movement by offering a range of elevations and substrates with natural gradients. Such exhibits allow for a more natural representation of the animals and can provide greater environmental complexity, providing long-distance viewing opportunities for both animals and visitors.

9.3.5 Training areas

As all captive individuals need health and welfare management, the provision of a suitable space for safe and practical training is essential for animal and keeper safety, but will vary greatly by species size, enclosure complexity, and necessary safety considerations (Lauderdale, Shorter, et al., 2021). Depending on the species, and the husbandry and management required, some species and individuals are worked "hands-off," while others require a very "hands-on" approach, with frequent husbandry tasks including health checks (Figure 9.4), pen cleaning, and pool maintenance required. For larger species, such as killer whales and walrus, providing this level of care and management can be difficult, with the need for separation or rotation of key individuals. This can be further complicated if the separation involves movement between water and land areas. Good enclosure design can prevent such issues by providing back-to-back pens, shared water systems, and moveable training areas, reducing the required training space and increasing the useable animal space.

Access and training for veterinary procedures and interventions is also a key consideration for all marine mammals, particularly due to their complex physiology. For example, the use of anaesthesia is not recommended for many marine mammals, particularly cetaceans, due to an inappropriate elicitation of their "dive reflex" and slow rates of recovery, making it an extremely risky veterinary procedure (Higgins & Hendrickson, 2013). As a result, all animals should be trained for such activities as part of their daily management pattern, removing the need for anaesthesia. This requires not only well-trained animals, but also easy access to workable safe areas, which will allow for the safe assessment of individuals and collection of samples if required. For example, the safe collection of oral samples from

Figure 9.2 The new naturalistic enclosure "Point Lobos" at Yorkshire Wildlife Park (UK), housing California sea lions.

well-trained species, including the killer whale, false killer whale (*Pseudorca crassidens*) and elephant seal (*Mirounga*), can be used to determine oral health and determine the presence of bacterial infections, such as *Helicobacter* (Goldman et al., 2002).

9.4 BEHAVIOURAL BIOLOGY AND MARINE MAMMAL WELFARE

The captive environment can restrict the performance of highly motivated behaviours due to difficulties in replicating the wide-ranging and varied foraging strategies of marine mammals. Ranging and foraging in particular are some of the most highly motivated behaviours in marine mammals and have the greatest potential to impact individual animal welfare. These two behaviours are discussed below, with other equally important and ecologically relevant behaviours, such as sociality, beyond the scope of this chapter. A case study on sociality in killer whales is provided at the end of this chapter, highlighting how complex, and difficult to replicate, this can be.

9.4.1 Ecological relevance of ranging behaviour

Species with large home ranges, and those which migrate or move between different areas during seasonal prey fluctuations, may display restricted behavioural repertoires in captive settings due to limited space availability or exhibit appropriateness. As highlighted by Anzolin, De Carvalho, Vianajr, Normade, and Souto (2004), and many others, attempting to replicate the natural environments of marine mammals in captivity is incredibly challenging, and has resulted in a huge diversity of captive facilities, leading to differing welfare standards and consequences. Despite the best efforts of captive population managers to provide complex enclosures and space to encourage movement, the captive environment is still restrictive, with natural ranging and foraging behaviours often constrained.

Ranging and movement motivations appear to be the most influential factors in potential behavioural restrictions in captive marine mammals (Clubb & Mason, 2007), and as such, methods to

Figure 9.3 The large naturalistic "Project Polar" enclosure at Yorkshire Wildlife Park (UK), housing four male polar bears with paddock rotation. Note the presence of two bears on the mound towards the back of the enclosure, showing the scale of the available space and the access to water.

alleviate or minimise these restrictions are key to improving captive behavioural welfare. For example, reduced swimming space and poor pool design can result in injuries from accidental incidents in species with specific movement patterns in the wild. Other traumas can also occur in captive environments, such as dental trauma, commonly seen in captive pinnipeds, and mandibular fractures (Higgins & Hendrickson, 2013). Terrestrial rangers, like the polar bear, are susceptible to experiencing abnormal repetitive behaviour (ARBs) in some captive settings (Clubb & Mason, 2003), likely due to an instinctive drive to move to locate high-value resources (e.g., a mate) and to hunt. Observations from the 1970s found that 82% of captive polar bears performed some form of stereotypic behaviour, dropping to 55% in 2004 (Mason, Clubb, Latham, & Vickery, 2007). Stereotypic behaviour (a type of ARB) in polar bears may include pacing (a repetitive form of walking) and head swinging. Similarly, repetitive circle swimming is displayed by aquatic mammals including seals, sea lions and walrus (Grindrod & Cleaver, 2000), and "back and forth" movements and muzzle hitting on an object or the waterline has been noted in West Indian manatees, *Trichechus manatus* (Anzolin et al., 2004).

Natural home range size, daily travel distance and time spent performing movement behaviours (including walking and swimming) can be species-specific risk factors for behavioural restriction. Seasonal differences in the intensity of these movement behaviours make it challenging to allow for their natural performance within a captive environment (due to differing spatial needs caused by behaviour change throughout the year). Similarly, wild seasonal fluctuations in prey type and availability further influence behavioural patterns. This can be seen in wild female polar bears, which show increases in daily movements with seasonal progression due to increasing prey scarcity (Kroshko et al., 2016; Hamilton et al., 2017). Although such fluctuations in available food may not occur in a captive setting, the seasonal motivation to change movement behaviours may remain.

Figure 9.4 Off-show polar bear training wall at Yorkshire Wildlife Park (UK), where bears voluntarily present requested areas of their body for veterinary care and health checks.

The definition and different types of "captive" should also be explored and discussed in greater detail, as any form of captive perimeter, whether in a closed tank or pool or in a naturalistic sea pen, has the potential to alter and restrict behaviour patterns. For example, 50% of rehabilitated manatees which were temporarily housed in constrained, natural corrals performed circle swimming, compared to 85% of manatees in an oceanarium, where four types of ARBs were regularly displayed (Anzolin et al., 2004). Therefore, determining the species-specific motivations for ARB performance is critical to the improvement of captive marine mammal welfare, helping to guide future management practices and enclosure design. The performance of ARBs is often linked to species-specific motor functions, which have a relevance in a wild context at a specific time and are often necessary for survival. Therefore, establishing natural wild behavioural repertoires is of great importance, as these behaviours can become heightened within a captive setting due to an inability to fulfil the behavioural role, resulting in redirected actions or even complete inaction. For example, apathy or lack of interest in the environment can stem from specific feeding regimes or lack of opportunities to contrafreeload (work for a reward). Understanding the origins and purposes of natural behaviours can therefore be critical to help replicate them in a captive setting.

Regardless of the standard of housing and management, behavioural restriction is highly suggestive of the captive environment and this can be difficult to assess and interpret across species. For example, polar bears have one of the largest annual ranges of any terrestrial carnivore, and will move large distances around a captive enclosure, often in the form of stereotypic pacing. Although observing a pacing polar bear may suggest poor welfare, with the behaviour appearing functionless, the pacing may be providing an outlet for redirected natural patrolling instincts, albeit it in a repetitive manner, with the animal using the edge of the enclosure as the edge of their home range. Similar explanations have also been suggested for dolphins and repetitive swimming within their pools (Clark, 2013). Contradictory to welfare perceptions, such stereotypic movement could be providing an outlet for natural behaviours, redirecting it into another action, indicative of positive coping strategies. However, this may not apply to all ARBs, as the replication of wild swimming behaviours may be difficult in closed water sources, preventing polar bears reaching their natural speeds of up to 32 km/hr (Derocher, 2012).

Intraspecific variation is also a key consideration when interpreting behavioural patterns, as ranging behaviours can differ across individuals, sexes etc. with large. This can be seen in polar bears, where well-fed males appear to move slower and travel half the distance of females (Hunter, 2011). For far-ranging species, like killer whales, this can also differ by social grouping, with some populations ranging much farther than others depending on the type of prey they consume. The importance of hunting efficiency and prey availability is also seen in polar bears, where a single female has been noted to remain completely within a 60 km territory for a whole year if sufficient prey is available (Derocher, 2012). In comparison, a different female has been recorded as swimming continuously for 232 hours, covering 687 km, highlighting the vast intraspecific variability (Hunter, 2011). Activity motivation may therefore be reduced when this "drive" is not as heightened. This has been demonstrated by Clubb and Mason (2007), who showed that 20 well-fed captive polar bears were only active for 15% of a 24-hour period, suggesting far less activity and motivation than their wild counterparts.

In reality, attempts to mimic home range distances and territory changes are completely unachievable, specifically when they are so variable, even within a species. It may be more relevant for highly mobile species (e.g., killer whales and dolphins) to improve environmental complexity and thus, space utility, rather than focus on simple metrics like distances travelled, with pool depth and horizontal dimensions having previously been shown to be relevant to naturalistic behavioural performance. Species with low-ranging behaviours, like fur seals and most sea lions, may be easier to care for by providing alternative mechanisms for the performance of specific motor functions, such as increased social play by replicating natural social groups and providing training opportunities. Similarly, captive sea otters spend less time swimming in their shallower pools and show significantly short dives compared to their wild counterparts, despite the ecological relevance of these behaviours, yet ARBs are not routinely observed in these species (Hempstead & Larson, 2019).

In summary, the overall motivations for performing ecologically relevant behaviours (e.g., movement behaviours) should be considered in marine mammal management and enclosure design. For example, swimming is primarily undertaken by most species to fulfil the need to find prey, requiring variable distances and speeds dependent on species, prey, social structure, and environment. Despite the cessation of the need (i.e., to hunt for food) to perform this behaviour, motivation will persist, and enclosures should promote welfare-positive movement patterns. However, not all wild behaviours manifest in the same way in all populations or individuals, and our understanding of their motivation is limited. For example, swimming for social play or group coherence is a different challenge and is later considered with reference to the killer whale (Box 9.1). Promoting such behaviours in captivity is therefore difficult.

9.4.2 Ecological relevance of foraging behaviour

Although the ecological relevance of highly motivated behaviours can be incredibly variable, foraging behaviour presents a unique challenge in captive marine mammal management. Meeting the involuntary behavioural demands triggered by complex physiological foraging motivations may is far beyond the scope of what any captive diet can provide. Hunger, for example, is triggered by involuntary and voluntary responses to varying environmental conditions and utilises circadian rhythms, light, and temperature changes, many of which are often beyond our control in a captive setting. Despite the regularity of captive diets, attempts to mimic wild foraging behaviour will always be challenging, as physiological and environmental conditions show great seasonal variability, particularly with respect to prey availability and abundance. Furthermore, highly complex and motivated foraging behaviours consist of numerous individual stages, including pursuit, capture, and consumption, each of which can be incredibly complex itself, specifically where social foraging is concerned. When considered part of a movement and ranging behaviour, foraging can take up a substantial amount of daily wild time budgets and can show huge complexity.

BOX 9.1: Case study: killer whales in captivity

Several species of cetacean are regularly housed in zoological facilities, notably bottlenose dolphins, beluga (*Delphinapterus leucas*) and killer whales. The practice of keeping this latter species in captivity has been the focus of a great deal of recent public and scientific controversy, with arguments ranging from this species' longevity under human care to philosophical arguments about the nature of killer whale cognition and sentience, and the ethical issues they imply. Here, we will briefly outline some of the features of killer whale behaviour in the wild that should be kept in mind in these debates. We stress that these largely outline hypotheses and areas for further investigation; a lack of robust empirical studies of killer whale welfare under human care makes firm, evidence-based conclusions difficult.

One of the most astonishing aspects of killer whale behavioural ecology is their social structure. Wild killer whales form long-lasting (sometimes lifelong) social bonds with their mother and other maternal kin. Unlike other well-known matrilineal species, killer whales of both sexes remain in their mother's group. This tends to result in populations being composed of distinct social groups, referred to as "matrilines," with little to no dispersal. Matrilines are stable units, however, multiple matrilines will regularly associate, forming open, multilevel societies. In the wild, these relationships exist in a dynamic, flexible social environment, with groups undergoing frequent fission and fusion, and spreading out over several km^2.

This social structure presents several challenges for zoological facilities. First, the lifelong mother-offspring social bond may represent a major hurdle. In captivity, there are several situations in which animals may need to be moved between facilities, which has the potential to sever what would otherwise be lifelong bonds. In addition, limited space may make it more difficult for individuals to disperse to diffuse aggressive interactions. More generally, it is unclear whether the necessarily small groups held in captive facilities allow killer whales to form the kinds of large, multifaceted social networks that they have in the wild, and whether this difference in social structure has implications for individuals' wellbeing. The largest number of killer whales currently (as of 2022) held at any single facility are the nine individuals at SeaWorld San Diego, which is approximately the size of a large matriline. However, matrilines are almost never completely socially isolated, and typically interact with many other matrilines, sometimes forming large, semi-stable social units which may contain over 40 individuals. It is currently accepted best practice to house cetaceans in groups that are compatible, however, it is unclear whether this fully allows for adequate social structure and stimulation. Comparative studies of relevant social interactions, such as physical contact and synchrony (Weiss, Franks et al., 2021), between captive and wild groups could help answer these questions.

Beyond the role of adequate space and freedom of movement in maintaining social cohesion, the degree of movement allowed to captive whales may itself be a cause for concern. Wild killer whales exhibit extremely large home ranges and long-range daily movements (over 100 km/d, Matthews et al., 2011), along with vertical movements over hundreds of metres in the water column (Wright et al., 2017). These types of movement are clearly impossible in captive settings, and outstrip any other animal regularly held in captivity. In terrestrial predators, species with larger home ranges are more likely to exhibit stereotypic behaviour indicative of poor welfare (Kroshko et al., 2016). It is unclear, however, to what degree this relationship holds true in cetaceans. Again, a lack of empirical studies is a major hindrance in ensuring the welfare of animals currently under human care.

A lack of empirical studies on killer whale welfare in captivity limits our ability to understand the degree to which these hypothesised issues are in fact impacting killer whale welfare. Direct behavioural studies of killer whale welfare are sorely lacking. In other taxa, knowledge gaps about welfare can be filled by our understanding of similar species. Our ability to do this with killer whales is limited by the extremity of killer whale physical size, behaviour, and social structure. Studies in the wild have made it clear that, within zoological facilities, there is simply nothing quite like a killer whale. Large scale, robust, and unbiased research is desperately needed to ensure the welfare of the killer whales that remain in captivity (Figure 9.8).

Foraging processes may be less heightened for species that do not have to travel far to find food, or for those which require less effort to source food, but spend more time on processing and consumption, as with the herbivorous sirenians, dugong (*Dugong dugong*) and manatee (Anzolin et al., 2004). Fulfilling the motivations of wild foraging repertoires can be extremely difficult in a captive setting, specifically when the simple presentation of "prey" significantly reduces the handling and processing time. Processing and consumption time are important for captive marine mammals, as often the food is presented so that it requires minimal animal processing time (e.g., when keepers hand feed). In captivity, walrus foraging time reduced when fed whole fish and clams as these were swallowed without processing, and as such, this increased time available for circle swimming (Kastelein, Jennings, & Postma, 2007). The presentation of food can also be manipulated to provide variability in terms of the type and regularity of food, e.g., fed whole and non-whole (shelled, cut or fillet), resulting in variability in the manipulation needed by the individual animal. Attempts to replicate the daily intake rate of some prey types are further complicated with difficulties in sourcing specialist prey items, such as crustaceans, if not already widely available.

Variation in prey type also has wider ecological relevance, with species often dependent upon seasonal prey availability, requiring greater ecological flexibility – something frequently overlooked in captive settings that often present a more restrictive diet. The restriction is not only behaviourally problematic; as wild dietary seasonal variation aids in thermal regulation, growth, and ageing processes, it also impacts on the quantity and quality of an individual's food choices. Providing variation can decrease frustration-related behaviours, reducing the potential gain from performing hunting related ARBs. Presenting limited diets may not necessarily fulfil the motivational needs of animals, but further work is needed to better understand how the lack of energy required for captive foraging relates to ARBs, and how it can be managed to minimise their performance.

Accessing varying prey types may be difficult for zoos as not all wild dietary items are practically or ethically accessible, and as such, the diet of certain species who would naturally prioritise food choices dependent on specific physiological requirements, may be hindered. For example, sourcing costs and food quality will vary across the global, with large discrepancies found between feeding low value commercially available fish against high value invertebrates (Hempstead & Larson, 2019). Seasonal fluctuations and increased costs will further impact on the availability of dietary items, with one or two main commercially available food types available, despite most wild diets consisting of at least four or five key species (Reeves & Stewart, 2003).

Limited variation and scheduled feeding plans can become predictable and do not always encourage natural spatial and temporal feeding patterns, which would mimic wild foraging times and patterns. For example, nocturnal feeding schedules are rare, despite being important for some species, like Galapagos sea lions, where diurnal schedules may not meet their foraging demands (Kooyman & Trillmich, 1986). The body mass:intake regime of larger marine mammals like the walrus or polar bear, may pose a further challenge as attempts to meet such foraging opportunities may require solitary feeding to minimise aggression between conspecifics, which has the potential to complicate the social dynamics of stable bachelor groups if not managed correctly. Whole carcass feeds for the polar bear could permit extended consumption time but this has its own limitations for fully aquatic species, with potential water quality, animal separation concerns, and visitor perception concerns.

The captive environment can become a rich foraging ground, with plentiful and regular supplies of high-quality food, especially if attempts are made to replicate wild foraging opportunities and patterns. The ecological relevance of foraging patterns, and their underlying motivational need, may be poorly understood for some species, as the performance of some foraging activities may be unnecessary in a captive setting. Furthermore, attempts to replicate expected daily wild foraging time requirements could be problematic, as individuals with reduced overall activity may have less "drive" or need to forage and thus may prefer some predictability in feeding regimes. What appears key is to not necessarily replicate the exact wild foraging patterns of marine mammals, but to provide a varied and unpredictable diet/feeding schedule that enables the performance of a similar time budget (where possible) with the same ecological relevance of wild foraging opportunities.

9.5 SPECIES-SPECIFIC ENRICHMENT FOR MARINE MAMMALS

Attempts to meet the ecological and behavioural needs of marine mammals are challenging, specifically when intraspecific variation and highly motivated behaviours (such as movement and foraging) are considered. Methods to mitigate or reduce ARBs or encourage welfare-positive behaviours via enrichment can positively influence animal welfare. The aim of enrichment is to occupy animals in a specific way, giving them increased opportunities to exhibit natural behaviours and to reduce any reliance on ARBs (Mason et al., 2007). If carried out correctly, with the species' ecological needs considered, wild behavioural repertoires can be performed in captivity for some species through novel feeding regimes or training programmes. The overarching desire for enrichment is to alter the motivational need to perform ARBs (Mason et al., 2007) by providing opportunities to perform some aspects of wild-type movement and foraging actions.

9.5.1 Enrichment types

Successful methods to reduce ARBs are listed in Table 9.2.

9.5.2 Training as enrichment

For highly intelligent species like cetaceans, complex cognitive processing can be promoted through training programmes, increasing problem solving skills and providing an outlet for highly instinctive motivations, resulting in the reduction of undesirable ARBs. Most zoos housing marine mammals carry out multiple daily training sessions, with many utilising some form of training programme for enrichment purposes (Lauderdale, Walsh, Mellen, Granger, & Miller, 2021). Operant conditioning methods as a form of associative learning, where animals learn to associate a specific command with a behaviour and a consequence, permit safe and calm health checks, in addition to advanced veterinary care, while ultimately providing a complex outlet for highly motivated behaviours. Training marine mammals requires patience with high value novel rewards needed to maintain animal interest and engagement. Sea lion species, including California and Steller sea lions (*Eumetopias jubata*) appear to gain inherent value from engaging in training activities (Kastelein & Wiepkema, 1988) and may be somewhat "easier" to accommodate than more behaviourally complex species, such as bottlenose dolphins. Simple training requirements vary in reward value for pinnipeds – fish tails being low value to a whole fish being high value. Some species/individuals appear to benefit greatly from the opportunity to "work" with their conspecifics in tasks, which may include their keeper, and are then able to fulfil some form of foraging activity, albeit in a different context. This instigates ecologically relevant activity and movement patterns, and social play, that are key affirmation behaviours for social pinnipeds, e.g., for sea lions.

To keep training as enriching and engaging as possible, behavioural complexity can be manipulated by varying activity, including simple target touch responses through to complex behavioural repertoires of flips, jumps, and long dives, as evidenced in Figures 9.5 and 9.6. Training sessions can form a large part of the daily management repertoires of marine mammals, whether in haul out areas or in water (Figure 9.7). Beyond the scope of this chapter, but worth acknowledging, is the improvement within the animal-keeper bond which can occur from positive experiences and voluntary engagement. Animal and human safety is essential when working with dangerous taxa, like most marine mammals, but as training has inherent value, it can be very advantageous. However, most training programmes use some form of semi-predictable schedules and can become routine, therefore variation and change of training sessions is important (Lauderdale, Walsh, et al., 2021).

9.5.3 Enrichment considerations

Despite the positive examples shown, care needs to be taken when implementing all forms of enrichment. Not all individuals will be able to manipulate enrichment in the same way (e.g., calves and elderly individuals) and there is the potential for the enrichment to increase competitive or agonistic behaviour. Species-specific relevance, ecological importance, and group practicalities should all be considered to help maximise its efficacy. Although frequently used, enrichment does not necessarily eliminate ARBs, nor allow replication of expected time budgets or provide outlets for highly motivated

Table 9.2 Examples of successful enrichment used for captive marine mammals, with reference sources provided for further information.

Ecological relevance	Behavioural relevance	Enrichment type	Species tested	Impact
Foraging	Increasing foraging opportunities	Fish in ice blocks, plastic bottles, or buoys	Harbour seal[1] Sea otter[2]	Reducing circling swimming
		Fish pull (fish tied to a fishing line and dragged)	Harbour seal[1]	Reducing circling swimming by 50%
		Roughage (fibrous indigestible bones & shells)	Sea otter[2]	Increased foraging time & efficiency
	Increasing foraging & consumption time	Sunken objects filled with food with drilled holes	Pacific walrus[3]	Increasing time from 5–10 m to 12–86 m for calves
	Increasing foraging complexity	Sea urchins, mussels & clams	Sea otter[2]	Increased food manipulation
	Reducing predictability	Timing, position, handler & varied food type	Harbour seal[1] West Indian manatee[4]	Reduced ARBs
Movement	Increasing investigative behaviour	Unfamiliar scents (vinegar, fragrance)	Polar bear[5]	Interrupted pacing bouts & increased sniffing
Proximity and choice	Environmental control	Increasing distance with conspecifics or public (off-show access)	Polar bear[6]	Increased pacing & increased social play, swimming
Social play	Increasing opportunities for social play and activity	Novel objects (hose pipe, mirror)	Harbour seal[1]	Increased social play
Socio-cognition	Increasing opportunities for cognitive manipulation & problem solving	Underwater maze	Bottlenose dolphin[7,8]	Males used, with two solving the puzzle without any training
Unpredictability	Increasing investigation	Varied enrichment presentation and training routines	Bottlenose dolphin[8,9] Steller sea lion[10]	Increased use of deeper enclosure depths & increased enrichment interaction with reduced ARBs

[1] Grindrod and Cleaver (2000)
[2] Hempstead and Larson (2019)
[3] Kastelein et al. (2007)
[4] Anzolin et al. (2004)
[5] Wechsler (1992)
[6] Ross (2006)
[7] Clark, Davies, Madigan, Warner, and Kuczaj (2013)
[8] Weiss, Ellis, and Croft (2021)
[9] Lauderdale and Miller (2020)
[10] Kastelein and Wiepkema (1988)

Figure 9.5 California sea lions being trained through operant conditioning at Yorkshire Wildlife Park, UK.

Figure 9.6 A South American sea lion having an oral health check at Dudley Zoo, UK.

behaviours (Clubb & Mason, 2007). However, if utilised correctly, enrichment will play a critical role in marine mammal husbandry, improving welfare, and providing individuals more opportunities to perform natural behaviour.

9.6 USING BEHAVIOURAL BIOLOGY TO ADVANCE MARINE MAMMAL CARE

Despite the use of positive behavioural training regimes for marine mammals, and common-place naturalistic enclosure design to promote welfare, we must acknowledge the concerns about the overall suitability of marine mammals in a captive environment. A complex ecology and specific behavioural requirements make captive care particularly challenging. It is possible that some species are not suitable for captivity regardless of the efforts made to replicate natural systems and processes. Specific assessments (e.g., of welfare) are required to determine how suitable a species is for the captive environment. Assessments on a broader taxonomic level will not provide adequate evidence, as significant

Figure 9.7 A California sea lion being trained at Blackpool Zoo (UK). Note the keeper is in the water and is utilising the same space within a free contact training system (Photo: Lauren Shields).

and unique differences often exist between closely related species, for example, the bottlenose dolphin and the killer whale. Behavioural flexibility may be key here; species with smaller home ranges or generalist foragers are more suitable and more adaptable to the constraints of a captive environment. Evidence of the performance of positive welfare behaviours and of enrichment success for the species highlighted does provide a positive outlook to move forward, however.

9.7 CONCLUSION

It is of no surprise that marine mammals are somewhat difficult to accommodate due to their extensive ecological diversity and natural histories, but where this is done well, they are an excellent asset to any zoo's animal collection. From the evidence presented here, the great diversity of these taxa (often the feature that creates our initial interest) makes them extremely specialised and therefore their captive needs vary greatly. Meeting these needs, whether this be by housing in natural water sources or within a traditional "closed" facility like an artificial tank or pool, will pose species-specific (and individual animal) challenges. The ability to replicate aspects of the wild for marine mammals may be possible by creating state-of-the-art facilities, use of extensive enrichment, and bespoke training programmes that encourage the performance of highly motivated behaviours. Other aspects, however, such as dietary and ranging/movement needs, may be more challenging. A reliance on seasonal variation, which has caused the evolution of complex behaviour patterns, is another challenging aspect of their ecology. The social needs of cetaceans, as presented in the killer whale case study, are also extremely complex both inter- and intraspecifically. Maintenance of naturalistic social groups for some species may be unachievable in the zoo. However, species such

Figure 9.8 Two wild, young male, killer whales, J45 and J38, from different matrilines travelling together to illustrate the "open social network" concept (Weiss, Ellis, & Croft, 2021).

as the California sea lion appear to thrive under human care, and consequently species-specific considerations are key to upholding good welfare. If managed appropriately using collection plans considerate to species that can thrive in captivity, ex-situ populations provide extensive opportunities to showcase the highly specialised adaptations and behaviours of marine mammals to the wider public.

ACKNOWLEDGEMENTS

Thanks go to Dr Andrew Mooney for the thorough proofread of this chapter and the subject-specific expertise in crafting key examples and ideas.

REFERENCES

Anzolin, D., De Carvalho, P.S.M., Vianajr, P.C., Normade, I.C., & Souto, A.D.S. (2004). Stereotypical behaviour in captive West Indian manatee (*Trichechus manatus*). *Journal of the Marine Biological Association of the United Kingdom*, *94*(6), 1133–1137.

Clark, F. (2013). Marine mammal cognition and captive care: A proposal for cognitive enrichment in zoos and aquariums. *Journal of Zoo and Aquarium Research*, *1*(1), 1–6.

Clark, F., Davies, S.L., Madigan, A.W., Warner, A.J., & Kuczaj, S.A. (2013). Cognitive enrichment for bottlenose dolphins (*Tursiops truncatus*): Evaluation of a novel underwater maze device. *Zoo Biology*, *32*(6), 608–619.

Clubb, R., & Mason, G. (2003). Captivity effects on wide-ranging carnivores. *Nature*, *425*, 473–474.

Clubb, R., & Mason, G. (2007). Natural behavioural biology as a risk factor in carnivore welfare: How analysing species differences could help zoos improve enclosures. *Applied Animal Behaviour Science*, *102*, 303–328.

Couquiaud, L. (2005). Chapter 4: Types and functions of pools and enclosures. *Aquatic Mammals*, 31(3), 320–325.

Derocher, A.E. (2012). *Polar bears: A complete guide to their biology and behaviour*. Baltimore, USA: The John Hopkins University Press.

Goldman, C.G., Loureiro, J.D., Quse, V., Corach, D., Calderon, E., Caro, R.A., Boccio, J., Heredia, S.R., Di Carlo, M.B., & Zubillaga, M.B. (2002). Evidence of *Helicobacter* sp. in dental plaque of captive dolphins (*Tursiops gephyreus*). *Journal of Wildlife Diseases, 38*(3), 644–648.

Grindrod, J.A.E., & Cleaver, J.A. (2000). Environmental enrichment reduces the performance of stereotypic circling behaviour in captive common seals (*Phoca vitulina*). *Animal Welfare, 10,* 53–63.

Hamilton, C.D., Kovacs, K.M., Ims, R.A., Aars, J., & Lyderson, C. (2017). An arctic predator–prey system in flux: Climate change impacts on coastal space use by polar bears and ringed seals. *Journal of Animal Ecology, 86,* 1054–1064.

Hempstead, C., & Larson, S. (2019). Short note sea otter (*Enhydra lutris*) diet diversity in zoos and aquariums. *Aquatic Mammals, 45*(4), 374–379.

Higgins, J.L., & Hendrickson, D.A. (2013). Surgical procedures in pinniped and cetacean species. *Journal of Zoo and Wildlife Medicine, 44*(4), 817–836.

Hunter, L. (2011). *Carnivores of the world.* (2nd edition). Princeton, USA: Princeton University Press.

Integrated Taxonomic Information System (ITIS) (2021). *Integrated taxonomic information system – Search results.* Retrieved 30 October 2021 from https://www.itis.gov/servlet/SingleRpt/SingleRpt

Kastelein, R., Jennings, N., & Postma, J. (2007). Feeding enrichment methods for Pacific walrus calves. *Zoo Biology, 26,* 175–186.

Kastelein, R.A., & Wiepkema, P.R. (1988). The significance of training for the behavior of Steller sea lions (*Eumetopias jubata*) in human care. *Aquatic Mammals, 14,* 39–41.

Kooyman, G.L., & Trillmich, F. (1986). Diving behaviour of Galapagos sea lions. In Gentry, R.L., & Kooyman, G.L. (Eds.), *Fur seals: Maternal strategies on land and at sea.* Princeton, USA: Princeton University Press, pp. 209–219.

Kroshko, J., Clubb, R., Harper, L., Mellor, E., Moehrenschlager, A., & Mason, G. (2016). Stereotypic route tracing in captive Carnivora is predicted by species-typical home range sizes and hunting styles. *Animal Behaviour, 117,* 197–209.

Lauderdale, L.K., & Miller, L.J. (2020). Efficacy of an interactive apparatus as environmental enrichment for common bottlenose dolphins (*Tursiops truncatus*). *Animal Welfare, 29,* 379–386.

Lauderdale, L.K., Shorter, K.A., Zhang, D., Gabaldon, J., Mellen, J.D., Walsh, M.T., Granger, D.A., & Miller, L.J. (2021). Habitat characteristics and animal management factors associated with habitat use by bottlenose dolphins in zoological environments. *PLoS One, 16*(8), e0252010.

Lauderdale, L.K., Walsh, M.T., Mellen, J.D., Granger, D.A., & Miller, L.J. (2021). Environmental enrichment, training, and habitat characteristics of common bottlenose dolphins (*Tursiops truncatus*) and Indo-Pacific bottlenose dolphins (*Tursiops aduncus*). *PLoS One, 16*(8), e0253688.

Mason, G., Clubb, R., Latham, N., & Vickery, S. (2007). Why and how should we use environmental enrichment to tackle stereotypic behaviour? *Applied Animal Behaviour Science, 102*(2007), 163–188.

Matthews, C.J., Luque, S.P., Petersen, S.D., Andrews, R.D., & Ferguson, S.H. (2011). Satellite tracking of a killer whale (*Orcinus orca*) in the eastern Canadian Arctic documents ice avoidance and rapid, long-distance movement into the North Atlantic. *Polar Biology, 34*(7), 1091–1096.

Nowak, R.M. (2003). *Walker's marine mammals of the world.* Baltimore, USA: The John Hopkins University Press.

Owen, M.A., Swaisgood, R.R., Slocomb, C., Amstrup, S.C., Durner, G.M., Simac, K., & Pessier, A.P. (2015). An experimental investigation of chemical communication in the polar bear. *Journal of Zoology, 295,*36–43.

Reeves, R.R., & Stewart, B.S. (2003). Marine mammals of the world: An introduction. In Nowak, R.M. (Ed.), *Walker's marine mammals of the world.* Baltimore: The John Hopkins University Press.

Rendell, L., Cantor, M., Gero, S., Whitehead, H., & Mann, J. (2019). Causes and consequences of female centrality in cetacean societies. *Philosophical Transactions of the Royal Society B, 374*(1780), 20180066.

Rose, N.A., Snusz, G.H., Brown, D.M., & Parsons, E.C.M. (2017). Improving captive marine mammal welfare in the United States: Science-based recommendations for improved regulatory requirements for captive marine mammal care. *Journal of International Wildlife Law and Policy, 20*(1), 1–35.

Ross, S. (2006). Issues of choice and control in the behaviour of a pair of captive polar bears (*Ursus maritimus*). *Behavioural Processes, 73*(2006), 117–120.

Ugaz, C., Valdez, R.A., Romano, M.C., & Galindo, F. (2013). Behavior and salivary cortisol of captive dolphins (*Tursiops truncatus*) kept in open and closed facilities. *Journal of Veterinary Behavior, 8*(2013), 285–290.

Wechsler, B. (1992) Stereotypies and attentiveness to novel stimuli: A test in polar bears. *Applied Animal Behaviour Science, 33*, 381–388.

Weiss, M.N., Ellis, S., & Croft, D.P. (2021). Diversity and consequences of social network structure in toothed whales. *Frontiers in Marine Science, 8*, 688842.

Weiss, M.N., Franks, D.W., Giles, D.A., Youngstrom, S., Wasser, S.K., Balcomb, K.C., Ellifrit, D.K., Domenici, P., Cant, M.A., Ellis, S., Nielson, M.L.K., Grime, C., & Croft, D.P. (2021). Age and sex influence social interactions, but not associations, within a killer whale pod. *Proceedings of the Royal Society B, 288*(1953), 20210617.

Wright, B.M., Ford, J.K.B., Ellis, G.M., Deecke, V.B., Shapiro, A.R., Battaile, B.C., & Trites, A.W. (2017). Fine-scale foraging movements by fish-eating killer whales (*Orcinus orca*) relate to the vertical distributions and escape responses of salmonid prey (*Oncorhynchus* spp.). *Movement Ecology, 5*(3), 1–18.

10

The behavioural biology of marsupials and monotremes

MARIANNE FREEMAN
University Centre Sparsholt, Winchester, UK

10.1 INTRODUCTION TO MARSUPIAL AND MONOTREME BEHAVIOURAL BIOLOGY

There are over 350 species of monotreme and marsupials and new species (e.g., the greater glider species, *Petauroides minor*, and *P. armillatus*) are still being describedh today (McGregor et al., 2020). Species of monotremes and marsupials have been kept in captivity for over 200 years, with evidence of kangaroos being on display in London in 1790 and wombats in Paris in 1803 (Jackson, 2003). Despite this long history of captive management, we are still uncertain about the many behavioural requirements of such unique and varied species and how to meet these requirements in the zoo environment. The mesoeutherians or marsupials are split into two main clades; American (consisting of 127 species of opossums and seven species of shrew opossum) and Australian (of which 210 are, except for one – the monito del Monte (*Dromiciops gliroides*) – all found across Oceania). The order Monotremea contains five species; platypus (*Ornithorhynchus anatinus*), short-beaked echidna (*Tachyglossus aculeatus*) and three species of long-beaked echidna, found in Papua New Guinea. The eastern long-beaked echidna (*Zaglossus bartoni*) has occasionally been kept in captivity, while the other two species (*Z. bruijni* and *Z. attenboroughi*) are Critically Endangered and difficult to locate in the wild.

DOI: 10.1201/9781003208471-12

Due to such vast diversity of species and lifestyles, this chapter will focus on the behavioural biology of the more commonly kept monotremes (platypus and short-beaked echidna), on the unique *Dasyuridae* (the carnivorous marsupials) and the more ubiquitous captive species of the order Diprotodontia (including the Phalangerforms, Vombatiforms, and Macropodiforms). Their diverse range of ecological niches from fossorial (such as the marsupial moles, *Notoryctidae*) to semi-aquatic (e.g., platypus) to arboreal species (e.g., *Petaurus* spp.), highlights the need to consider appropriate habitat and biologically relevant resources when undertaking species-specific behavioural management. However, their distinctive but varied reproductive strategies of, not only oviparity and neonatal births, but semelparity (a single reproductive event before death) in species such as the marsupial mouse (*Antechinus stuartii*), *also* require species-specific knowledge to care for and conserve this unique group of mammals in captivity.

10.2 MARSUPIAL AND MONOTREME ECOLOGY AND NATURAL HISTORY RELEVANT TO THE ZOO

Knowledge of the natural history of a species is imperative to assist zoos in improving the care and welfare offered to the captive collection, ultimately resulting in being able to maintain self-sufficient captive breeding programmes. Our understanding of the natural history of many monotreme and marsupial species is still limited, especially in the smaller more elusive species, however, research is constantly building on previous knowledge and current literature from the wild should be a first port of call for anyone looking to improve captive management. Some examples of the diversity of species and behavioural ecology that is relevant to the zoo can be seen in Table 10.1.

Both the platypus and short-beaked echidna are two species where knowledge of their ecology has made huge advances in recent years. Despite very different geographical ranges and habitats, these two species have elements of their behavioural biology that align. Both are solitary with home ranges that overlap with at least one other individual. The short-beaked echidna home range is dependent on sex and habitat, with vegetation dictating cover and invertebrate foraging availability (Rismiller & Grutzner, 2019). Both species forage for invertebrate prey (Table 10.1); platypus can spend up to 29 hours continuously foraging during the winter (Bethge, Munks, Otley, & Nicol, 2009). Both species also dig burrows though platypus are loyal to theirs; a platypus will spend up to half of their day in burrows dug up along the banks (Bethge et al., 2009). Short-beaked echidnas will shelter in irregular self-dug burrows and some subspecies will form nests where they remain with their young while they suckle, and others form pouch-like folds in their skin to carry early neonates (Wallage et al., 2015). In all individuals, offspring are left in filled in burrows after around 50 days while the female searches for food (Rismiller & Grutzner, 2019). The platypus females raise their young in nests within their burrows which they will block themselves into, after the first month she will emerge to feed, with the offspring finally emerging after ~130 days, juvenile dispersal from the natal area occurs in their first year (Holland & Jackson, 2002). Short-beaked echidna have a relatively long weaning period of up to seven months and a comparatively long lifespan for such a small animal, with captive individuals living into their 50s.

One factor that faces many of the marsupials is the ecological pressure of an unpredictable environment, whether that is coping with varying seasonal fluctuations in temperature, harsh arid environments with irregular rainfall or unexpected bush fires and they have developed behavioural adaptations to respond to this instability. Carnivorous marsupials are from the order Dasyuromorphia which includes the family *Myrmecobiidae* a monophyletic group consisting of the numbat and *Dasyuridae*, comprising over 70 species. Many of the smaller more nocturnal species have traditionally been less successfully kept in captivity (Jackson, 2003). Their relatively short lifespans and high post reproductive senescence are key challenges that faces the captive manager of such species. The majority of these smaller dasyurid species can be found in harsh arid deserts, several, including the northern quoll (*Dasyurus hallucatus*), kaluka (*Dasykaluta rosamondae*), and marsupial mouse (*Antechinus stuartii*), are semelparous, where the male dies after the first breeding season (Hayes et al., 2019).

The smaller species of dasyurids are insectivores as they do not have the high energetic requirements of larger carnivore species. The largest of the dasyurids, the Tasmanian devil, meets a body weight

Table 10.1 Examples of behavioural adaptations of selected monotreme and marsupial species appropriate to captivity (taxonomic orders represented in bold).

Species	Distribution and Red List status (as of February 2022)	Behavioural adaptations
Monotremata		
Short-beaked echidna	Widespread across Australia, Tasmania and New Guinea occurring in deserts, grasslands, and alpine regions. Least Concern	They can sense subterranean invertebrate prey through vibrations picked up in their lower jaw and mechanoreceptors in their front feet to help them locate and dig in the soil for food.
Platypus	Riparian species occupying freshwater habitats on the eastern coast of Australia. Near Threatened	Well-adapted bill sifts through riverbed substrate for invertebrates but are elusive, easy alarmed animals, diving under the water if disturbed. They will eat less mobile prey if given a choice so a varied diet in captivity encourages foraging.
Dasyuromorpha		
Tasmanian devil (*Sarcophilus harrisii*)	Found in Tasmanian and recently introduced to New South Wales, Australia. Dry sclerophyll woodland habitat. Endangered	Solitary but will socialise when feeding on scavenged livestock or wild mammal carcasses. Will also predate on prey up to the size of wombats.
Spotted-tail quoll (*Dasyurus maculatus*),	Widespread across eastern Australia from north Queensland to Tasmania preferring rainforests or closed eucalyptus forests. Near Threatened	Quolls, as arboreal animals can prey on species such as gliders (*Peturus* spp.) and other medium sized marsupials in addition to a diet of invertebrates and vegetation.
Numbat (*Myrmecobius fasciatus*)	Restricted to two isolated populations in southwest Western Australia. Endangered	The numbat principally feeds on termites that are scooped up with a long tongue. Numbats do not possess a pouch instead, skin folds cover developing young.
Diprotodontia		
Long-nosed potoroo (*Potorous tridactylus*)	South-eastern Australia in habitats ranging from rainforest and wet sclerophyll forest to open forest and scrub; Near Threatened	*Potoroidae* are specialised fungivores with a good sense of smell to help locate the subterranean food sources.
Common brushtail possum (*Trichosurus vulpecula*)	Widespread across Australia in eastern northern and western regions, Tasmania, and naturalised in New Zealand. Equally wide range of habitats from semi-arid, forest, or urban areas. Least Concern	Primarily herbivorous but a generalist feeder. Prehensile tail and adapted hind feet for their arboreal lifestyle.
Sugar glider (*Petaurus breviceps*)	Queensland to New South Wales found in coastal forests. Least Concern	Specially adapted patagia allow airborne travel over long distances when moving through the canopy.

(*Continued*)

Table 10.1 (Continued) Examples of behavioural adaptations of selected monotreme and marsupial species appropriate to captivity (taxonomic orders represented in bold).

Species	Distribution and Red List status (as of February 2022)	Behavioural adaptations
Koala (*Phascolarctos cinereus)*	Eastern and south-eastern South Australia in dry eucalyptus forests preferring riparian habitats when in semi-arid areas. Vulnerable	Specialist folivores consuming *Eucalyptus* leaves and feeds young on pap (maternal faeces) to help in weening.
Common wombat (*Vombatus ursinus)*	Open fields, alpine grassland, and woodlands in southern and eastern Australia and Tasmania. Least Concern	Semi-fossorial grazers with varying complexity burrows. Olfactory communicate is used and frequent cube shaped faeces are left around their territory.
Red kangaroo *(Macropus rufus)*	The occupy open habitats ranging from arid deserts to scrubland and grassland across Australia except for the rainforest areas and southern and eastern coast. Least Concern	Terrestrial grazers living in groups of up to ten individuals. They are reasonable swimmers.
Goodfellow's tree kangaroo (*Dendrolagus goodfellowi*)	Rainforest of New Guinea. Endangered	Arboreal macropods that can move their hindlimbs independently for branch stability.
Red-necked wallaby *(Macropus rufogriseus)*	Coastal scrub and sclerophyll forest of eastern Australia. Introduced to New Zealand and several countries in Europe. Least Concern	A group-living, primarily grazing macropod that practices reconciliation after high-intensity conflicts.

that suits a lifestyle of predation on small mammal species. However, some individuals can grow large enough to justify the energetic costs of feeding on prey that is greater than half their body weight and Tasmanian devils have been known to predate on juvenile wombats and wallaby species (Dickman & Jones, 2003). Partly through competition for the limited and unpredictable resources, many species in this order are solitary, though some, such as the marsupial mouse, nest in groups of several individuals.

Diprotodontia is the largest marsupial order, the majority of which are herbivores, though the some of the possums (i.e., those of the *Burramyidae* and *Petauridae*) and the musky rat kangaroos (*Hypsiprymnodon moschatus*) are omnivores and the potoroids are fungivorous.

Three suborders of diprotodontids are:

- Phalangerforms, these range from social species, such as the sugar glider that live in colonies to solitary species, such as the bear cuscus (*Ailurops* spp.).
- Vombatiforms, two families of koala and wombat (*Vombatus* and *Lasiorhinus* spp.), both have a posterior pouch opening, likely descended from the ancestral vombatoid, but useful for a fossorial, digging species such as a wombat.
- Macropodiforms, with a name translating to large foot, all share a similar locomotion of propulsion via leaping from their two large hind feet. This group includes the kangaroos, wallabies, potoroos, and rat kangaroos.

Understanding the habitats that these species occupy, their modes of locomotion and ecological pressures helps guide captive husbandry and allow caregivers to consider the wider range of behavioural repertoires that these species should be exhibiting in captivity.

10.3 ENCLOSURE CONSIDERATIONS FOR MARSUPIALS AND MONOTREMES BASED ON BEHAVIOURAL EVIDENCE

The variety of breeding strategies can be a key factor to consider when planning enclosure designs for monotremes and marsupials. These strategies can dictate nesting resources, social housing parameters, and feeding opportunities. Behaviour biology in terms of locomotion and circadian rhythm will also be important factors to plan the design to accommodate each individual species' needs.

10.3.1 Nesting behaviours

Short-beaked echidna have proven difficult to breed in captivity and successful breeding was only achieved when housed in enclosures at least 72 m^2 and nesting opportunities in the form of burrow boxes were provided (Rismiller & Grutzner, 2019). In addition, heat lamps at shelter sites allow for more efficient energy conservation strategies (Wallage et al., 2015). In platypus, females control breeding, through changes in temporal activity, exercising flight behaviour and resisting encounters (Thomas et al., 2018b). Thus, opportunities for avoidance behaviours should be considered in enclosure design, such as space, hiding places, and multiple ponds. Courtship behaviour of the platypus has been most observed in slow water flow and mating behaviour has been noted to take place in water areas with partially submerged logs or ramps, either for the female to hold on to or to use as a barrier if needed to escape (Thomas et al., 2018b).

Platypus nest after mating and the burrow to this nest is then plugged, a behaviour hypothesised to prevent flooding of the nest, maintain humidity in the burrow, protection from predators or to confine young offspring (Thomas et al., 2018a). For greater rearing success, opportunities for burrowing (with scope cams to monitor) within substrate of at least 50 cm deep, the inclusion of suitable nesting materials (such as rush leaves, eucalyptus leaves, grasses, reeds, and willow) and clay-based soil for sealing the entrance should be provided. Offspring will also use separate burrows from the female before natal dispersal, so additional space is recommended (Thomas et al., 2018a).

Nesting behaviour is common in some marsupial species too, several dasyurid species construct quite elaborate burrow systems (i.e., western quolls (*Dasyurus geoffroii*) and crest-tailed mulgara (*Dasycercus cristicauda*)). Nest are dome shaped and secured within a burrow or some other form of protective structure like a rocky overhang (Dickman & Jones, 2003). Furnishings, branches, rocks, hollow logs, and planted tussocks can all help encourage nest building. Wild gliders, such as stripped possums (*Dactylopsila trivirgata*), collect leaves and carry them in their tail back to tree hollows that they use for nests, to replicate this, provision of fibrous bark materials and a nest box is needed. The potoroids, as shy, elusive animals need plenty of areas to shelter, resting diurnally in squats, under tussocks and thickets. They may even use vacant burrows of other species to shelter in. Provision of planting and furnishings can allow opportunities for nesting and shelter. Their close relatives, the brush-tailed bettong (*Bettongia penicillate*), require elevated nest boxes, while the burrowing bettong (*B. lesueur*) can have complex burrowing systems.

10.3.2 Locomotion

Wombats are a subterranean marsupial species, having vast burrow systems that they occupy. The consideration of this behavioural need is important, wombats can dig up enclosures and if not provided with the opportunity to burrow, behavioural problems and welfare issues can ensue (Jackson, 2003). Deep enclosure walls (1 m) are necessary to prevent escape. If housed indoors the provisions of substrate/leaf litter will allow for digging and this facility can also provide opportunities for a reverse lighting schedule to house this nocturnal species. Digging helps keep claws short and sandy-loam soil can also be given for dust bathing too. With a low metabolism, seeking shelter in burrows can be a less energetic method for temperature regulation for the species.

Arboreal marsupials have the advantage of three-dimensional space to increase resource accessibility and occupy an environment less open to predation, however, this also requires more complex navigation and movement. Quadrupedal locomotion is the predominant form, though in some species there is an aerial phase of movement. The climbing and jumping ability of arboreal species should be considered when planning enclosure perimeters, edge planting or furnishings, to prevent escapes. Double door entrances are worthwhile for some of the smaller, more agile species. The provision of a branch network is essential for arboreal marsupials. Smaller marsupials are less subjected to branch movement but larger arboreal species such as quolls (*Dasyurus* spp.), bear cuscus and tree kangaroos (*Dendrolagus* spp.) have adopted larger stride gaits to increase stability on moving branches. These larger species require stronger horizonal branches for their quadrupedal locomotion but provision of structures that have some lateral or vertical movement can help maintain similar natural locomotion patterns and climbing abilities to that of wild arboreal marsupials.

Taller enclosures of around 4 m^3 are needed for the gliders (Jackson, 2003). The behavioural adaptation of gliding is made possible using the patagia (a membrane of skin between the fore and hind limbs). This behaviour is formed through a lift off stage of either dropping or springing from one location, the animal can maintain velocity and have some control over their movement while in free fall (though in shorter glides this is harder to achieve). On landing, the animal will stretch forward their limbs and pull back their head ready to grip, allowing space for head momentum after braking. So large gaps between structures is needed to encourage gliding with suitable take-off and landing point. Spatial memory is thought to be important for the gliders as they traverse through the complex arboreal network and thus a modular branching network could keep them mentally challenged.

Koalas require at least 3 m tall supports with natural forks for sitting (Figure 10.1), this is particularly important as koalas sleep and rest for 18–20 hours a day and moving between trees happens infrequently throughout the day. Koalas move faster on ground (more like macropods) than in the canopy, but would prefer not to descend, unless essential, opting to leap up to 1 m from tree-to-tree

Figure 10.1 Koala resting in a tree fork at Zoo Zurich (Photo: Clinic for Zoo Animals, Exotic Pets and Wildlife, University of Zurich).

(Gaschk, Frère, & Clemente, 2019). Vertical structures of no more than 3 m apart and joined allows lateral movement. For all arboreal animals, trees or other vertical structures should be rough to prevent slipping when climbing. Tree kangaroos need nest boxes at least 2 m off the ground, they can move along horizontal branches with quadrupedal motion demonstrating independent hind limb movement, but they can also hop along the branches too, rarely jumping from tree-to-tree and instead preferring to move between trees overground.

The macropods are predominantly bipedal and move with a hopping behaviour, using the hind limbs for propulsion and their tail for counterbalance. At high speeds, this is one of the most energy-efficient modes of locomotion. In captivity, large enclosures with limited furnishings and straight fencing are recommended to allow them range to move. In addition, they can display thigmotaxis (a tendency to stick close to and hop along the fence line when frightened) so enclosures should have minimal corners and without tree/furnishings along the line of the fence. Pentapedal locomotion is used when grazing – their tail gives support and propulsion, akin to a fifth limb. Provision of an "A" framed wooden shelter and tussock vegetation can offer sheltering opportunities for many of the wallaby species but rock wallabies (*Petrogale* spp.) and wallaroos (*Macropus* spp.), with their rock-hopping behaviour, should have rock piles to allow this behaviour to flourish. Sunny cliff caves with multiple openings and ledges are a preferred habitat choice of wild rock wallabies. Larger macropods will need open areas and mature trees for shelter and dust baths or well-drained soil to roll in. Species, such as red kangaroos and red-necked wallabies can tolerate being in walk-through exhibits (see Section 10.4).

10.3.3 Feeding behaviour

Each of the different feeding strategies needs to be taken into consideration when planning the enclosures. For the smaller carnivorous marsupials, opportunities to hunt can be encouraged with live invertebrate feeds scattered in the leaf litter. For larger dasyurids, whole prey carcass feeds encourage feeding behaviours. Tasmanian devils gorge themselves in the wild, eating ~90% of the carcass (Pollock et al., 2021). To replicate wild feeding behaviour and ensure individuals are not overfed, starve days can be implemented into their feeding schedule. Provision of large bones can be a form of gnawing enrichment. Cuscus and brushtail possums (*Phalangeridae*) have a diet that is made up of fruits, foliage, and other vegetation. Feeding these arboreal marsupials on branches can encourage activity. Gliders are predominantly exudivores that undertake bark stripping feeding behaviour to reach sap and any invertebrates that they encounter. Feeding structures that replicate bark with insects and gum inserted within encourage natural feeding behaviours of stripping bark (Murray, Waran, & Young, 1998). The exception to this type of diet is the greater glider (*Petauroides* spp.) which has a niche like that of the koala, feeding on different varieties of eucalyptus leaves. Pots in several different locations within the enclosure, with horizontal or forked branches nearby, are required to hold cut eucalyptus stems and multiple feeding opportunities should be provided to prevent any aggression in group-housed animals and provide a choice of the different eucalyptus species. Koalas have specific eucalyptus preferences, and these can vary throughout the year. Koalas and other marsupial folivores will sniff leaves before ingesting and this has been suggested as a possible behaviour linked to formylated phloroglucinol compounds (FPC) and that koalas may learn to identify suitable species choices based on either pre- or post-ingestion behavioural consequences, the latter of which has been noted in other marsupial species (Moore & Foley, 2000).

10.3.4 Social behaviour

Sociality in the macropods varies on a scale from solitary (tree kangaroos), solitary with some communal feeding (pademelons (*Thylogale*), hare wallabies (*Lagorchestes*) and nailtail wallabies (*Onychogalea*)), monogamous pairs (rock wallabies, quokka (*Setonix brachyurus*)) to gregariousness, of which the degree relates to body size. Generally, a single male with up to five females is a suitable grouping ratio with the exception of the tree kangaroo, whereby any males should only be introduced to the female for mating and then removed (Jackson, 2003). Mixed species exhibits can be possible though even in seemingly compatible species displacement and agonistic behaviours may occur (Rendle, Ward, & McCormick, 2018).

Despite being predominantly solitary in the wild many dasyurids can be group housed in captivity, though when females have young many species can become aggressive so the removal of the male at this time is usually advised. If a Tasmanian devil female is the bigger of a pair, breeding will likely be unsuccessful. In the wild, Tasmanian devils will naturally congregate to feed on large prey items and through a series of ritualised communication that includes yawning and other postures, vocal communication, and cloacal dragging, will eventually settle the individuals into a dominance hierarchy (Pollock et al., 2021). Injuries can occur but this is thought to be mainly due to intrusion in dens or aggression from unreceptive females. It is suggested that group-housing Tasmanian devils can help prepare them for intraspecific competition should they be reintroduced into the wild (Skelton & Stannard, 2018).

Most of the possums are social and optimally kept in large enclosures, in groups, with some exceptions, phalangerids are generally solitary, as are greater gliders. Leadbeater's possum (*Gymnobelideus leadbeateri*) are monogamous and so a dominant pair ensures group cohesion. Pygmy possums (*Burramyidae*) can be rotated on a "round-robin" system with females housed alone and alternatively housed with a different male on/off, every three days, (Jackson, 2003). This works well for quolls too, who become aggressive to repeated mating attempts by the same male and need to be mixed with new males. Female-female competition over limited resources in platypus may impair breeding so social groups should be considered carefully (Thomas et al., 2018b).

10.3.5 Circadian rhythms

Many marsupials, such as dasyurids are strictly nocturnal but can be kept on a reversed lighting system to allow visitors to experience their active behaviours. Some species are fully active all night long and some experience bursts of activity, though little has been published on dasyurids activity budgets and nocturnal behaviours. It does appear in some species that as well as photoperiod activated circadian rhythms, light rain also stimulates foraging activity (Dickman & Jones, 2003).

10.4 BEHAVIOURAL BIOLOGY TO ADVANCE MARSUPIAL AND MONOTREME WELFARE

Managing animals in captivity, as a one-plan approach to conservation, involves the need to retain wild behaviours to ensure that animals that are returned to the wild can cope with an unpredictability that is seldom experienced in captivity. While in captivity, the goal is that of optimal welfare for the individuals and in recent years this has moved to a focus on positive welfare indicators, by bringing in opportunities for play, exploration, and communicating through positive emotional signals. However, the foundation of animal welfare is still to ensure that animals are not overly suffering from the conditions that they are in, which could lead to adverse effects on their longevity and ability to breed successfully.

Hing, Narayan, Thompson, and Godfrey (2014) noted the main stressors in captivity were capture and handling, which at extreme levels can lead to fatalities from issues like capture myopathy. Through reintroduction research, the behavioural stress effects of captivity appear short lived. However, increased time in captivity correlated with the vulnerability of released Tasmanian devils to vehicle strikes (Grueber et al., 2017) and so small amounts of stress can be beneficial and help prepare animals for the wild environment.

10.4.1 Visitor effects

The effects of the zoo visitor on marsupial welfare is a topic that has begun to receive a lot of attention, with research spanning the macropods and koala. Macropods are often housed in walk-through exhibits, though certain species are known to be unsuitable for walk-through interactions and with larger macropods, there is also the possibility of injury both to the visitor and animal. Individuals that are hand-fed can show feed aggression, and public chasings can result in collisions with perimeter, so roped-off areas, for refuge, are recommended (Jackson, 2003).

Despite these obvious welfare issues other more subtle behaviour consequences have been recorded, such as avoidance behaviour, signs of suppressed feeding, reduced interactive behaviours, increased social proximity and inactive behaviours, in conjunction with the presence of visitors (Sherwen & Hemsworth, 2019). However, "visitor effect" is not consistent with all macropods or even across all groups and proximate hypotheses, such as conditioning to visitor feeds or competition from other species in a mixed exhibit may affect the response to visitors. Additionally, considering the ultimate hypotheses of evolutionary pressures could explain differences in visitor effects of species that are more affected by avian predators or are island species with no predation pressure. Elucidating visitor effects needs to be undertaken by controlled experimentation to reduce extraneous variables and isolate cause and effect to find the optimal way of managing these species. Even isolating the specific sensory trigger of visitor effects can be difficult, visitors may be a visual stimulus for increased vigilance, or the accompanying noise may be the factor or a combination of both. Koalas responded to visitors with an increase in vigilance which was also present with auditory stimuli only (Sherwen & Hemsworth, 2019). Koalas, evolutionarily more susceptible to stress and with sensitive mechanoreceptors, are likely to not only be susceptible to noise and public presence (due to the vibrations created) but also to impacts of enclosure renovations that occur nearby and within the wider zoo environment too.

10.4.2 Seasonality

Animals experience seasonal fluctuations in the wild and visitor impacts may be season variations that captive animals can habituate to. The seasonal variation may also be beneficial to an animal's health and welfare as seen in some placental animals when fluctuation in their diet, behaviour, and environment is experienced. It can also improve breeding success with increases in food availability stimulating breeding physiology. In fact, species of dasyurid live longer and have higher reproductive rate if the environment (particularly rainfall) is unpredictable (Collett, Baker, & Fisher, 2018).

Torpor is a common behavioural adaptation to deal with unpredictable environmental changes and fluctuations in food availability. The process of experiencing torpor can stimulate some species, such as pygmy possums, to breed. Not only can it help by affording increased survival during times of low food availability and fluctuations in seasons, but it is also thought to play a functional part in coping with unpredictable events such as droughts, floods, or forest fires. Unlike full hibernation, torpid animals can rouse from signals, such as smoke, offering the chance to escape, and then reduce their activity levels with low post-fire food availability. This response is so well developed that captive animals will be seduced into torpor from the smell of charcoal and ash regardless of the abundance of food availability (Geiser, 2021).

While daily torpor is a behavioural adaptation in several of the orders of marsupial and monotreme, a few species undergo full hibernation, such as the feathertail glider (*Acrobates pygmaeus*) as well as the Monito del Monte (the only hibernating mammal in South America) and pygmy possums, in which eastern pygmy possums (*Cercartetus nanus*) can hibernate for a year. Captive-bred feathertail gliders differ from wild gliders in morphology and lengthened activity periods and shorter, shallower torpor but with slower rewarming resulting in individuals that struggle with hypothermia and rewarming (Geiser, 2021). On the other hand, in arid zones dasyurids torpor may last two times that of captive animals. Behavioural adaptations are important to maintain if individuals should be released back to the wild and opportunities to adapt to more wild type conditions should be maintained in captivity. The mountain pygmy possum (*Burramys parvus*) emerges out of hibernation in time for the next breeding season. As does the short-beaked echidna and males will emerge earlier than females to co-inside with this.

10.4.3 Breeding behaviours

One thing that is unique among marsupials is their developmental/parental care strategies. Knowledge of the breeding behavioural ecology can help to improve captive management. Some species are semelparous with, either; abrupt mortality or facultative male die off, others are seasonal or continuous breeders, and there are facultative breeders, remaining in anoestrous if the conditions are poor. Semelparity can be an issue to consider when planning welfare and quality of life assessments, as high stress situations can increase mortality in some semelparous species, while others do not undergo male die off in captivity where resources are abundant (Hing et al., 2014). Breeding strategies have evolved to help cope with the unpredictable. Many marsupial species undergo embryonic diapause and in most macropods the blastocyst ceases cell division leaving an embryo in a suspended state until conditions are suitable to continue with development. From a behaviour point of view, this means that as soon as offspring are born, the female can be ready to breed again, and the diapause process is then controlled by signals from lactation or seasonal photoperiods and biological clocks. Planned breeding needs to consider oestrous cycles and ovulation periods or their mechanisms.

Behavioural observations can help to elucidate these periods, which might be very time-sensitive, especially in the more solitary species. Some species will spontaneously ovulate, others require specific behaviours to induce ovulation. Koalas are seasonal breeders that have induced ovulation (where ovulation is stimulated through the presence of male semen) and show behavioural indicators of oestrous, including an increase in activity, hiccoughing, decrease in appetite, and bellowing. Despite the varying ecological niches, the breeding behaviour in dasyurids are very similar. Behavioural indicators of oestrus can be strong in these species overriding the need for more invasive hormonal testing (Jackson, 2003). Signals include increased activity in the marsupial mouse and Tasmanian devil females, which can be aggressive to male encounters, can become submissive during oestrous. However, wombats have comparatively shorter oestrous cycles of just a few days and behavioural responses are more individualised. Wombat males will chase females in a figure of eight, which ends with a bite to her rump before mating occurs. Care must be taken when introducing individuals for breeding or if pair housed, observation made for any possible injuries.

Breeding is difficult in many species dasyurids, some species (i.e., phascogales and antechinus) are monoestrous, timed by internal biological clocks and photoperiod changes. Artificial light timings have worked well for some species like the fat-tailed dunnart (*Sminthopsis crassicaudata*); a polyoestrous species that can be brought into the breeding season with a short-day light cycle for three weeks, twice a year, but this is less effective for other species. Keeping these species outside of their native range appears to allow the best chance of successful breeding. Though, with advances in UV lighting, improvements to light technology may hold the answer to these captive breeding issues, especially considering that vision in some marsupial species relies more on UV sensitivity.

10.4.4 Communication and cognition

All behaviour, ultimately, is some form of communication and understanding the sensory perception of marsupials can help us to determine the best care. Marsupials must communicate, whether for cooperation, during breeding encounters, territorial disputes, with offspring (which includes communication with very underdeveloped and dependent young) and be vigilant for predators. When considering any effect of the zoo environment on animal behaviour, a species' perceived range of sensory abilities must be considered. If trying to undertake training for captive management or reintroduction we should also consider the interspecific communication between human and animal. If we are training animals, ensuring that targets are suitably distinct in colour for species with a limited colour range is equally as important as ensuring that the environment allows the individuals to undergo their full range of communication. It is important to understand the communication cues that animals are giving to ensure we react appropriately and pre-empt needs but an awareness of their understanding of the trainer is also important, sensory perception can be useful here but any ability to follow social referencing cues can imply higher-level awareness of others. Though limited currently to one study, kangaroos have so far demonstrated some level of awareness by directing their gaze to humans during an unsolvable task (McElligott, O'Keeffe, & Green, 2020).

Olfactory communication is frequently employed, the flehmen response (mouth open lip curled baring the gum and wrinkling the nose) is commonly observed when a male kangaroo investigates the female as a behavioural indicator of female oestrous. Other species have unique breeding calls, cloacal scent mark, sternal scent mark, or urination scent mark to send communication signals. Scent marking is an important form of communication in wild wombats; pheromones through their dung are often used as a method of communication in an otherwise solitary species. In controlled captive experiments, the presences of male wombat faeces resulted in more avoidance of other wombats and more use of burrows and hides compared with control groups, suggesting this is a likely stress response (Descovich, Lisle, Johnston, & Phillips, 2012).

To ensure positive welfare, opportunities for play and exploration should be available. Play, as functional training for predatory skills, has been observed in the larger dasyurids (marsupial mice, quolls, and crest-tailed mulgara) juveniles (Dickman & Jones, 2003). Social learning of breeding behaviour and exploration of the environment is important, play can also be driven by high competition for resources. Juveniles can observe competitive displays, as the adult male kangaroos do, where one opponent will lean back on his tail while rubbing handfuls of grass on his chest and so will learn this behaviour for the future. If this display is not enough to reduce competition, aggressive interactions may result with abdominal kicks and wrestling with the forelimbs and in imitation, juveniles can undertake play-fighting too.

While there have been limited studies into the cognition of marsupials, those that have been undertaken shine light on a group of species that possess complex cognitive abilities. Large brain-to-body size ratios in the possums, such as the honey possum (*Tarsipes rostratus*) and western pygmy possum (*Cercartetus concinnus*), are thought to be indicative of the spatial awareness and understanding needed to undertake complex exploration of the habitat in search of very unpredictable food sources. Being aware of these cognitive needs can help design suitably challenging enrichment programmes.

10.5 SPECIES-SPECIFIC ENRICHMENT FOR MARSUPIALS AND MONOTREMES

A well-constructed enrichment programme should consider the key goals that are intended for that species, or individual, whether aiming for a more "wild-type" activity level, increasing behavioural diversity of natural behaviours that are missing from the captive repertoire, reducing abnormal behaviours, or even preparing a species for reintroduction. Programmes should be devised to cover a range of categories that can be offered to the animals on an interchangeable and unpredictable schedule. This should include physically, sensory, nutritionally, socially, and cognitively enriching aspects of their environment.

For the fossorial species, the provision of substrates that can be used in digging is important in maintaining these behaviours, which are innate or learnt aspects of their biology, but also to stimulate their mental capacities, keeping them occupied and in good physical health. Equally, the provision of enclosure structures that allow the arboreal species to climb, jump, or glide with a range of resources spread around to encourage this activity. For smaller marsupial species that have high daily activity levels that cannot fully be met in captivity, running wheels can be provided to allow active behaviours, though their use should be monitored to prevent this becoming an abnormal repetitive behaviour.

We can look to eutherian equivalent taxa to get tried and tested enrichment ideas such as scent trails and as well as the provision of raw bones for the larger carnivorous marsupials and stimulating the mechanoreceptor vibrissae with intermittent vibrations can maintain these animals' natural foraging abilities (especially if this is followed by a daily feed). Feed should be provided in enriching and varied presentations, such as hanging a whole carcass, bungee feeders for resistance feeding, scatter feeds, mechanical insect feeders, subterranean feeds for those that forage for earthworms and hypogeal fungi, browse for the arboreal species, and for the exudivores, specialised devices that can encourage natural foraging abilities of peeling bark back with their teeth and licking out the sap from underneath (see Figure 10.2).

Figure 10.2 Glider enrichment example adapted from Murray et al. (1998). Credit: Heather Tanton.

Figure 10.3 Ground cuscus (*Phalanger gymnotis*) target training (photo: Erika Oulton/RZSS).

Varying feeding times and feeding towards the end of the day encourages nocturnal activity patterns and can encourage more natural behaviours. Planting foliage can provide shelter and active climbing opportunities. Training, itself, can be enriching (Figure 10.3) and can also be considered social enrichment for those social species or hand-raised individuals that have imprinted on humans. Training can also be used to help individuals cope with predators, find prey, or avoid toxic prey should they be reintroduced to the wild in the future.

10.6 USING BEHAVIOURAL BIOLOGY TO ADVANCE MARSUPIAL AND MONOTREME CARE

Our understanding of monotreme and marsupial husbandry has improved in recent years, though many areas are still under researched. Box 10.1 details some of the key behavioural biology research that still needs to be undertaken and applied to captive management practices. Videos of behaviour patterns and livestream webcams (see Box 10.2) can be valuable resources for more information and research of this group of animals.

BOX 10.1: Areas for future research for marsupials and monotremes

More research is needed to:

- Determine suitable enrichment programmes including enrichment that is needed to maintain specific behavioural adaptations that help cope with the natural environment.
- Understand the cognitive abilities of these species, which will improve our knowledge of their evolution and can be applied to captive management through improved enrichment and training programmes.
- Assess nocturnal behaviours. With most marsupial species being nocturnal, 24-hour behaviour patterns should be investigated, and considerations made as to how to apply findings into management to improve their welfare.
- Evaluate UV lighting, especially for species housed outside of native ranges, and understand effects on health, reproduction, and communication.
- Determine the effects of training on monotremes and marsupial behaviour and welfare.
- Further evaluate any effects of visitor presence, and other zoo-specific stimuli on the behaviour and welfare of monotremes and marsupials.
- Identify and understand appropriate positive welfare indicators specific to these animals.

BOX 10.2: Further information about marsupial and monotreme care

The following videos, links, and live-stream webcams can be useful resources to inform and research monotreme and marsupial behaviour:

- https://www.publish.csiro.au/zo/zo14069 – see supplementary material for video of breeding behaviour in short-beaked echidna
- https://www.publish.csiro.au/am/am21006 – see supplementary material for prey processing behaviours in Tasmanian devils
- San Diego Zoo Safari Park platypus webcam: https://sdzsafaripark.org/cams/platypus-cam
- RZSS Edinburgh Zoo koala webcam: https://www.edinburghzoo.org.uk/webcams/koala-cam/#koalacam
- Wild rock wallaby webcam: https://www.nationalparks.nsw.gov.au/brush-tailed-rock-wallaby-cam
- Koala vocalisations: https://wildambience.com/wildlife-sounds/koala/
- Tree kangaroo locomotion: https://www.youtube.com/watch?v=E27v1CNLhKA
- Glider behaviour: https://www.youtube.com/watch?v=sZ-PzBRSPsw
- Husbandry and best-practice guidelines for several marsupials species: Australasian Society of Zoo Keeping (aszk.org.au) and YFRW-EAZA-Best-Practice-Guidelines-FINAL2-February-2022.pdf

10.7 CONCLUSION

Knowledge on wild ecology of marsupials and monotremes is vitally important to animal husbandry and welfare, and ultimately conservation success of these evolutionary unique animals. Many have adapted to cope with unpredictable environments, adopting specialised behaviours to help them to survive. Captive management requires careful consideration of these behaviours when designing enclosures and planning daily care, taking account of their reproductive, social, and feeding behaviours, and the daily activity patterns of each individual species. While great progress has been made, further research is needed to continue to improve our understanding and provide better care in areas such as cognitive abilities, which in turn can improve enrichment opportunities, providing optimal husbandry and promoting positive welfare.

REFERENCES

Bethge, P., Munks, S., Otley, H., & Nicol, S. (2009). Activity patterns and sharing of time and space of platypuses, *Ornithorhynchus anatinus*, in a subalpine Tasmanian lake. *Journal of Mammalogy, 90*(6), 1350–1356.

Collett, R.A., Baker, A.M., & Fisher, D.O. (2018). Prey productivity and predictability drive different axes of life-history variation in carnivorous marsupials. *Proceedings of the Royal Society B: Biological Sciences, 285*(1890), 20181291.

Descovich, K.A., Lisle, A.T., Johnston, S., & Phillips, C.J.C. (2012). Space allowance and the behaviour of captive southern hairy-nosed wombats (*Lasiorhinus latifrons*). *Applied Animal Behaviour Science, 140*(1), 92–98.

Dickman, C.R., & Jones, M.E. (2003). *Predators with pouches: The biology of carnivorous marsupials*. CSIRO Publishing, Collingwood, Australia.

Gaschk, J.L., Frère, C.H., & Clemente, C.J. (2019). Quantifying koala locomotion strategies: Implications for the evolution of arborealism in marsupials. *Journal of Experimental Biology, 222*(24), jeb207506.

Geiser, F. (2021). *Ecological physiology of daily torpor and hibernation*. Springer, Cham, Switzerland.

Grueber, C.E., Reid-Wainscoat, E.E., Fox, S., Belov, K., Shier, D.M., Hogg, C.J., & Pemberton, D. (2017). Increasing generations in captivity is associated with increased vulnerability of Tasmanian devils to vehicle strike following release to the wild. *Scientific Reports, 7*(1), 1–7.

Hayes, G.L.T., Simmons, L.W., Dugand, R.J., Mills, H.R., Roberts, J.D., Tomkins, J.L., & Fisher, D.O. (2019). Male semelparity and multiple paternity confirmed in an arid-zone dasyurid. *Journal of Zoology, 308*(4), 266–273.

Hing, S., Narayan, E., Thompson, R.C., & Godfrey, S. (2014). A review of factors influencing the stress response in Australian marsupials. *Conservation Physiology, 2*(1), cou027.

Holland, N., & Jackson, S.M. (2002). Reproductive behaviour and food consumption associated with the captive breeding of platypus (*Ornithorhynchus anatinus*). *Journal of Zoology, 256*(3), 279–288.

Jackson, S.M. (2003). *Australian mammals: Biology and captive management.* CSIRO Publishing, Collingwood, Australia.

McElligott, A.G., O'Keeffe, K.H., & Green, A.C. (2020). Kangaroos display gazing and gaze alternations during an unsolvable problem task. *Biology Letters, 16*(12), 20200607.

McGregor, D.C., Padovan, A., Georges, A., Krockenberger, A., Yoon, H.-J., & Youngentob, K.N. (2020). Genetic evidence supports three previously described species of greater glider, *Petauroides volans, P. minor*, and *P. armillatus. Scientific Reports, 10*(1), 1–11.

Moore, B.D., & Foley, W.J. (2000). A review of feeding and diet selection in koalas (*Phascolarctos cinereus*). *Australian Journal of Zoology, 48*(3), 317–333.

Murray, A.J., Waran, N., & Young, R. (1998). Environmental Enrichment for Australian Mammals. *Animal Welfare, 7*, 415–425.

Pollock, T.I., Hocking, D.P., Hunter, D.O., Parrott, M.L., Zabinskas, M., & Evans, A.R. (2021). Torn limb from limb: The ethology of prey-processing in Tasmanian devils (*Sarcophilus harrisii*). *Australian Mammalogy, 44*(1), 126–138.

Rendle, J.A.J., Ward, S., & McCormick, W.D. (2018). Behaviour and enclosure use of captive parma wallabies (*Macropus parma*): An assessment of compatibility within a mixed-species exhibit. *Journal of Zoo and Aquarium Research, 6*(2), 63–68.

Rismiller, P.D., & Grutzner, F. (2019). *Tachyglossus aculeatus* (Monotremata: tachyglossidae). *Mammalian Species, 51*(980), 75–91.

Sherwen, S.L., & Hemsworth, P.H. (2019). The visitor effect on zoo animals: Implications and opportunities for zoo animal welfare. *Animals, 9*(6), 366.

Skelton, C.J.A., & Stannard, H.J. (2018). Preliminary investigation of social interactions and feeding behavior in captive group-housed Tasmanian devils (*Sarcophilus Harrisii*). *Journal of Applied Animal Welfare Science, 21*(3), 295–303.

Thomas, J., Handasyde, K., Parrott, M.L., & Temple-Smith, P. (2018a). The platypus nest: Burrow structure and nesting behaviour in captivity. *Australian Journal of Zoology, 65*(6), 347–356.

Thomas, J.L., Parrott, M.L., Handasyde, K.A., & Temple-Smith, P. (2018b). Female control of reproductive behaviour in the platypus (*Ornithorhynchus anatinus*), with notes on female competition for mating. *Behaviour, 155*(1), 27–53.

Wallage, A., Clarke, L., Thomas, L., Pyne, M., Beard, L., Ferguson, A., Lisle, A., Johnston, S., Wallage, A., Clarke, L., Thomas, L., Pyne, M., Beard, L., Ferguson, A., Lisle, A., & Johnston, S. (2015). Advances in the captive breeding and reproductive biology of the short-beaked echidna (*Tachyglossus aculeatus*). *Australian Journal of Zoology, 63*(3), 181–191.

11

The behavioural biology of flightless birds

LISA WARD
Wild Planet Trust, Paignton, UK

LINDA HENRY
SeaWorld San Diego, San Diego, USA

11.1 INTRODUCTION TO THE BEHAVIOURAL BIOLOGY OF RATITES AND PENGUINS

Ratites constitute a range of species that, at a glance, can seem limited in biological and ecological diversity. However, species can be found in a range of habitats and ecological niches, with some considerable physiological and behavioural variation. When discussing ratites, this chapter will include example species from Struthionidae, Apterygidae, Casuariidae, Dromaiidae, Rheidae, and Tinamidae. The tinamous are a family with a complex taxonomic history, but more recently have been included within the ratite radiation (Phillips, Gibb, Crimp, & Penny, 2010). They are also popular in zoological collections and as such have been included in this chapter. The size range of ratites is dramatic, from the formidable red-necked ostrich (*Struthio camelus camelus*), the world's largest bird at an impressive 210–275 cm tall and 110–156 kg mass, to the comparatively diminutive dwarf tinamou (*Taoniscus nanus*) at 14 cm tall and weighing around 43 g.

The commonality between this group is the inability to fly and the associated lack of a keel on the sternum. The exception are the tinamous, who have a keeled sternum and some ability to fly, though not particularly powerfully and rather reluctantly (Winkler, Billerman, & Lovette, 2020).

DOI: 10.1201/9781003208471-13

Other common features and adaptations to a life on land that impact their behavioural ecology include:

- A stocky build, that among other things, allows for better regulation of body temperature and fat storage
- A strong beak and jaw for grazing and browsing
- Keen eyesight (except for the *Apteryx* spp.) particularly in species living on open terrain
- Powerful legs and feet, allowing for great speed and dexterity and/or the ability to travel great distances
- Vestigial wings (tinamous have functioning, but often small wings) that are used for display and temperature regulation. However, wings can be almost completely absent (e.g., emu, *Dromaius novaehollandiae*).

Despite these similarities, there is also species-specific variation, which must be considered when managing these taxa in captivity. For example, the lesser rhea (*Pterocnemia pennata*) will inhabit open terrain and spend a large part of its day foraging for green matter. Lesser rheas are social, living in groups or small harems during the breeding season. In contrast, the southern cassowary (*Casaurius casaurius*) inhabits dense forest and lives a very solitary lifestyle. They primarily consume fruits and invertebrates and can be very territorial. The species is aggressive towards conspecifics, except for a few weeks during breeding season. The behavioural biology of these two species means the management will vary greatly in terms of enclosure design, group structure, diet and food presentation, and breeding management.

Without the constraints of flight, penguins, like ratites, have gained a robust body style. Penguins are heavy-bodied, densely feathered, colonial marine birds that forage only at sea. However, penguins must come on land for nesting and moult. Of the 18 penguin species, only 10 are commonly found in zoos and aquariums representing the genera *Aptenodytes, Eudyptes, Spheniscus*, and *Eudyptula*. Shared characteristics relative to behavioural ecology among penguins include:

- Fusiform body shape with legs situated posteriorly rendering an upright posture on land
- Large keel structure to support "aquatic" flight with flipper-like wings lacking primary feathers
- Long-lived with a unique and physiologically taxing "catastrophic" moult (i.e., all feathers lost at the same time)
- Visual acuity in both air and water
- Monomorphic, but with well-developed visual and vocal behavioural displays used to communicate nest territories, allow mate and chick recognition, maintain group cohesion and in predator defence (Jouventin & Dobson, 2017)
- Both territorial and social, with synchronised colonial reproductive strategies and at-sea social behaviour
- Biparental investment in nesting, incubation, and chick-rearing
- Chick-rearing strategy, whereby chicks form conspecific aggregations (i.e., crèches) in the post-guard nestling stage
- Well-adapted for annual periods of weight gain and fasting
- Photoperiod-dependent for the timing of annual biological activities

Though superficially similar, penguin species differ in body size, habitat, thermoregulatory adaptations, and reproductive biology. The size range in penguins varies from the second-largest species, the king penguin (*Aptentodytes patagonicus*), 80–90 cm tall, 14–16 kg, to the smallest little blue penguin (*Eudyptula minor*), 33 cm, 1–1.2 kg. Habitats vary from open, arid shorelines, to icy or rocky shores and steep hillsides, to densely vegetated islands. Chiefly adapted for heat retention in cold ocean waters, individual species possess varying adaptations for thermoregulation on land based on latitude distribution. Reproductive biology ranges from the king with a 14–16-month breeding cycle to chinstrap (*Pygoscelis antarctica*) and rockhopper (*Eudyptes chrysocome*) with a four-month breeding cycle (Williams & Busby, 1995). The equal importance of both water and land features, social needs, as well as thermoregulatory requirements presents challenges to ex-situ management and enclosure design while their social nature provides opportunities for mixed-species colonies.

11.2 RATITE AND PENGUIN ECOLOGY AND NATURAL HISTORY RELEVANT TO THE ZOO

For many ratites, breeding strategy and social structure are complex, and are key behavioural features that should be considered when informing decisions on management to improve welfare and reproductive success. For greater rhea (*Rhea americana*), group size depends on food availability, predator abundance and season, and will impact an individual's behaviour. Out of breeding season, groups are a mixture of sexes and ages, and larger groups form when food is abundant. When in larger groups, individual vigilance behaviours (such as raising their heads to scan the environment) reduce but overall vigilance within the group increases. Leading up to breeding season, smaller harem groups emerge. Males then become more vigilant and aggressive when defending a harem and a nest site (Carro & Fernández, 2008). Therefore, group size and structure are clearly a key feature that will affect behaviour and breeding success.

The full complexities of ratite social structure can be hard to replicate in captivity. With the aggressive and solitary southern cassowary, remaining solitary in the captive environment when not breeding is not only easy to achieve, but also important for safety and welfare. However, recreating the fluid social structure of more gregarious species such as greater rhea or common ostrich (*Struthio camelus*) may not always be practicable. Focusing on the smaller, seasonal group sizes seen in wild populations will be a positive start; ostrich, rhea, and tinamou do well when housed as a harem with one male and multiple females, throughout the year. Smaller group sizes are not uncommon for these species while still allowing the expression of natural behaviours. Furthermore, consideration can still be made to offer seasonal variation, such as separating females during incubation or introducing females to additional males for polyandrous (one female mating with multiple males) species. Table 11.1 provides details of habitat type, feeding strategy, and social and breeding behaviours of example ratite species commonly housed in ex situ populations. More information on the biology and ecology of these select species, and that of other ratitie species too can be found in Davies (2002).

By contrast, penguins' colonial social structure is easy to support in managed settings, given sufficient minimum colony size to support species-typical social activities and mate choice. The social welfare of an ex-situ Humboldt colony (*Spheniscus humboldti*), including age, colony size, and the number of breeding pairs were the best predictor of breeding success (Blay & Côté, 2001; Marshall et al., 2016). In another survey of zoos, Schweizer et al. (2016) found there was a positive association between king penguin population density and yearly egg productivity. The same survey found a negative relationship between uneven sex ratio and egg productivity. Such findings may be true for other penguin species. The *AZA Penguin Care Manual* currently recommends a minimum colony size of ten individuals (Schneider et al., 2014) but most exhibits exceed this minimum. Blay and Côté (2001) found a group size of ten Humboldt pairs to be successful. While *Spheniscus* species are recommended for single-species habitats to prevent hybridisation, other species may flourish in mixed-species exhibits, benefitting from the combined social stimulation. Intraspecific behavioural synchrony was observed in a mixed colony of chinstrap and gentoo (*Pygoscelis papua*) penguins, whereby each species formed separate colonies within the enclosure (Foerder, Chodorow, & Moore, 2013).

Penguin species vary in their responses to caregivers and enclosure settings both at the species and individual levels. Recognising species-specific differences in behaviour and individual personality (boldness versus shyness) is the first step to mitigating adverse outcomes. Little blue penguins tend to be secretive and nocturnally active, complicating exhibiting this species under usual zoo and aquarium visiting hours. Chinstraps tend to be shy, selecting spaces remote from zoo-visitor activity, but author experience suggests that king, gentoo, and macaroni (*Eudyptes chrysolophus*) penguins often appear to choose visitor attentions, although research also shows aversive reactions in free-living settings when tourists are present. Humboldt penguins tend to be shy and do not readily habituate to tourist presence, whereas Magellanic penguins (*Spheniscus magellanicus*) are shown to habituate to regular daily visits. Penguins in stressful settings may show no outward behavioural changes, so observation alone may be unreliable when assessing acceptance of exhibit conditions or visitor presence (Seddon, Ellenberg, Higham, & Lück, 2008).

Penguins are visual predators, diving to varying depths to forage, beyond the photic zone in some species, e.g., king penguins (Williams & Busby, 1995). Penguins are carnivores; fish, krill, and cephalopods comprise the principal free-living diet. Food provision in the managed setting consists of fresh-frozen, thawed fishes, and krill. Fish are often hand-fed, but penguins benefit from a variety of food provision strategies. Self-feeding from trays allows choice and control and increases opportunities for less competitive geriatric or juvenile birds to access food. Fish broadcast in pools (even including live fish where feasible) encourages species-typical

Table 11.1 A summary of the habitat, feeding strategy, and social and breeding behaviours of example species of penguin and ratite commonly housed in zoological collections.

Penguins			
Species	Range, Habitat & Red List status (as of February 2022)	Diet & feeding	Social & Breeding Behaviour
King penguin (*Aptenodytes patagonicus*)	Primarily sub-Antarctic, nesting on islands between 45–55° South latitude. Rocky, island shorelines during breeding. Travel to the Polar Front to forage during summer nesting. Least Concern	Pelagic fish (primarily myctophids at the Polar Front), squid, and krill.	Serially monogamous. No nest, defending a small territory. Form courtship trios consisting of two males and a single female; female selects mate. Male takes first incubation bout for up to two weeks. An annual breeder that generally recruits one chick every other year due to unique early and late breeding resulting from over winter chick provisioning strategy. Migratory. Colonial.
Banded penguin: African (*Spheniscus demersus*)	South-western coast of Africa; Benguela bioregion. Sandy, rocky beaches and sparsely vegetated islands. Endangered	Schooling pelagic fish such as anchovy and sardine, also cephalopods, mackerel, and hake.	Monogamous and philopatric. Colonial. Digs nests in burrows or under vegetation.
Banded penguins: Magellanic (*S. magellanicus*) & Humboldt (*S. humboldti*)	Primarily southern coastal South America: Magellanic on the east coast to Punta Tombo; Humboldt on the west coast. Humboldt habitat is arid, rocky, or desert coastlines and islands, sparsely vegetated. Magellanic habitat similar: rocky or sandy, beaches with sparse vegetation. Magellanic: Least Concern Humboldt: Vulnerable	Pelagic schooling fish such as anchovies, silversides and sardines, squid, and krill.	Species display similar behaviour. Monogamous (extra-pair paternity reported in Magellanic). Colonial. Digs nest in burrows or under vegetation.

Brush-tailed penguins: *Pygoscelis* spp. (gentoo, chinstrap)	Chinstraps are circumpolar. Gentoos are sub-Antarctic. Species overlap in the South Shetland Islands. Antarctic ice pack, breeding in Ice-free rocky shorelines and adjacent hills Gentoo: Least Concern Chinstrap: Least Concern	Primarily feed on krill but will also forage fish such as myctophids.	Species display similar behaviours and phenology. Chinstrap is migratory in winter. Monogamous (though both extra-pair copulations and extra-pair paternity is reported in Adelie and gentoo). Colonial. Builds nest of collected stones.
Crested penguins: *Eudyptes* spp. (macaroni, rockhopper)	Circumpolar, sub-Antarctic (45–65Os) islands. Vegetated islands, rocky shorelines, and cliffs. Macaroni: Vulnerable Southern rockhopper *E. chrysocome*: Vulnerable; Northern rockhopper *E. moseleyi* Endangered	Krill and fish (myctophids), some squid.	Species display similar behaviours and phenology; highly synchronous egg lay within colonies. Lay two eggs but generally only one egg/chick is reared. Monogamous. Colonial and territorial. Migratory.
Little blue penguin: *Eudyptula minor*	New Zealand, Tasmania, and southern Australia. Dunes, coastal forests, rocky coastlines, even travelling up streams and short distances. Least Concern	Anchovy, pilchards, some krill, and cephalopods.	Secretive and nocturnal. Monogamous and philopatric. Colonial. Digs nests in burrows or under vegetation, or nests in caves.
Ratites			
Common ostrich (*Struthio camelus*)	North Africa, south of the Sahara and Southern Africa. Arid to semi-arid, open desert, savanna, or grassland. Least Concern	Succulents, seeds, grass, and leaves Grazes on low level vegetation.	Form small flocks consisting of one male, a dominant female, and multiple other females. All eggs are laid in a communal nest, but only the dominant pair will incubate and rear young.
Greater rhea (*Rhea americana*)	Southern and southeastern South America. Grassland or savanna, occasionally grassy wetland. Near Threatened	Grasses but also seeds, fruits, and invertebrates. Nomadic, moving regularly to find a food source.	Form mixed groups of two to 20 birds outside of breeding season. Primarily polygynous, with some polyandry. During breeding season males separate into harems with multiple females. All eggs are laid in a communal nest. Males take on all incubation and rearing. Females may search for further mates.

(*Continued*)

Table 11.1 (Continued) A summary of the habitat, feeding strategy, and social and breeding behaviours of example species of penguin and ratite commonly housed in zoological collections.

Penguins			
Species	**Range, Habitat & Red List status (as of February 2022)**	**Diet & feeding**	**Social & Breeding Behaviour**
Emu (*Dromaius novaehollandiae*)	Australia. Grassland, savanna, and forest. Least Concern	Opportunistic feeders, taking seeds, leaves, berries, and invertebrates. Nomadic, moving to utilise new food sources.	Often solitary, pairs will form for a few months leading up to breeding season. Monogamous, though females display some polyandry. Incubation and rearing are carried out by the male. Female's leave after laying and may mate with other males.
Southern cassowary (*Casaurius casaurius*)	Indonesia, Papua New Guinea, Northeastern Australia. Rainforest, woodland, swamps, and mangroves. Least Concern	Highly frugivorous, will also eat invertebrates and small vertebrates. Spend a third of their day foraging, usual in the early mornings and late afternoons.	A solitary species, pairs form briefly during breeding season. Polyandrous. Incubation and rearing are carried out by the male. Females may search for an additional male to breed with.
North Island Brown kiwi (*Apteryx mantelli*)	New Zealand. Dense forest and shrubland. Vulnerable	Primarily an insectivorous species, prioritising earthworms, then other invertebrates and some fruits and seeds. Forage using scent, sound, and touch and are quite territorial.	A nocturnal and largely solitary species. Monogamous. Pairs have very vocal breeding displays. Pairs may have overlapping territories. Incubation and rearing are carried out by the male.
Elegant crested tinamou (*Eudromia elegans*)	Argentina and southern Chile. Savanna, shrubland, and grassland. Least Concern	Fruit, seeds, and leaves. Also consumes invertebrates in the summer months Searches for food at ground level, occasionally jumping up to reach fruits.	Social, often found in groups of 5 – 10 birds but can form large feeding groups in winter. Vocalisations are an important part of social interactions. Polygamous with polyandry also common. Incubation and rearing are carried out by the male.

foraging and swimming behaviour. Table 11.1 provides some details of the types of habitats, feeding strategy, and social and breeding behaviours of example species housed in human care.

11.3 RATITE AND PENGUIN ENCLOSURE CONSIDERATIONS BASED ON BEHAVIOURAL EVIDENCE

Using knowledge of an animal's natural history can create an environment that goes some way to allowing them to express natural behaviours and reduce stress. Figure 11.1 highlights some features that can be included within an enclosure to encourage natural behaviours. Most ratite species exhibit a nomadic lifestyle influenced by food availability. For the emu, food availability and intake are highly dependent on season and weather. Therefore, the ability to adapt to change and to regularly locate new food sources is a key behavioural feature. While depletion of a food source is not an issue in captivity, it could be replicated to encourage naturalistic behaviours. For example, regularly relocating feeding areas would increase foraging behaviour and subsequently time spent walking.

Given the behavioural and physiological adaptations for running at speed or travelling long distances, access to sufficient space to express these behaviours will improve mental and physical fitness. Whether the behaviour is motivated by territorial defence, avoidance, breeding, or simply inquisitiveness, in a captive environment these fast and often agile birds need access to some degree of flat, open space to run and explore, that is unhindered by barriers or other hazards.

Enclosure complexity requires consideration and can be achieved by providing a variety of features akin to those found in the wild. For forest species such as the North Island brown kiwi (*Apteryx mantelli*), this may include heavily planted areas of trees and bushes, foraging sites, some open space, changes in topography, variation in substrates and multiple options for breeding and roosting sites. Awareness of species-specific behaviours will help indicate how species choose to use their

Figure 11.1 Examples of enclosure features that encourage natural behaviour. Top left – plants and rocks can provide complexity and visual barriers for elegant crested tinamou (Photo: R. Ward). Top right – large, open spaces allow for safe movement for common ostrich, such as running. Mixed species can increase behavioural repertoires (Photo: M. Hepher). Bottom left – low-level planting provides areas for greater rhea to nest or rest. Bottom middle – sandy areas provide bathing and nesting opportunities for common ostrich (Photo: M. Hepher). Bottom right – forest species such as the southern cassowary benefit from complex, heavily planted enclosures with some open space.

environment. For example, the greater rhea will utilise loose shrubs and grasses of certain heights to provide shelter, avoiding dense shrubs or long grasses and preferring to forage in areas that allow for easy escape from danger.

While ratite nests are not flamboyant or complex, attention should be given to what type of environment the species chooses to nest in. Enclosure furnishings can replicate wild preferences, features such as well positioned logs or grasses will allow birds the opportunity to choose a favoured nest site. Male tinamous will dig a nest bowl next to dense clumps of grass, so knowing which types of plants to use is beneficial.

Understanding natural behaviours helps provide a safe environment for both animals and keepers. Some ratites can be aggressive to other animals including conspecifics. Larger species, notably the ostrich and cassowary species, can also be a risk to staff. The opportunity to segregate aggressive birds, both from staff and other animals, may be vital. Mixed species exhibits can work for some ratites and has the benefit of providing the opportunity for a more varied range of behaviours to be expressed. For example, the common ostrich is often housed with various ungulates and this can work well, but enclosures must be designed with welfare and safety in mind. When studying wild populations, Sauer (1970) recorded that while 75% of behaviours between ostrich and other species were avoidance or ignoring, 16% of interactions were agonistic. To reduce the likelihood of escalated agonistic behaviours in captivity, safe areas for birds should be provided, as should unhindered access to shelter, food, water, and nesting sites. Nesting areas are particularly important when considering that ratites are ground-nesting birds as other species could cause injury to adults, eggs or chicks, or there may be aggression towards other species during breeding season. Such factors can be accounted for with good planning and enclosure design.

Bathing is an important behaviour readily expressed in many ratite species. While bathing is often associated with maintenance such as hygiene and temperature regulation, the motivation and causation for bathing is hard to establish. Research on dust bathing in domestic fowl found removing the ability of a bird to dust bathe led to an increase in dustbathing behaviour when it was once again made available (Olsson & Keeling, 2005). This suggests that whatever the primary motivation, bathing behaviour is high value and therefore opportunity to express this behaviour is beneficial to welfare. For species in dry habitats, this is often in the form of dust bathing. For species that inhabit damp areas, they often bathe in water. Emus and cassowaries, particularly, are keen swimmers when the opportunity arises.

Penguins straddle a life requiring both water and land provision to express the full complement of species-typical behaviours. Penguins forage at sea but are dependent on dry land for reproduction and moult. Most penguins are gregarious, so group size is important to promote social behaviours. Enclosures should provide adequate land and water space to accommodate the colony size, the behavioural range and size of housed species, and support social preferences.

Pools provide three-dimensional interactive space for penguin behavioural expression including foraging, "porpoising," diving, synchronised swimming, and bathing. Overall square footage, surface area, and depth should be sufficient to support the population. Current enclosures include depths up to 7–8 m, but a variety of depths ranging from a minimum of 3 m are beneficial. Built-in water currents or waves promote healthy swim behaviours and provide environmental variability. Chlorine must be avoided, and the surface of the penguin's pool should be surface skimmed to remove debris (e.g., fallen leaves) that can collect on the penguin's plumage and alter time budgets (Blay & Côté, 2001). Water quality is important to maintaining healthy penguin plumage, forage behaviour, and access to drinking. Penguins have been maintained in both saltwater and freshwater pools, but Reisfeld et al. (2013) found Magellanic penguin swimming behaviour was greatly enhanced in saltwater when compared to freshwater. Woodhouse et al. (2016) identified saltwater pools as one factor associated with decreased probability of cataract in macaroni penguins, and African penguins face a greater likelihood for abnormal moult when housed with freshwater pools (Golembeski, Sander, Kottyan, Sander, & Bronson, 2020).

The water-land interface should comprise both gradual beaches and vertical edges to meet penguins' variability in pool access and egress behaviour: *Pygoscelids* swim up towards the vertical edges to leap out while kings and *Spheniscus* use beaches to walk out. Little blue and *Eudyptes* make use of both types. Geriatric birds of all species may require beach or ramp access for safe entry and exit.

Land areas must be large enough to accommodate both dry rest, habitat elements ("furniture") and nesting space. Substrates should be variable in texture and topography with sufficient drainage. Concrete substrates were associated with hatching success in Humboldt penguins (Blay & Côté, 2001) but are not considered appropriate substrates for penguin foot health. Artificial snow is used as a substrate in some climate-controlled exhibits. Species-specific nesting space set away from the pool should be large enough to accommodate colony needs, especially in mixed-species habitats. Dry rest areas can be seasonally altered to serve as nesting spaces where exhibit footprint is limited.

Nesting strategies vary. King penguins, which prefer flat spaces for nesting territory, also need areas away from nests for copulations. Many penguins such as *Pygoscelis* use loose-rock substrates which provide sufficient materials for substrate and to build their nests in colonial groupings. *Eudyptes* will choose rocky areas or elevated spaces for nesting. *Spheniscus* dig burrows in dirt substrates for nesting or make use of provided nest boxes. In a survey of British zoos, Humboldt penguin chick productivity was highest when burrow boxes were provided with sand and gravel as the nesting substrate (Blay & Côté, 2001).

The post-guard nestling stage, when chicks form aggregations or crèches, is an important consideration in exhibit design. Twelve of the 18 penguin species are reported to form crèches during this growth stage, when both parents must forage at sea to sufficiently provision the chick(s) (Wilson, 2009). Species include king, chinstrap, gentoo, African, rockhopper, and macaroni. In human care, parents will naturally leave chicks alone during this phase rendering the chicks vulnerable to aggression. For zoos and aquariums, the crèche phase presents both a challenge and an opportunity. The challenge is to secure chicks from conspecific interference or aggression while maintaining visual and vocal contact with the colony. This is best done in the primary habitat in a contained area or corral near the rookery. Plexiglas or similar clear material helps maintain visual contact. The advantage is that the crèche phase is a natural segue for chicks to learn the routines of life in human care as well as social-colonial interactions. In the corral they will no longer have access to parental feedings and chicks readily habituate to hand-feeding from caretakers within a few days. Sufficient space to accommodate this transitional phase is a vital provision in penguin exhibit design and to chicks' developmental welfare.

Exhibit design relative to behavioural outcomes have been marginally studied in penguins. In general, most research agrees that exhibit size is important for animal welfare through positive impacts on behaviour, social preferences, and choice within the environment (Browning & Maple, 2019). In Humboldt penguins, Marshall et al. (2016) compared behaviour and breeding success to enclosure parameters. They found that total enclosure area per penguin (land plus pool) best-predicted pool use, followed by land area, and by pool surface area, suggesting the importance of land–water ratios. They recommended up to 21 m^2 per bird. Hatching success was positively associated with land–water ratio. Blay and Côté (2001) found a correlation between increased hatching success in Humboldt penguins with increasing pool size. Enclosure design can maximise land–water ratios by stacking elements vertically (Figure 11.2), e.g., two-thirds land over a full exhibit pool with one-third air-water interface and variety in land elevations to maximise nesting, hides, and dry rest spaces.

Many penguin exhibits are incorporating pathways for visitors directly within the enclosure (Figure 11.3, lower right), but care should be taken to understand species-specific flight distances and response to visitor noise and proximity (Edes et al., 2021). Chiew et al. (2021) found that visitors positioned above (described as "looming") and near to little penguins in a pool may have been perceived as a threat, resulting in penguins moving away from that area. By contrast, gentoo penguin pool usage and behavioural diversity were associated with higher numbers of visitors (Collins, Quirke, Overy, Flannery, & O'Riordan, 2016; Edes et al., 2021). An important aspect of enclosure design is choice and control regarding conspicuousness to visitors. Land spaces and pools should incorporate areas that are out of guest view. Visitor interaction areas should be situated so that penguins have a choice of whether to engage or avoid visitor presence.

Other factors to consider are whether exhibits are situated outdoors, or if species needs require the indoor provision of specific temperature and lighting (Figure 11.3). While *Spheniscus* can be exhibited in outdoor settings in local climates similar to their home range, there is an increasing concern for mosquito-borne illnesses where indoor enclosures could provide better protection. Some species are

Figure 11.2 A penguin exhibit shown from two angles illustrating stacked habitat elements to maximise both water and land space in a limited footprint. Here, the land space is a simulated ice-edge overhang of complex dry rest space (~ 325 m^2) with rock elevations and substrate variety (e.g., river rocks, artificial snow), underlaid by a saltwater pool (~372 m^2) that extends 12 m under the dry rest and with a long (~30.5 m) air–water interface of variable width, along with the visitor viewing window, featuring both steep edge and beach exits. These photos also show exhibit design flexibility for winter (upper photos, snowy dry rest) and summer (lower photos) depicting a rookery configuration within the same habitat footprint.(Photo top: SeaWorld San Diego).

high latitude, requiring specialised lighting conditions to meet physiological needs. Visitor activities such as camera flash, video, and smartphone lighting may affect lighting regimes or startle penguins under winter lighting conditions. Even temperate latitude species require appropriate photoperiod provision. All light sources should provide appropriate colour-rendering to meet the needs of the penguin visual system. Schull et al. (2016) found that king penguin beak spot colouration (ultraviolet brightness) conveys an honest signal of fitness that may be important in mate choice.

In all exhibits where visitors are in direct proximity to penguins, there is a risk of foreign objects entering the penguin habit. Small items, such as coins, are often ingested, with increased frequency prior to breeding and moult (unpublished data). Penguins have been observed to ingest seashells around the time of egg-laying (Vanstreels et al., 2020; Boersma, Rebstock, & Stokes, 2004) and this may influence the ingestion of coins which may appear similar in character to mollusc shells. Coins containing zinc represent a heavy-metal health hazard. Exhibit design that prevents visitors from direct pool contact or a system of routine monitoring and retrieval may be warranted. Husbandry measures vary with the size and design of the enclosure but as water birds, penguin exhibits should have adequate drainage, including built-into nest areas and burrows, for ease of maintenance, cleaning, and sanitising.

11.4 RATITE AND PENGUIN BEHAVIOURAL BIOLOGY AND WELFARE

In addition to species-specific behaviours, an understanding of individual behaviours and personality traits can also help when monitoring health and

Figure 11.3 Penguin enclosure examples. Top left – Humboldt exhibit with a window apparatus that can be raised to provide outdoor access as weather permits. Top right –central-habitat pool access. Lower left – an expansive outdoor habitat with shallow and deep-water pools. Lower right –an immersive walk-through habitat.

welfare. For example, a confident common ostrich will hold their head and body high, and their tail erect, whereas a more submissive ostrich will hold their head and tail lower (Bertram, 1992). Knowing what normal behaviour is for an individual, as well as understanding group dynamics, make it easier to monitor changes that could indicate a welfare or health concern.

Knowledge of ratite nutritional requirements is varied and limited in many species; however, behavioural biology can help determine broader needs. Ratites display considerable changes to feeding throughout the year, owing to their reproductive strategies. Understanding these changes is vital, as it can help improve health and reduce the risk of nutritional diseases such as obesity. Male emu appetite increases pre-breeding, then during incubation they do not leave the nest. Instead, they rely on pre-established fat and energy reserves, losing up to a fifth of their body weight. Female ratites often lay multiple eggs, in multiple clutches over each season. Therefore, consideration must be given to the nutritional availability throughout the year based on the species and the individual circumstances, to ensure birds remain physically fit.

Ratites can be difficult to catch, restrain, and transport due to physical and behavioural traits such as speed, gait, nervousness, and aggression. Placing ratites in situations where they are stressed, or fearful, increases the likelihood of serious injury or death. Many zoos have found ratites to be highly responsive to husbandry training techniques that improve their management and reduce the risk of health or welfare issues (Figure 11.4). A tinamou group living in a large mixed-species exhibit trained to recall to a smaller holding area will allow for better visual health checks or easier manual restraint, thus reducing the risk of stress and injury to all the animals within that enclosure.

As with ratites, penguins benefit from an understanding of individual behaviours and personality traits to inform health and welfare monitoring. Research on king penguins identified six comfort (self-maintenance) behaviours that comprised up to

Figure 11.4 Examples of ratite husbandry training. A & B: Crate training a cassowary at Chester Zoo (Photos: Z. Newman). C: target training an ostrich at Paignton Zoo. D & E: Training an emu to voluntarily accept injections and greater rhea station training at Copenhagen Zoo (Photos: A. Pederson).

22% of time spent ashore during breeding or fasting (Viblanc, Mathien, Saraux, Viera, & Groscolas, 2011). Penguins exhibiting poor plumage may not be allocating sufficient time to self-maintenance that may be an indication of subclinical pathologies, warranting veterinary intervention. Caregivers should be familiar with individual bird habits to better identify aberrant behaviours such as sedentary or reduced activity, or self-isolating from the main colony, and identify causation. Seasonal variations owing to breeding and moult may complicate such observations so routine annual behavioural monitoring to establish both species and individual behaviour patterns can promote welfare for penguins. Where colony numbers are high, remote monitoring using remote frequency identification devices (RFID) or time-depth recorders (TDR) can help manage data collection and facilitate analysis of behaviours indicative of positive, or negative, welfare (Fuller, Heintz, & Allard, 2019; Kalafut & Kinley, 2020).

Penguins have been successfully trained to perform husbandry behaviours such as foot exams, blood sampling, and voluntary exhibit shifting. However, penguins may also be accepting of handling especially in birds showing affiliative behaviours towards caretakers. Where colony size makes training behaviours unfeasible, non-voluntary restraint for exams or blood sampling is relatively easy and can be completed relatively quickly, allowing the bird to return to the colony.

Captive penguins are susceptible to Aspergillosis, avian malaria, and pododermatitis. Aspergillosis usually manifests as a secondary infection in the air sacs. While indoor enclosures benefit from air filtration, Aspergillosis can still occur. In outdoor exhibits, the *Aspergillus* fungus is ubiquitous. Penguins with robust immune function should not be susceptible to Aspergillosis. However, chronic stress and melatonin suppression can contribute to reduced immune function. Behavioural observations to identify primary chronic stressors may reduce infections. Monitoring faecal glucocorticoid metabolites may help to clarify environmental or situational stressors. Avian malaria is a vector-borne blood parasite for which penguins have no natural resistance. Behavioural indicators of infection are isolation from colony social behaviour as well as inappetence and reduced alertness. Pododermatitis is another health and welfare concern in penguins. This condition is often associated with sedentary behaviour and poor substrate variety. Enrichment provision to engage social swimming behaviour and modification of substrates should alleviate or eliminate mild infections.

Abnormal moult has been recently identified as a welfare concern is some penguin species. The factors affecting moult are complex but subtle light/dark confusion may result from exposure to artificial light inputs at night, adversely affecting moult outcomes (Figure 11.5). Adherence to consistent photoperiodic timing is essential to penguin health and welfare. More research is needed to better identify the factors associated with aberrant moult.

11.5 RATITE AND PENGUIN SPECIES-SPECIFIC ENRICHMENT

Knowledge of an animal's behavioural repertoire allows keepers to provide husbandry routines that encourage natural behaviours and reduce the risk of these behaviours being lost in captive populations. Research (on multiple species) highlights that movement makes up a large percentage of a wild ratite's activity budget, and the causation for this behaviour is largely foraging. Bertram (1992) observed that ostriches spent 29% of their day travelling and 33% feeding. A feeding routine that encourages regular movement, through increased frequency or regular relocation of feeding areas for example, can reduce monotony and encourage foraging. de Azevedo, Lima, Cipreste, Young, and Rodrigues (2013) found abnormal behaviours in greater rhea (such as pacing and coprophagy) reduced when food was offered scattered or hung across their enclosure. Foraging and walking behaviours increased significantly, not only during the enrichment phase, but after it ended. Food was still offered in their normal feeding station during the enrichment phase, but individuals chose to forage, suggesting foraging is of high value and should be prioritised when looking to improve captive ratite welfare.

Providing safe sources of food within the enclosure, such as fruiting trees or palatable grasses, can provide excellent enrichment. Piles of logs or leaf litter that encourage invertebrates for insectivorous species can be beneficial, as can areas of planting and small water bodies that encourage flying insects that species such as greater rhea may enjoy catching.

Further examples of enrichment items that can be temporarily added to the environment to increase novelty include:

- Water bodies to allow swimming behaviours for emu and cassowary
- Sprinkler systems for birds to shower under
- Bubble machines for species that like to catch flying insects
- Foraging boxes that food can be hidden in to encouraging digging or probing
- Items that can be safely hung from trees or shrubs for pecking and manipulation e.g., browse, strong boomer balls or willow balls with food inside

Primary enrichment for penguins often comes from their exploration of complex and dimensional habitat designs that incorporate species-specific habitat elements such as a large pool, water currents, underwater caves, periodic wave production, timed-release underwater forage options, rocky shorelines,

Figure 11.5 Potential sources of light bleed from spaces adjacent to penguin enclosures that may affect photoperiod. Top left – from exterior doors or windows. Top right – from visitor area pathway or graphics lighting. Bottom – from holiday or building lights.

hides, elevations for climbing, and snow or ice elements. Penguins spend most of their annual cycle at sea in social groupings. Penguins engage in synchronous swim behaviours with conspecifics. Penguins are naturally curious, and zoos and aquariums continue to innovate enrichment items to challenge the behavioural repertoire of penguins in human care (Figure 11.6).

Figure 11.6 Enclosures for penguins should be designed to support their curious nature, encouraging species-typical behaviours both in and out of the water. Right photos show underwater complexity to encourage swimming. Lower left features a waterfall element to encourage grooming. Upper left illustrates penguins on a voluntarily public walk as a form of enrichment.

Successful enrichment approaches include:

- Variable feeding schedules and hand-feeding locations
- Frozen krill balls to promote foraging behaviour
- Random provision of high-value fish paired with an auditory cue to call birds to the water; fish can alternately be tossed to opposite ends of the pool to promote foraging competition as birds race between fish amounts, often eliciting porpoising behaviour
- Water features (sprinklers, waterfalls, mists, streams) are popular among penguins, especially *Eudyptes* species, for bathing and plumage maintenance
- Manipulatives such as river rocks, large sticks, pine needles, plastic tubing, strands of pampas grass or similar artificial grasses replicating the collection of nest materials (it is not recommended to use organic enrichment in indoor habitats due to the risks of fungal introduction)
- Floating elements, including balls, kelp (natural or artificial), and ice shapes to encourage active swim behaviour
- Bubbles to elicit curiosity and to increase locomotor activity
- Non-floating elements such as small artificial sea stars or coral that stay on the pool bottom but can be moved and manipulated under water to provide challenge and encourage repetitive and lengthy dives to satisfy curiosity
- Live fish can be fed to replicate predator-prey behaviour (if allowed by national legislation)

All enrichment items should be evaluated for safety from ingestion, physical injury, or introduced pathogens.

11.6 FUTURE DIRECTIONS IN RATITE AND PENGUIN BEHAVIOURAL BIOLOGY

Our understanding of the full behavioural repertoire for these flightless birds is limited. Much of the husbandry and management recommendations for ratites and penguins is based on anecdotal avicultural experience. It is recommended that current practices, while successful and ongoing, still be reviewed and supported with research data. The ongoing well-being of the species in human care depends on a commitment to an evidence-based, scientific approach to best practices. Box 11.1 lists opportunities for research that, while not exhaustive, if explored could benefit our understanding of these uniquely adapted birds.

BOX 11.1: Recommended areas of investigation that the authors believe will contribute to an improved understanding of ratite and penguin behaviour, as well as strengthen best practices for these species in managed care.

Ratites

- Vocalisation:
  - Vocalisation is important for certain species of ratite, such as the tinamous. However, there is some research that suggests vocalisations are more commonly used in other species than previously understood. Gaining a more thorough understanding of ratite vocalisation and particularly its importance on social or breeding behaviours, could improve reproductive success.
- Nutrition:
  - Improving knowledge of the nutritional impact of seasonal variation in wild populations and how to apply this to captive populations. Not only would this allow for improved replication of wild behaviours, but it may increase breeding success and reduce nutritional disease.
- Group dynamics:
  - It would be beneficial to understand more about the relationships of individuals within social ratite species and the importance of social structure on their behaviour and welfare.
  - Research using Social Network Analysis as a tool will be useful in understanding group dynamics and may help advise better management decisions, such as ideal group size or the importance of individual relationships within group dynamics (see Rose & Croft, 2015).
- Mate selection:
  - When looking at long-term breeding plans, it would be beneficial to understand if there was any behavioural value in providing mate choice to captive ratites. For example, would allowing solitary species such as cassowary the option to "choose" their own mate improve breeding success. For polygamous species such as the greater rhea, could collections house larger mixed-sex groups and allow the birds to form their own harems.
- Enrichment:
  - There is a requirement for a better understanding of the value and effectiveness of different enrichment techniques and items, and the impacts these have on individual and group behaviours.

Penguins

- Nutrition:
  - Procuring fish for penguins in zoos and aquariums is becoming increasingly difficult. Fish species that have historically provided nutritional support are no longer sustainable or available. As alternate fish sources are sourced and incorporated into captive penguin diets, more

research is needed on the sustainability and nutritional components of alternate fish species to best meet penguin nutritional needs. Microplastics and heavy-metal contamination is becoming more prevalent in fish forage. Research is needed to determine the degree to which these elements may be affecting welfare outcomes for penguins in zoos and aquariums.

- Visitor effects:
  - As zoological facilities drive more engaging penguins exhibits, bringing visitors into direct contact with penguins, more research is needed to determine if such designs are a source of chronic stress. As mentioned in the text, penguins may be behaviourally stoic in the presence of negative stimuli but may show physiological changes such as increased corticosterone levels or increased heart rates. Measuring faecal glucocorticoid metabolites (FGCM) is a non-invasive means of assessing FGCM as an indicator of stress in penguins relative to visitor activities or presence.
- Photoperiod and Artificial Light at Night (ALAN):
  - As zoological facilities incorporate more evening lighted events, the impact of light trespass into adjacent penguin habitats, or light exposure from visitor area activities (e.g., night-time dinner parties) expose penguins to light inputs that may serve to suppress melatonin expression and subsequently impact proper immune function. Outdoor penguin habitats in urban areas may suffer from light pollution in the form of sky glow or direct street or pathway lighting. More research is needed into the impacts of ALAN on penguin health and welfare.
- Behavioural monitoring:
  - Zoos and aquariums are committed to environments where penguins can thrive. However, empiric data are needed to support exhibit designs and outcomes. RFID and TDR technologies are just starting to be used with ex-situ penguins to assess individual's behaviours as indicators of well-being. This type of research should be expanded across zoos and aquariums.
- Extra-pair copulations (EPC), extra-pair paternity (EPP), intraspecific brood parasitism (IBP), and studbook data:
  - Studbooks have been compiled to document parentage of penguins in managed care and guide population decisions to maintain genetic diversity. However, increasing evidence suggests that both in-situ and ex-situ penguins (so far *Pygoscelis* and *Spheniscus* species), while socially monogamous, may be genetically polygamous, resulting in confusion about the genetic parentage of nestlings. From the standpoint of studbook management as well as behavioural interest, this bears further investigation.

11.6.1 Further information

Further information on ratites from the Association Zoos & Aquaria (AZA), including species fact sheets, enrichment features and annual newsletters can be found here:

Avian Scientific Advisory Group (2020). *Struthioniformes Taxon Advisory Group.* https://www.avianscientific.org/struthionformes

Further details on ratite behaviour and ecology can be found in these papers:

Barri, F.R., Roldán, N., Navarro, J.L., & Martella, M.B. (2012). Effects of group size, habitat and hunting risk on vigilance and foraging behaviour in the lesser rhea (*Rhea pennata pennata*). *Emu, 112*, 67–70.

Brooks, D.M. (2015). Behaviour, reproduction and development in little tinamou (*Crypturellus soui*). *The Wilson Journal of Ornithology, 127*(4), 761–765.

Cunningham, S.J. & Castro, I. (2011). The secret life of wild brown kiwi: studying behaviour of a cryptic species by direct observation. *New Zealand Journal of Ecology, 35*(3), 209–219.

Deeming, D.C. & Bubier, N.E. (1999). Behaviour in natural and captive environments. In D.C. Deeming (ed.). *The ostrich: Biology, production and health.* CABI International, Wallingford, UK, pp. 83–104.

Martella, M.B., Renison, D., & Navarro, J.L. (1995). Vigilance in the greater rhea: effects of vegetation height and group size. *Journal of Field Ornithology, 66*(2). 215–220.

Pérez-Granados, C. & Schuchmann, K. (2021). Vocalisations of the Greater Rhea (*Rhea americana*): an allegedly silent ratite. *Bioacoustics, 30*(5), 564–574.

Sales, J. (2006). Digestive physiology and nutrition of ratites. *Avian and Poultry Biology Reviews, 17*(2). 41–55.

Further information helpful to advancing penguin husbandry include:

AZA Penguin Taxon Advisory Group. (2014). Penguin (Spheniscidae) Care Manual. Silver Spring, MD: Association of Zoos and Aquariums. Available to download from: https://www.avianscientific.org/penguin

Borboroglu, P.G. & Boersma, P.D. (2015). *Penguins: natural history and conservation.* University of Washington Press, Seattle, USA.

Lee, L., Tirrell, N., Burrell, C., Chambers, S., Vogel, S., & Domyan, E.T. (2018). Genetic tests reveal extra-pair paternity among Gentoo penguins (*Pyogoscelis papua ellsworthii*) at Loveland Living Planet Aquarium: Implications for ex situ colony management. *Zoo Biology, 37*(4), 236–244.

Marasco, A.C.M., Morgante, J.S., Barrionuevo, M., Frere, E., & de Mendonça Dantas, G.P. (2020). Molecular evidence of extra-pair paternity and intraspecific brood parasitism by the Magellanic Penguin (*Spheniscus magellanicus*). *Journal of Ornithology, 161*(1), 125–135.

de Mendonça Dantas, G.P., Gonzaga, L.G., da Silveira, A.S., Werle, G.B., da Cruz Piuco, R., & Petry, M.V. (2020). Extra-pair paternity and intraspecific brood parasitism in the Gentoo Penguin (*Pygoscelis papua*) on Elephant Island, Antarctica. *Polar Biology, 43*(7), 851–859.

Müller-Schwarze, D. (1984). *The behavior of penguins: Adapted to ice and tropics*. State University of New York Press, New York, NY.

11.7 CONCLUSION

Ratites and penguins are unique taxa that have evolved species-specific behavioural traits to survive in their natural environment. Using knowledge of their wild ecology, biology, and behaviour to establish husbandry practices that are evidence-based ultimately improves the welfare of those species housed in zoological facilities. Ratites have not received much attention in the peer-reviewed literature, but it is hoped that ex-situ bird managers, as well as in-situ biologists will take up the challenge. Penguins are among the most well-studied sea birds with numerous books and literature on a spectrum of topics including behaviour, but still new questions arise. It is a necessary process to continually research, establish, and implement practices, then research again as we strive to conserve these iconic and charismatic species.

REFERENCES

Bertram, B.C.R. (1992). *The ostrich communal nesting system*. Princeton University Press, Princeton, USA.

Blay, N., & Côté, I.M. (2001). Optimal conditions for breeding of captive Humboldt penguins (*Spheniscus humboldti*): A survey of British zoos. *Zoo Biology, 20*(6), 545–555.

Boersma, P.D., Rebstock, G.A., & Stokes, D.L. (2004). Why penguin eggshells are thick. *The Auk, 121*(1), 148–155.

Browning, H., & Maple, T.L. (2019). Developing a metric of usable space for zoo exhibits. *Frontiers in Psychology, 10*, 791.

Carro, M.E., & Fernández, G.J. (2008). Seasonal variation in social organisation and diurnal activity budget of the greater rhea (*Rhea americana*) in the Argentinean pampas. *Emu – Austral Ornithology, 108*(2), 167–173.

Chiew, S.J., Butler, K.L., Fanson, K.V., Eyre, S., Coleman, G.J., Sherwen, S.L., Melfi, V., & Hemsworth, P.H. (2021). Effects of the presence of zoo visitors on zoo-housed little penguins (*Eudyptula minor*). *New Zealand Journal of Zoology, 49*, 1–22.

Collins, C.K., Quirke, T., Overy, L., Flannery, K., & O'Riordan, R. (2016). The effect of the zoo setting on the behavioural diversity of captive gentoo penguins and the implications for their educational potential. *Journal of Zoo and Aquarium Research, 4*(2), 85–90.

Davies, S.J.J.F. (2002). *Ratites and tinamous.* Oxford University Press, Oxford, UK.

de Azevedo, C.S., Lima, M.F.F., Cipreste, C.F., Young, R.J., & Rodrigues, M. (2013). Using environmental enrichment to reduce expression of abnormal behaviours in greater rhea *Rhea americana* at Belo Horizonte Zoo. *International Zoo Yearbook, 47*, 163–170.

Edes, A.N., Baskir, E., Bauman, K.L., Chandrasekharan, N., Macek, M., & Tieber, A. (2021). Effects of crowd size, composition, and noise level on pool use in a mixed-species penguin colony. *Animal Behavior and Cognition*, *8*(4), 507–520.

Foerder, P., Chodorow, M., & Moore, D.E. (2013). Behavioural synchrony in two species of communally housed captive penguins. *Behaviour*, *150*(12), 1357–1374.

Fuller, G., Heintz, M.R., & Allard, S. (2019). Validation and welfare assessment of flipper-mounted time-depth recorders for monitoring penguins in zoos and aquariums. *Applied Animal Behaviour Science*, *212*, 114–122.

Golembeski, M., Sander, S.J., Kottyan, J., Sander, W.E., & Bronson, E. (2020). Factors affecting abnormal molting in the managed African penguin (*Spheniscus demersus*) population in North America. *Journal of Zoo and Wildlife Medicine*, *50*(4), 917–926.

Jouventin, P., & Dobson, F.S. (2017). *Why penguins communicate: The evolution of visual and vocal signals*. Academic Press, Oxford, UK.

Kalafut, K., & Kinley, R. (2020). Using radio frequency identification for behavioral monitoring in little blue penguins. *Journal of Applied Animal Welfare Science*, *23*(1), 62–73.

Marshall, A.R., Deere, N.J., Little, H.A., Snipp, R., Goulder, J., & Mayer-Clarke, S. (2016). Husbandry and enclosure influences on penguin behavior and conservation breeding. *Zoo Biology*, *35*(5), 385–397.

Olsson, I.A.S., & Keeling, L.J. (2005). Why in earth? dustbathing behaviour in jungle and domestic fowl reviewed from a Tinbergian and animal welfare perspective. *Applied Animal Behaviour Science*, *93*(3–4), 259–282.

Phillips, M.J., Gibb, G.C., Crimp, E.A., & Penny, D. (2010). Tinamous and Moa Flock Together: Mitochondrial Genome Sequence Analysis Reveals Independent Losses of Flight among Ratites. *Systematic Biology*, *59*(1), 90–107.

Reisfeld, L., Moraes, K., Spaulussi, L., Cardoso, R.C., Ippolito, L., Gutierrez, R., Silvatti, B., & Pizzutto, C.S. (2013). Behavioral responses of Magellanic Penguins (*Spheniscus magellanicus*) to saltwater versus freshwater. *Zoo Biology*, *32*(5), 575–577.

Rose, P.E., & Croft, D.P. (2015). The potential of social network analysis as a tool for the management of zoo animals. *Animal Welfare*, *24*(2), 123–138.

Sauer, E.G.F. (1970) Interspecific behaviour of the South African ostrich. *Ostrich*, *40*(1), 91–103.

Schneider, T., Olsen, D., Dykstra, C., Huettner, S., Sirpenski, G., Sarro, S., Waterfall, K., Henry, L., Dubois, L., Jozwiak, J., & Diebold, E. (2014). *Penguin (Spheniscidae) care manual: Published by the association of zoos and aquariums in association with the AZA animal welfare committee*. Association of Zoos and Aquariums, Silver Springs, MD, USA.

Schull, Q., Dobson, F.S., Stier, A., Robin, J.P., Bize, P., & Viblanc, V.A. (2016). Beak color dynamically signals changes in fasting status and parasite loads in king penguins. *Behavioral Ecology*, *27*(6), 1684–1693.

Schweizer, S., Stoll, P., von Houwald, F., & Baur, B. (2016). King penguins in zoos: Relating breeding success to husbandry practices. *Journal of Zoo and Aquarium Research*, *4*(2), 91–98.

Seddon, P.J., Ellenberg, U., Higham, J.E.S., & Lück, M. (2008). Effects of human disturbance on penguins: The need for site-and species-specific visitor management guidelines. In J.E.S. Higham, & M. Lück (eds). *Marine wildlife and tourism management: Insights from the natural and social sciences*. CABI Publishing, Wallingford, UK, pp. 163–181.

Vanstreels, R.E.T., Pichegru, L., Pfaff, M.C., Snyman, A., Dyer, B.M., Parsons, N.J., Roberts, D.G., Ludynia, K., Makhado, A., & Pistorius, P.A. (2020). Seashell and debris ingestion by African penguins. *Emu – Austral Ornithology*, *120*(1), 90–96.

Viblanc, V.A., Mathien, A., Saraux, C., Viera, V.M., & Groscolas, R. (2011). It costs to be clean and fit: Energetics of comfort behavior in breeding-fasting penguins. *PLoS One*, *6*(7), e21110.

Williams, T.D., & Busby, J. (1995). *Bird families of the world. 2. The penguins: Spheniscidae*. Oxford University Press, Oxford, UK.

Wilson, D. (2009). Causes and benefits of chick aggregations in penguins. *The Auk*, *126*(3), 688–693.

Winkler, D.W., Billerman, S.M., & Lovette, I.J. (2020). Tinamous (Tinamidae) (v. 1.0). In S.M. Billerman, B.K. Keeney, P.G. Rodewald, & T.S. Schulenberg (eds). *Birds of the world*. Cornell Lab of Ornithology, New York.

Woodhouse, S.J., Peterson, E.L., & Schmitt, T. (2016). Evaluation of potential risk factors associated with cataract in captive macaroni (*Eudyptes chrysolophus*) and rockhopper penguins (*Eudyptes chrysocome*). *Journal of Zoo and Wildlife Medicine*, *47*(3), 806–819.

12

The behavioural biology of waterbirds

PAUL ROSE
University of Exeter, Exeter, UK
WWT, Slimbridge Wetland Centre, Slimbridge, UK

ANDREW MOONEY
Trinity College Dublin, Dublin, Ireland

JOANNA KLASS
Woodland Park Zoo, Washington, USA

12.1 INTRODUCTION TO WATERBIRD BEHAVIOURAL BIOLOGY

The waterbirds covered in this chapter include representative species of the Orders Anseriformes (ducks, geese, swans, and screamers), Phoenicopteriformes (flamingos), Ciconiiformes (storks), Gruiformes (cranes, rails, trumpeters, finfoots, fluff-tails, and the limpkin, *Aramus guarauna*) and the Pelecaniformes – pelicans, herons, ibis, spoonbills, the shoebill (*Balaeniceps rex*), and the hamerkop (*Scopus umbretta*). Table 12.1 provides details of example species from across these taxonomic orders. This diverse collection of birds represents some of the most ancient of avian species, the Anseriformes (Kuhl et al., 2021), and includes species that inhabit every one of the Earth's continents. Species can be very generalised in their ecological needs, coping well with the anthropogenic environment (e.g., mallard, *Anas platyrhynchos* or grey heron, *Ardea cinerea*) or can be very specific in their required habitat, existing only in a narrow ecological niche (e.g., lesser flamingo, *Phoeniconaias minor*, or torrent duck, *Merganetta armata*). Many of these waterbird species have long legs and/or webbed feet as adaptations to life in a wetland environment. Species can swim or dive underwater, and may collect feed from the water's surface, from the water column or from the wetland substrate.

DOI: 10.1201/9781003208471-14

Table 12.1 Representative species of waterbird with selected ecological information useful for planning husbandry and management in the zoo.

Species	Distribution and Red List status (as of September 2021)	Social behaviour	Biometric information
White-headed duck (*Oxyura leucocephala*)	Spain, North Africa, Western and Central Asia. Large wetland systems with a preference for areas with dense emergent vegetation, particularly during nesting. Utilises fresh, alkaline, brackish, and saline bodies of water. Endangered	Gregarious during winter and on moulting grounds. Flexible social system. Flock size varies throughout the year, averaging 500 individuals or less. Polygynous mating system that sees smaller flocks with a predominance of adult males during breeding.	Up to 50 cm long. Up to 70 cm wingspan. Up to 900 g mass.
Red-breasted goose (*Branta ruficollis*)	Breeds on the Taymyr, Yamal, and Gydan Peninsulas of the Siberian tundra. Winters in Southeast Europe and Southwest Asia, mainly along the north and west Black Sea coasts. Vulnerable	Highly sociable. Form large flocks during migration and winter. Family groups overwinter together. Monogamous. Territorial during nesting. Multiple pairs may nest in close proximity to one another.	Up to 56 cm long. Up to 135 cm wingspan. Up to 1.7 kg mass.
Trumpeter swan (*Cygnus buccinator*)	Breeds across North America. Winters in southern Alaska, western British Columbia, the Pacific Northwest, and in pockets across central North America. Year-round in the American Northwest and southern Alaska. Least Concern	Gregarious and sociable in winter and during migration. Forms strong monogamous pair bonds. Highly territorial during nesting. Cygnets will stay with adults until next nesting season.	Largest species of swan in the world. Up to 1.58 m long. Up to 2.1 m wingspan. Up to 13.6 kg mass.
Southern screamer (*Chauna torquata*)	Bolivia, northern Argentina, Paraguay, and Brazil. Common in the Pantanal. Endemic to South America. Least Concern	Monogamous pairs are territorial and aggressive during breeding. Sociable and gregarious during winter and nonbreeding.	Largest screamer species. Up to 90 cm long. Up to 1.7 m wingspan. Up to 5.0 kg mass.
Great white pelican (*Pelecanus onocrotalus*)	Southern Europe into the Middle East, Asia, and sub-Saharan Africa. Least Concern	Nests colonially, often in large numbers, social fishing activity increases foraging efficiency, migrates in groups.	Amongst the heaviest of all flying birds. Up to 180 cm long. Up to 360 cm wingspan. Up to 15 kg mass.
Chilean flamingo (*Phoenicopterus chilensis*)	Range restricted to alkaline wetlands suitable for feeding and breeding. Predominantly Argentina, Bolivia, Chile, and Peru. Near Threatened	Highly sociable, with specific inter-bird social bonds within groups. Nests colonially and a large flock size is important for synchronised group courtship display.	Up to 130 cm long. Up to 140 cm wingspan. Up to 2.3 kg mass.

Saddle-billed stork (*Ephippiorhynchus senegalensis*)	Across sub-Saharan Africa. Requires habitat areas that contain larger expanse of open water when compared to other storks of a similar nature. Least Concern	Most commonly found in pairs (potential life-long pair bond) or alone. Loose flocks have been observed but birds remain distant. Territorial and limited social display due to solitary or pair bonded lifestyle.	Amongst the tallest of all storks. Up to 142 cm long. Up to 270 cm wingspan. Up to 7.5 kg mass.
Scarlet ibis (*Eudocimus ruber*)	Tropical South America and the Caribbean (e.g., Trinidad & Tobago). Least Concern	Colonial and sociable, including with other waterbird species. Roosts and nests arboreally and forages together in shallow wetlands.	Up to 63 cm long. Up to 54 cm wingspan. Up to 1.4 kg mass.
Goliath heron (*Ardea goliath*)	Wide distribution across sub-Saharan Africa with a smaller population in the Middle East and Asia. Predominantly a large fish specialist during hunting, goliath herons frequent undisturbed waterways and rarely travel over land. Least Concern	Solitary or found as a bonded pair. Fluid social behaviour when breeding, as can nest as a single pair or in large colonies (sometimes with other species).	Largest species of heron. Up to 152 cm tall. Up to 230 cm wingspan. Up to 5kg mass.
African spoonbill (*Platalea alba*)	Expansive, shallow waterways and wooded wetlands of sub-Saharan Africa and Madagascar. Least Concern	Colonial and sociable. Roosts and nests arboreally and forages together in shallow wetlands.	Up to 90 cm tall. Up to 135 cm wingspan. Up to 1.8 kg mass.
Hamerkop (*Scopus umbretta*)	Wide distribution across sub-Saharan Africa, into Madagascar, with a smaller population in the Middle East. Least Concern.	Flexible social system. Can be found alone, in pairs, or in sizeable flocks. Complex repertoire of social behaviours and vocalisations depending upon on current social environment.	Up to 56 cm tall. Up to 94 cm wingspan. Up to 0.5 kg mass.
Red-crowned crane (*Grus japonensis*)	Breeding range in Siberia, northern China, and Mongolia. Winters in Korea and central China. Resident in Hokkaido, Japan. Endangered	Strong pair bond that lasts for life (30 to 40 years in the wild). elaborate pair bonding and courtship ritual. Territorial when breeding but will flock together during winter.	The heaviest species of crane. Up to 158 cm tall. Up to 250 cm wingspan. Up to 10.5 kg mass.
Black crake (*Zapornia flavirostra*)	Wetlands that provide ample cover. Wide distribution across sub-Saharan Africa, avoiding arid locations. Least Concern	Aggressive and territorial. Found in pairs or pairs with young. Adults will drive away intruders or attack other birds within the vicinity of their nesting location.	Up to 23 cm long. Up to 25 cm wingspan. Up to 118 g mass.
Grey-winged trumpeter (*Psophia crepitans*)	Wetlands of tropical South America, specifically the northern stretches of the Amazon rainforest, Brazil, Colombia, Ecuador, French Guiana, Guyana, Peru, Suriname, and Venezuela. Near Threatened	Gregarious in and out of the breeding season. Multiple individuals may lay eggs in a nest and all help raise youngsters. Wide range of vocalisations and is a sentinel species within the rainforest.	Up to 56 cm long. Up to 1.3 kg mass. No wingspan information available.

Some species are terrestrial foragers, whereas others struggle to walk well on land. Species will nest in a variety of locations – in trees, reedbeds, or thickets, on the ground and exposed, or on floating nests on the water's surface. Multiple species are also long-distance travellers, migrating efficiently across hundreds of kilometres between suitable feeding and breeding sites, being in tune with seasonal changes to the environment around them.

Waterbirds present complex social and reproductive behaviours, many species perform ritualised courtship displays when attracting a breeding partner or strengthening an existing pair bond (Johnsgard, 1965). While the males of many duck species participate in large group displays to vie for female attention, e.g., common goldeneye (*Bucephala clangula*) and northern pintail (*Anas acuta*), whereas the flamingos all rely on both males and females partaking equally. All of these species-specific behaviours must be taken into consideration when developing a husbandry and management plan, if successful breeding is to occur, viable captive populations maintained, and animal welfare standards kept high.

12.2 WATERBIRD ECOLOGY AND NATURAL HISTORY RELEVANT TO THE ZOO

Literature on the ecology of free-living waterbirds is helpful for guiding captive management, specifically regarding inter-specific interactions and foraging activity (e.g., avoiding predation of smaller species from larger species in multi-taxa holdings), identification of biologically relevant resources (nesting sites, perching and loafing spaces, water area, volume, and depth to promote feeding, locomotory and social activities) and for determining an appropriate social structure for numbers held in a captive flock. As many species of waterbirds are staples of mixed-taxa aviaries (Klausen, 2014), using wild information on social interactions (between and within a species) helps to create a harmonious mix of birds, a social environment conducive to successful reproduction and increased opportunities for the enrichment and diversification of daily activity patterns (Rose, 2018). When waterbirds congregate together in the wild, the individual foraging niche for their species keeps them apart and reduces competition. Therefore, knowledge of how the birds feed and their temporal activity pattern for foraging can help in resource distribution; this in turn provides opportunities for increased behavioural diversity (foraging can occur as per wild habits) and unwanted aggression, displacement bullying or harassment of individual birds in enclosure areas of high-value enclosure zones is reduced or eliminated.

Understanding the causal factors for a behaviour can improve the husbandry of waterbirds and provide a more enriched environment and lifestyle in the zoo. Figure 12.1 uses the collective fishing behaviour of great white pelicans as an example of how behavioural information can underpin zoo husbandry. By understanding the stimulus that triggers the behaviour (e.g., the presence of other birds foraging or the potential presence of food in the water) the way in which the bird interacts with its environment can be assessed. The pelican may wish to perform this foraging behaviour achieve a specific goal (i.e., there is a need to feed at a specific time of day) or the pelican may be contrafreeloading – wishing to work for a reward even though that reward is easily accessible (McGowan, Robbins, Alldredge, & Newberry, 2010). Functional substitution is the creation of a specific resource from or feature of the animal's wild habitat that provides an outlet for an appetitive behaviour in the zoo. Appetitive behaviours are those actions whereby the performance of the activity is more important than the end result. In this case for the pelicans, providing opportunities for fishing is key. The birds do not necessarily have to catch fish, but be allowed to complete their foraging routine and therefore feel satiated once this behaviour pattern has been fully expressed. Measuring how much time the pelicans spend foraging, or attempting to forage, provides data on how important it is to the bird's daily time budget, and therefore how it can be encouraged as an activity in the zoo. This information further evidences how to alter an enclosure or husbandry routine to accommodate the pelican's natural behaviour pattern. Table 12.2 provides further examples of behavioural adaptations and how these can guide housing and husbandry – with varying placement and location of food around an enclosure one of the best ways of enabling a wild-type time budget in captive waterbirds.

Mating systems and reproductive strategies can be a significant driver for a large suite of behaviours

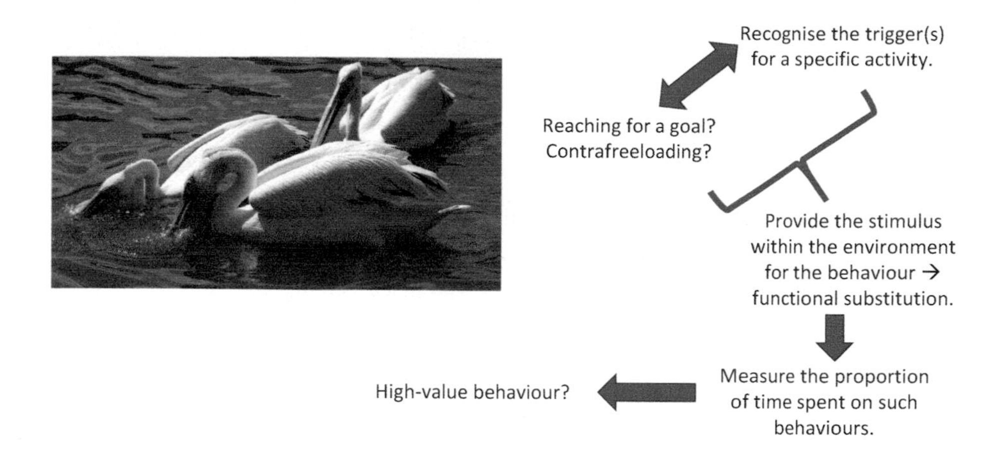

Figure 12.1 Adapted from Rose, Brereton, and Croft (2018), how understanding the trigger or stimuli for the performance of an adaptive behaviour can provide higher-quality welfare states; as animals living under human care can be provided with an environment that enables them to perform high value (e.g., highly motivated or with a fitness benefit) behaviour patterns. In this case, cooperative fishing in flocks of great white pelicans can be enabled in the captive environment by housing an appropriately sized social group, allowing the birds access to a waterbody with an expensive area and sufficient depth for birds to fish together, and by feeding the birds in the water, facilitating the foraging action.

in waterbirds, from social structure and flock composition to the determination of which food resources are being utilised. In many species of ducks, laying hens are driven to bulk up on calcium and protein-rich food sources, often seeking out diet items that they may typically ignore during non-nesting seasons. For example, a female Baikal teal (*Sibirionetta formosa*) housed in a zoological setting may begin to exhibit behavioural changes leading up to and during egg lay. While preparing to begin her clutch, she starts to exhibit more territorial behaviours and becomes more assertive during feeding times, perhaps even seeking out the caretaker when she otherwise would not. Instead of showing preference for floating pellets or seeds as she does in the winter, she shows an affinity for invertebrates and fish, coupled with an overall increase in appetite. It is important to note these dietary preferences and overall changes in behaviour as they are excellent indicators of a bird's needs and reproductive status. Such indicators also relay key information about an individual's health. When annual seasonal patterns in diet and behaviour are regularly recorded, such as with the female Baikal teal, their needs become predictable and apparent. If in one nesting season she fails to exhibit her typical diet pattern, it could be an indication of an underlying health issue or that her environment is insufficient for reproduction. This warrants an evaluation of current husbandry practices, including an enclosure assessment and potential veterinary examination.

Understanding species' natural histories and reproductive strategies can allow for incredible opportunities to increase their overall welfare state. Some waterbird species, such as the common eider (*Somateria mollissima*), will form multifemale crèches with their young to improve offspring survivability and reduce resource stress on individual females (Kilpi, Öst, Lindström, & Rita, 2001). Figure 12.2 shows a group of American common eiders (*S. m. dresseri*) that comprises several females, males, and their young. The adults' lives are highly enriched through the opportunity to express a natural repertoire of parenting behaviours, while the ducklings are able to learn valuable feeding and social skills directly from conspecifics. Opportunities for social learning and the transmission of important information can be provided within and between populations when such natural social groupings are maintained. For example, supporting the explanation of behavioural complexity during foraging for great white pelicans (Figure 12.1), this species has been documented as engaging in novel opportunities for social learning that result from problem-solving (Danel et al., 2020), demonstrating the

Table 12.2 Examples of behavioural adaptations of waterbirds and how this should be accommodated in zoo management for these species.

Species	Behavioural adaptation	Accommodating behaviour in husbandry and housing
Scaly-sided merganser (*Mergus squamatus*) 	Long, narrow bill with tooth-like lamellae to grasp fish and invertebrates. Legs are positioned wide and towards the posterior end of the body for diving.	Provide opportunities for pursuit hunting, such as offering live fish, scatter feeding fresh or thawed fish into the pool, and offering sinking enrichment devices with visible prey items to entice diving behaviour. Provide deep water with low turbidity and high visibility to encourage diving and sight hunting.
Snow goose (*Anser caerulescens*)	Stout, sturdy bill with well-defined lamellae for grubbing tubers, roots, and rhizomes from muddy marshes and tundra.	Provide partially submerged foraging areas, either via tubs or directly into the bird's pool, with sunken live plants, seeds, pellets, greens, or grass clippings. Take care to avoid making muddy areas in the bird's enclosure as these have the potential for spreading soil-transmitted diseases.
Yellow-billed stork (*Mycteria ibis*) 	Touch-sensitive bill used for fishing while wading in murky water.	Provide food within the bird's pool or similar to encourage foraging activity. Ensure storks can forage within enough space as a social group.
Lesser flamingo	Densely packed fine lamellar in bill to collect cyanobacteria and diatoms from top 4 cm of water column. Varied feeding strategies (walking, wading, swimming) based on weather conditions or water depth.	Provide fine ground flamingo pellet to encourage filtering behaviour. Algal and Daphnia blooms in the pools in the enclosure allow for natural foraging. Provide pools with a range of water depths that can accommodate all individuals when foraging.

Eurasian spoonbill (*Platalea leucorodia*)	Forages by walking in shallow water, sweeping bill from side-to-side to catch small fish, crustaceans, and other aquatic prey.	Provide space for sweeping foraging action within waterbodies in an enclosure. Ensure a flock of birds is maintained to promote social interactions.
Little egret (*Egretta garazetta*)	Bright yellow feet allow the egret to scare prey, making the prey visible for catching.	Flowing water and pools of different sizes and shapes, with planted banks and different substrates to encourage foraging activity.
Grey crowned crane (*Balearica regulorum*)	Wetland and grassland dwelling and will collect food by foraging in long grass.	Allow areas of long grass to grow and seed, which provides a naturalistic grassland habitat for birds to forage in (stripping seed heads from grass).
Pink-backed pelican (*Pelecanus rufescens*)	Perches, roosts, and nests off the ground, often in very close proximity to other individuals.	Provide perching at various heights and opportunities for nesting colonially off the ground.

Figure 12.2 American common eiders exhibiting crèche behaviour. Giving birds the opportunity to express behaviours such as incubating and rearing their young in a safe, appropriate setting provides tremendous benefits to their overall welfare (Photo: Ian Gereg).

advanced cognitive abilities of waterbird species and hence the importance of creating opportunities for positive challenge and social variation within their captive environments.

12.3 ENCLOSURE CONSIDERATIONS FOR WATERBIRDS BASED ON BEHAVIOURAL EVIDENCE

The majority of species in the taxonomic orders listed utilise wetland environments when conducting daily activity patterns. The way in which the bird interacts with the wetland can be judged from its size and shape, and the anatomy, physiology, and morphology of its bill, wings, and legs/feet. Specific sensory perception should also guide enclosure design and layout, and this is especially important when considering how to encourage foraging or social behaviours in different areas of the exhibit. The predatory nature of some waterbird species, e.g., marabou (*Leptoptilos crumenifer*), also needs to be considered when housing species together. Refuge areas, space to move away from larger species and sufficient feeding areas to reduce possible predation attempts are all required in a mixed-species enclosure that houses waterbird species from different ecological niches (Figure 12.3).

Wild evidence also provides support for the shape and size of resources within an enclosure so that captive waterbirds can display a naturalistic activity pattern and diversity of behaviours as documented in free-living individuals (Figure 12.4). Such wild information should be consulted during the enclosure planning and construction stages, as well as when furnishings (e.g., vegetation, substrates, waterways) are being placed in situ.

When designing an enclosure, it is important to consider its purpose (e.g., mixed-species enclosure, private parent-rearing pen, brooder and hand-rearing runs, etc.). For example, if the goal is to allow for multiple species to parent-rear their offspring in the same enclosure, species composition considerations must be thoroughly scrutinised to maintain quality welfare for all individuals in that space (e.g., avoid species that prey on eggs or young birds, select species that occupy different

Figure 12.3 Top left and right – mixing species from different niches and with different foraging strategies requires space and multiple resources present to reduce detrimental inter-species interactions. Bottom left – a range of perching and substrates provides overlapping but species-specific opportunities for behavioural diversity in a mixed-species waterbird aviary. Bottom centre – zoo signage explaining that marabou and lesser flamingos are in adjacent enclosures (to provide a Rift Valley diorama style exhibit); consider the view of the predator (stork) from the prey (flamingo) and how this may impact bird behaviour and welfare. Bottom right – even enclosures for mixed-species of the same genus of bird (example here are pelicans) will require consideration of specific resources utilised in different ways by each species in the enclosure.

nesting habitats and structures). Hartlaub's ducks (*Pteronetta hartlaubii*) are fiercely protective and attentive parents. If they are allowed to rear their ducklings in a mixed-species exhibit, care must be taken to ensure that there are ample resources available (e.g., multiple pools, shelters, perches, nest boxes) in order to reduce competitive pressure within the community and avoid potentially harmful interactions amongst inhabitants. If the family establishes a territory on one pool, it is critical to ensure that displaced birds have alternative choices to turn to.

Successful reproduction and expression of reproductive behaviours (i.e., courtship, mating, parenting) are key positive welfare indicators for waterbirds. If an individual fails to display species-appropriate reproductive behaviours, there may be one or more insufficient elements in their environment affecting their overall welfare. For example, in a species that exhibits seasonal monogamy (i.e., forms seasonal pair bonds) such as the gadwall (*Mareca strepera*), providing the space for multiple males to perform their courtship displays will allow for the female to select the best possible mate. The female bird will also need an enclosure that supplies her with a variety of nest site options. If the males do not have sufficient or ecologically suitable space to display, or if the female does not have the opportunity to choose her mate and/or a safe nest location, inappropriate behaviour or a lack of behaviour may be observed (e.g., the female does not lay eggs in a nest, males do not display, no copulation will occur). Waterbird

Figure 12.4 Examples of behaviour patterns of captive flamingos and evidence for habitat choices from the wild. Top centre – lesser flamingos in Tanzania's Rift Valley, illustrating the environment of the soda lakes flamingo inhabit. Top left – recreation of an alkaline wetland for captive Caribbean flamingos that has used natural evidence. Top right – stylised "tropical" enclosure flamingos based on anthropomorphic sentiments instead of natural evidence. Bottom left and centre – examples of different foraging and locomotion behaviours of flamingos – swimming and wading – used to collect food. Enclosure pools should be large enough to provide different water depths to allow for swimming to occur. Bottom right – natural setting for nesting in captive flamingos; an open space, with clear views, on an island, provides a safe refuge for raising young.

species have a multitude of nesting habits, mating systems, and habitat preferences, and these must be considered when designing an enriching enclosure. Figure 12.5 shows examples of various nesting structures utilised by Anseriformes species boasting a wide range of life-history strategies.

12.4 BEHAVIOURAL BIOLOGY AND WATERBIRD WELFARE

Many species of waterbirds provide visible cues to overall welfare state via their activity patterns and plumage/body condition. Understanding the behavioural syndrome (i.e., consistent behaviour patterns expressed by an individual, often termed personality) allows bird keepers to identify when an animal may be experiencing poorer welfare and therefore requires extra care or attention (Figure 12.6 top). Annual physical health checks of birds are important to determining individual bird mass, essential data for promoting good welfare in colonial or group-dwelling species (Figure 12.6 bottom). Regular review of records from weighing can see if individuals have lost mass and therefore are not coping well with a busy and competitive social environment. Changes to husbandry and group management, e.g., providing access to more high value resources such as foraging sites, perching and nesting areas, can reduce unnecessary aggression to specific individuals and encourage birds to regain condition. As so many species of waterbird are highly sociable, knowledge of the social dynamic of a flock and whether any form of social structure or

Figure 12.5 Top left shows a red-breasted goose nesting in an open-topped structure of logs and branches, allowing the bird direct sightlines around the nest much like it would have on the peninsulas of Arctic Russia (Photo: Ian Gereg). Top centre and top right are examples of cavity and snag nesting options lofted above water features to give the bird a sense of security, with a pink-eared duck (*Malacorhyncus membranaceus*) occupying the palm log in the right photograph (Photos: Pinola Conservancy). Bottom left is a mound built out of moss and other vegetation near the water's edge by the large, ground-nesting trumpeter swan (Photo: Ian Gereg). Bottom centre has a natural log in a mixed-species aviary with vegetation near the nest opening to provide partial cover (Photo: Pinola Conservancy). Bottom right shows a female white-headed duck defending a nest built in a bed of emergent iris (Photo: Joanna Klass). Stifftail species such as the white-headed duck are highly aquatic and maladapted to walking, typically seeking out nesting sites in or near the water.

hierarchy occurs within a group, is important for maintaining long-term good health and welfare.

Utilisation of behavioural husbandry techniques that train birds to voluntarily participate in their own routine health care can increase welfare exponentially, as control and choice over the immediate environment and how the bird engages with it is provided by the training (Melfi & Ward, 2020). Training eliminates the need to put the bird through a potentially stressful situation (e.g., netting, chasing, manual restraint) and grants the bird the ability to choose to participate (or not) in the training session. Natural histories should be taken into consideration when developing a training plan in order to create the best possible environment for reaching behavioural husbandry goals. Figure 12.7 shows a juvenile male hooded merganser (*Lophodytes cucullatus*) being trained to swim into a crate. Hooded mergansers can be skittish and, as excellent divers, spend a large portion of their time in the water. White-faced whistling ducks (*Dendrocygna viduata*) on the other hand have long legs and a slender build, perfect for perching and walking. This knowledge can be applied to their behavioural husbandry plans. Using crate training as an example, a walk-in crate is the perfect option for more terrestrial species and a swim-in crate more appropriate for many dabblers, divers, and sea ducks.

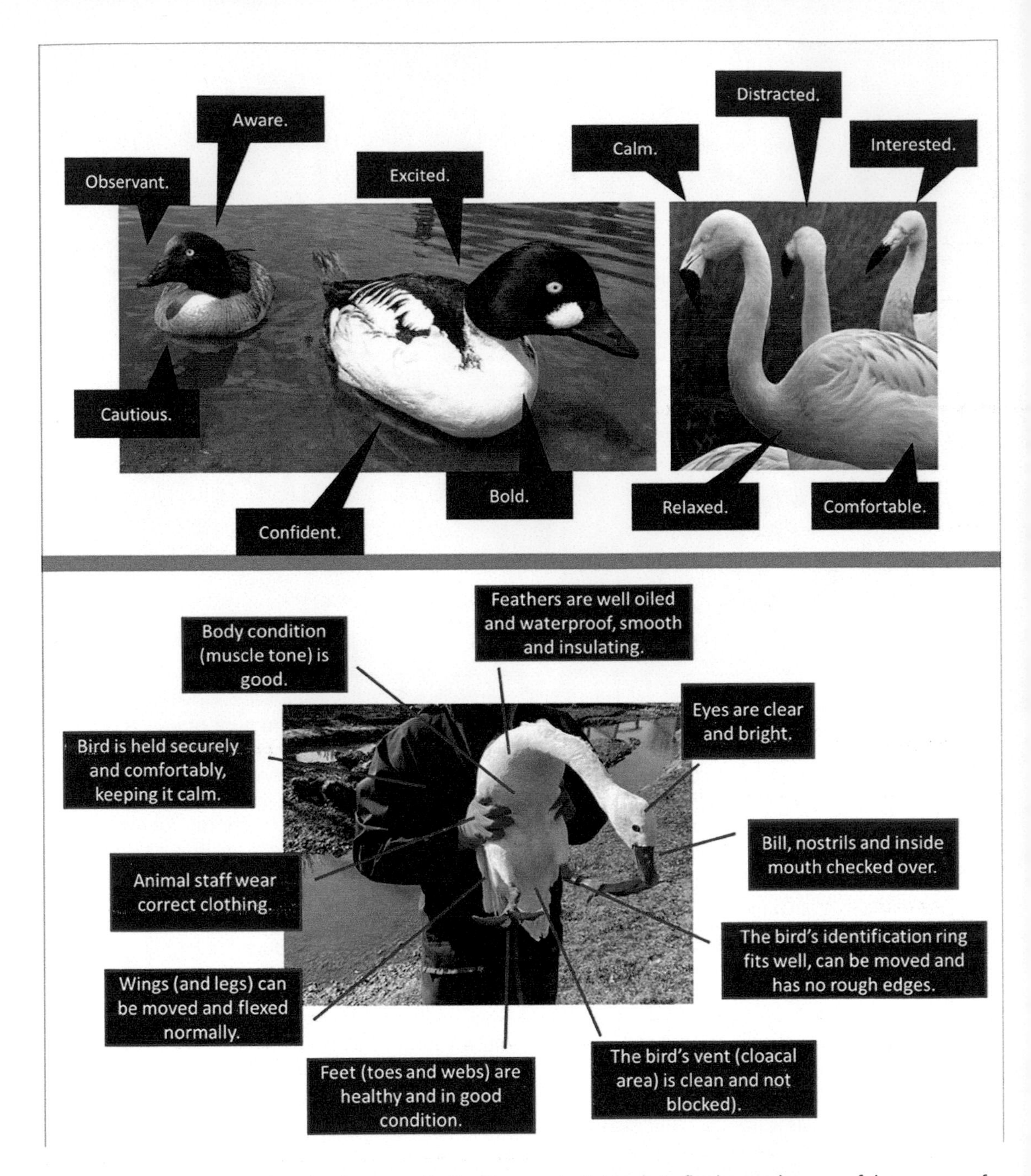

Figure 12.6 Top. Assessing the "personality" of known individuals in flocks can be a useful measure of bird welfare. Waterbirds comprise a range of charismatic species where individuals can display known consistent behavioural syndromes, akin to personality traits, that can be systematically recorded and compared over time to build up a picture of the usual behavioural characteristics of that individual. Any change in expected behavioural expression can be further investigated to determine if the cause is a potential health and welfare issue. Bottom. It is important to regularly use physical health checks to ensure that bird wellbeing is maintained at a high standard. Species-specific anatomical and physical adaptations, important to the correct performance of daily activity patterns (e.g., bill anatomy and morphology), can be assessed for condition and integrity so that no handicaps to bird behaviour are present (e.g., due to damage or trauma due to the environment around the bird).

Figure 12.7 Left, white-faced whistling ducks participate in a crate training session. Their long legs are well suited for both terrestrial and aquatic lifestyles, making a walk-in kennel a suitable option for voluntary crating behaviour (Photo: Tamlyn Sapp). Right, a juvenile male hooded merganser is being positively reinforced for a swim-in crate behaviour. This style of crate can be extremely useful in behavioural husbandry training for species that spend a large portion of their activity budget in the water, such as mergansers, stifftails, and eiders (Photo: Bill Robles).

12.5 SPECIES-SPECIFIC ENRICHMENT FOR WATERBIRDS

Suitable enrichment for waterbirds includes a wide range of options that can be nutritional, occupational, physical, sensory, and social in design and construct. These enrichment options include changes to enclosure design and layout, alterations to feeding regimes, enhancement of social groups, and provision of resources to encourage reproductive and nesting activities. Ideas for flamingo enrichment are outlined in Rose, Brereton, and Gardner (2016) to provide ways of maximising the opportunities that captive flamingos have for the display of the widest possible range of behaviours. Differing water depths that allow for feeding opportunities (from the surface, in mid-water and in the substrate), for example, enable flamingos to use a wider range of enclosure areas, reducing time spent inactive, and improving overall bird health.

Encouraging activity across species of waterbird can be helpful in improving health and wellbeing. Especially if environmental enrichment programmes focus on encouraging activity patterns similar to that seen in wild birds. Wild flamingos, for example, have a time-activity pattern that revolves around foraging, preening, and loafing (resting or sleeping), interspersed with courtship and nesting activities at the appropriate time of year. As heightened levels of inactivity are noted in captive flamingos, when comparing their behaviour to that documented in wild flocks (Rose, 2021; Rose et al., 2018), zoo enrichment can help encourage more time spent on foraging and therefore usage of wider areas of the enclosure. Foot health can be poor in many heavy species of waterbird (e.g., pelicans, large storks, and flamingos). Enriched enclosures that provide multiple opportunities for perching can alleviate pododermatitis ("bumblefoot"), which manifests itself as pressure points, nodules, cracks, and fissures on the plantar surface of the foot (Nielsen, Nielsen, King, & Bertelsen, 2012), causing pain and discomfort, as well as sites for potential bacterial infection. Increased movement improves blood flow to the foot, enhancing the immune response at the extremities and promoting chances of healing.

Enrichment for waterbirds can include:

- Bubblers to move the water column around, encouraging filter feeding and other foraging actions in the bird's pool.
- Water features, such as sprinklers, waterfalls, misters, hoses, and novel water tubs or dishes.
- Foraging trays for species that like to dabble in shallow water. Long, shallow seed trays filled with soil and water, vegetation and seeds can be easily installed, cleaned, and changed, to allow for occupational and nutritional enrichment.
- Offering live prey items, such as minnows, mussels, and invertebrates (mealworms, crickets, waxworms) to appropriate species.
- Collecting seed heads from water plants, such as reeds, rushes and sedges, and scatter feeding in the bird's pool.
- Diverse planting and opportunities for shelter makes shy and secretive species feel comfortable, increasing their activity and movements around an enclosure.
- Management of the social group to provide a biologically relevant age structure and flock demographic.
- Changes to substrate across the enclosure (e.g., sand, mud, clay, soil, bark, leaves, matting, river rock) allows for different environmental interactions to occur.
- Perching heights, shapes, and sizes (even for flight restrained birds, if close to the ground) allows for increasing behavioural diversity.
- Nesting sites and provision of nesting material.
- Allowing birds to choose their own partners or to invest in a long-term relationship.
- Allowing birds to incubate and rear their own young or serve as fosters to build pair bonds.
- Opportunities for grazing, up-ending, stamp-feeding or other species-specific foraging action to occur in a recreation of the important aspects of the bird's wild habitat.

Figure 12.8 details the opportunities for an enriched behavioural repertoire that can result from waterbirds being housed in an enclosure that is ecologically relevant and ecologically complex in its design

Figure 12.8 The overall enclosure provided for waterbirds can be enriching if the ecological needs of each species housed are considered in the design and construction of the exhibit, and used to guide husbandry and management protocols.

and structure, and social mixing. Further examples of how enrichment opportunities can be provided holistically, via small alterations to an enclosure (in this case for captive crane species) is explained in Rose and Young (2014).

12.6 USING BEHAVIOURAL BIOLOGY TO ADVANCE WATERBIRD CARE

This chapter has explained the importance of knowledge of species-specific adaptive traits and behaviour patterns when designing and implementing husbandry and management practices for a range of waterbirds in the zoo. Waterbirds, from a diverse array of habitats and biomes, require specialised care in the zoo; care that is firmly grounded in knowledge of their evolutionary history, habitat choices, and behavioural complexities.

An example of how behavioural biology can be useful to population management and future conservation potential, Box 12.1 illustrates the multifactorial influences over nesting attempts of captive flamingos being successful or not, and highlights why some factors (notably flock size) are key to increasing the chances of successful breeding.

BOX 12.1: Advancing population management of flamingos using behavioural evidence

Flamingos are amongst the most commonly kept bird species in zoos currently. Despite this ubiquity and popularity, the management and successful breeding of captive flamingos has proven problematic due to their unique breeding behaviour and natural history, making it difficult for zoos to create self-sustaining populations for any flamingo species. Although there are records of flamingos in captivity dating back to Ancient Rome, it was not until 1857 that the first captive breeding event was recorded at London Zoo (Edwards, 2012). Since then, captive flamingo populations have grown considerably, primarily as a result of importing wild-caught individuals, with flamingos now being found in at least two-thirds of all European Association of Zoos and Aquaria (EAZA) institutions. All six flamingo species can be found in zoos, however the greater flamingo (*Phoenicopterus roseus*), Chilean flamingo and American/Caribbean flamingo (*P. ruber*) are the commonest to be seen. Although it is impossible to replicate all aspects of flamingo natural history in zoos, species-specific behavioural evidence is key to the successful management of these populations and should be used to guide management strategies to promote successful reproduction.

Reproductive success in captive flamingos is determined by many factors, ranging from enclosure design to climate, but flock size has been consistently identified as the most important factor influencing the level of breeding success in captive flocks. The largest flocks show the greatest reproductive success. Wild flamingos occur in huge colonies, often numbering in the thousands to millions (depending on species) where they use coordinated group displays to facilitate pair formation and stimulate synchronous nesting. In captivity, the optimum flock size for reproduction has recently been identified. By utilising nearly 30 years of captive flamingo reproductive data, approximately 100 individuals for greater, Chilean, and American flamingos, and 150 individuals for lesser flamingos are needed for the best chances of reproductive success (Mooney, 2021), see Figure 12.9. However, current captive flock sizes average approximately 25 birds (varying with species). This new evidence helps to explain the poor rates of reproductive success seen in zoos to date, particularly for lesser flamingos. Although most zoos would like to increase their flock sizes, there are simply not enough surplus birds available and many zoos are reluctant to relocate their flamingos to other institutions, making it difficult to create larger flocks. To encourage reproduction, institutional flock sizes should be increased from the current average of 25 birds, to this 100-150 bird optimum. To achieve this goal, zoos with very small flocks should relocate their birds to other zoos with existing flocks, and successfully breeding flocks should be used to create surplus birds that can be made available to other institutions. This evidence shows that concerted efforts must be made by zoos to cooperate on creating self-sustaining flamingo populations to prevent the decline of these charismatic species in ex-situ facilities.

Wild flamingos may be considered as serially monogamous, potentially changing mates every year. However, captive flamingos can show high degrees of mate fidelity, spanning multiple breeding seasons across time. These long-lasting relationships are believed to be inversely related to flock size, with smaller flocks providing limited opportunities for mate choice. Consequently, the creation of larger flocks should also be seen as a priority from a behaviour and welfare perspective. In addition to flock size, the sex ratio of a flock can also influence the level of breeding success observed, particularly where a male bias occurs, as this is associated with higher rates of egg breakage and performance of behaviours that disrupt nesting. Similarly, the occurrence of same-sex relationships, trios, and quartets, which are observed more frequently in captive populations, could be linked to an uneven sex ratio within the flock. While there is clear evidence that captive flamingo flocks should consist of an even sex ratio, and this is naturally assumed to be the case, nearly 25% of all captive flamingos are yet to be sexed due to group management approaches used for flamingo record keeping. Identifying the true sex ratio of existing flocks should be a clear priority for all institutions, and zoos with uneven sex ratio flocks should relocate individuals in order to create an ideal ratio in their group. The creation of sex-specific colonies may also help to manage known biases in the population, which can be seen by steps being undertaken by the Association of Zoos & Aquariums (AZA) in North America for the lesser flamingo, where there is a known male bias.

The movement of individuals between institutions to even out sex ratios may have additional reproductive benefits, as there is evidence that introducing new individuals into an existing flock can stimulate reproductive activities and increase breeding success the following year. This reproductive benefit is in addition to that gained from increasing flock size itself, and is one management strategy that could be used to encourage reproductive behaviour in smaller flocks. Another issue observed in captive flamingos is the influence of wing conditions on the male's ability to always successfully balance on the female during copulation. Potentially up to 75% of pinioned male flamingos are unable to successfully reproduce due to flight restrain. Keeping male flamingos fully winged will prevent wide-scale infertility in zoo flocks. Changes to traditional flight restraint practices will involve upfront extra expenditure on enclosure infrastructure, but this is balanced out longer term by the easier and more sustainable management of the birds themselves.

The evidence presented here (Figure 12.9) shows that current management guidelines require wide-scale change to flamingo husbandry and management practices, alongside extensive institutional collaboration to achieve ex-situ population aims. However, such species-specific and evidence-based management strategies are necessary in order to achieve long-term sustainability for all captive flamingo populations moving forward if zoos wish to continue to display these charismatic and popular birds well into the future.

12.6.1 Further information to inform waterbird care

For more evidence of wild ecology and behaviour patterns, the following list of livestream webcams are excellent resources for research and investigation:

Wildfowl at WWT Slimbridge Wetland Centre, UK: https://www.wwt.org.uk/wetland-centres/slimbridge/experience/webcam/

Flamingos at Kamfers Dam, Kimberley South Africa: https://explore.org/livecams/african-wildlife/flamingo-cam

Wetland river systems such as the Mississippi: https://explore.org/livecams/raptor-resource-project/mississippi-river-flyway-cam

Hula Nature Reserve (Israel): https://www.camscape.com/webcam/hula-nature-reserve-webcams/

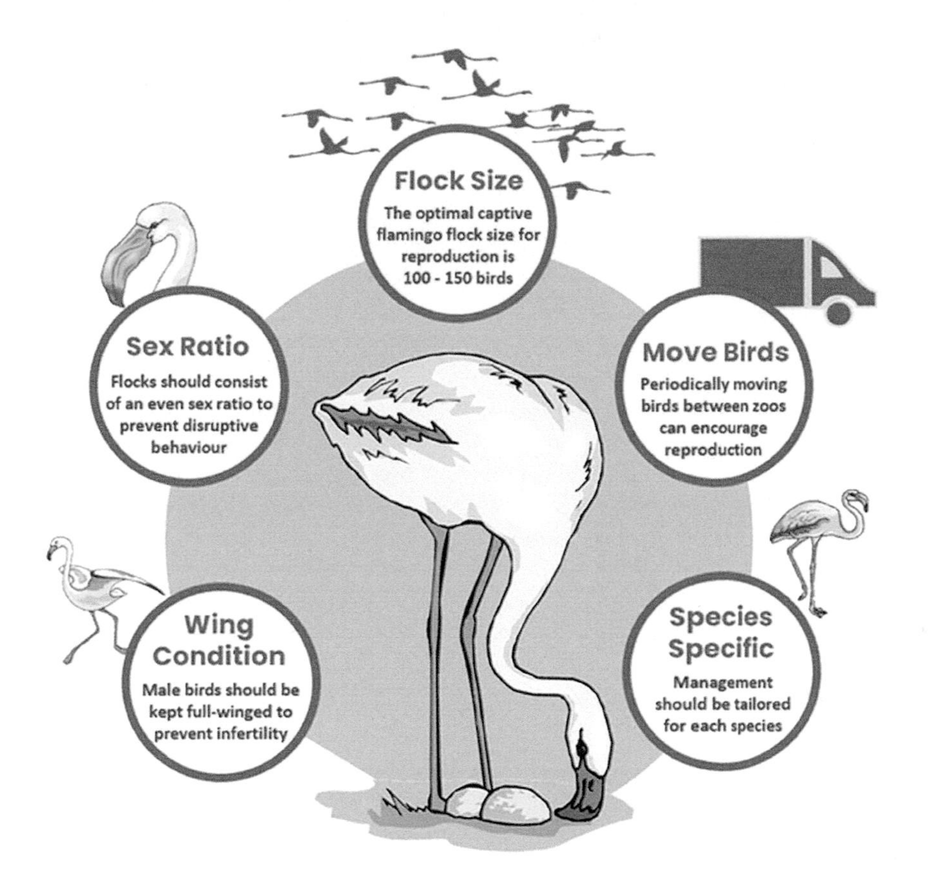

Figure 12.9 Species-specific evidence should be used to guide captive population management and promote population sustainability. For flamingos, flocks should consist of 100–150 birds, with some birds periodically moved between zoos to encourage reproduction. Birds should be kept full-winged and all individuals should be sexed to ensure the flock consists of an even sex ratio.

Blackwater National Wildlife Refuge (Maryland, USA): https://www.friendsofblackwater.org/waterfowl-cam.html

Further information on welfare assessment methods for Anseriformes, which can be adapted for other waterbird species, including practical methods for assessment of psychological, physical, and behavioural indicators of welfare are available in:

Rose, P. E. & O'Brien, M. (2020). Welfare assessment for captive Anseriformes: A guide for practitioners and animal keepers. *Animals, 10*(7), 1132.

This paper also provides useful information on flight restraint (e.g., feather trimming) and, from an ecological perspective, when this may be appropriate for specific species of duck, goose, swans, or screamer, dependent on wild behaviour patterns and habitat preferences.

12.7 CONCLUSION

Waterbirds are common residents of zoos and aquariums; with a diverse array of colours, shapes, sizes, and behaviour patterns, they are excellent species for zoos to exhibit. They engage visitors with messages on animal and habitat conservation, anatomy and evolutionary history, specialisations to extreme environments, human influences over wildlife (e.g., history of domestication or population declines) and courtship and breeding behaviours. When displayed in an ecologically relevant manner, waterbird species can become some of a zoo's most popular exhibits. Given the huge diversity of form, function, and taxonomy, it is essential that a "one size fits all" approach is avoided when housing and husbandry of zoo waterbirds is being developed – not all ducks simply "go on a pond." With careful consideration of wild ecology and natural history, zoo-housed waterbirds can lead enriched lives within sustainable and viable managed populations that complement the key aims of the modern zoo.

REFERENCES

Danel, S., Troina, G., Dufour, V., Bailly-Bechet, M., von Bayern, A.M.P., & Osiurak, F. (2020). Social learning in great white pelicans (*Pelecanus onocrotalus*): A preliminary study. *Learning & Behavior, 48*, 344–350.

Edwards, J. (2012). *London Zoo from old photographs, 1852–1914* (2nd ed.). London, UK: John Edwards.

Johnsgard, P.A. (1965). *Handbook of waterfowl behavior*. Lincoln, USA: University of Nebraska.

Kilpi, M., Öst, M., Lindström, K., & Rita, H. (2001). Female characteristics and parental care mode in the crèching system of eiders, Somateria mollissima. *Animal Behaviour, 62*(3), 527–534.

Klausen, B. (2014). A mixed-species exhibit for African water birds (including pelicans, flamingos, spoonbills and storks) at Odense Zoo, Denmark: Breeding success, animal welfare and education. *International Zoo Yearbook*, 41(1), 61–68.

Kuhl, H., Frankl-Vilches, C., Bakker, A., Mayr, G., Nikolaus, G., Boerno, S.T., Klages, S., Timmermann, B., & Gahr, M. (2021). An unbiased molecular approach using 3′-UTRs resolves the Avian Family-level Tree of Life. *Molecular Biology and Evolution, 38*(1), 108–127.

McGowan, R.T.S., Robbins, C.T., Alldredge, J.R., & Newberry, R.C. (2010). Contrafreeloading in grizzly bears: Implications for captive foraging enrichment. *Zoo Biology, 29*(4), 484–502.

Melfi, V.A., & Ward, S.J. (2020). Welfare implications of zoo animal training. In V.A. Melfi, N.R. Dorey, & S.J. Ward (Eds.), *Zoo Animal Learning and Training*, pp. 271–288. Hoboken, USA: Wiley-Blackwell.

Mooney, A. (2021). The value of ex situ collections for global biodiversity conservation in the wild. Doctoral dissertation, Trinity College Dublin, University of Dublin, Ireland.

Nielsen, A.M.W., Nielsen, S.S., King, C.E., & Bertelsen, M.F. (2012). Risk factors for development of foot lesions in captive flamingos (Phoenicopteridae). *Journal of Zoo and Wildlife Medicine, 43*(4), 744–749.

Rose, P.E. (2018). Ensuring a good quality of life in the zoo. Underpinning welfare-positive animal management with ecological evidence. In M. Berger & S. Corbett (Eds.), *Zoo animals: Behavior, welfare and public interactions*, pp. 141–198. New York, USA: Nova Science Publishers Inc.

Rose, P.E. (2021). Evidence for aviculture: Identifying research needs to advance the role of ex situ bird populations in conservation initiatives and collection planning. *Birds, 2*(1), 77–95.

Rose, P.E., Brereton, J.E., & Croft, D.P. (2018). Measuring welfare in captive flamingos: Activity patterns and exhibit usage in zoo-housed birds. *Applied Animal Behaviour Science, 205*, 115–125.

Rose, P.E., Brereton, J.E., & Gardner, L. (2016). Developing flamingo husbandry practices through workshop communication. *Journal of Zoo and Aquarium Research, 4*(2), 115–121.

Rose, P.E., & Young, P. (2014). Enriching enclosures for canes at WWT Slimbridge. *Ratel*, 41(1), 23–27.

13

The behavioural biology of parrots

JOHN E. ANDREWS
AZA Population Management Center at Lincoln Park Zoo, Chicago, IL, USA

13.1 INTRODUCTION TO PARROT BEHAVIOURAL BIOLOGY

Parrots described and discussed in this chapter include representatives of the Psittaciformes, an order containing around 350 species (Homberger, 2006). The order is further divided into six different family groups consisting of Cacatuidae (Cockatoos), Psittacidae (Afrotropical and Neotropical parrots), Psittrichasiidae (New Guinea and Madagascar Parrots), Psittaculidae (Old World Parrots), Nestoridae (New Zealand Alpine parrots), and Strigopidae (Kākāpō, *Strigops habroptila*). Table 13.1 describes some examples of parrots among these various families. Unique among bird lineages, parrots are mostly distributed among tropical and forested habitats along or around the equator, predominately in the Neotropical, African and Australasian tropical forest habitats (Munn, 2006). Most species are found in forested areas in lowlands or mountains, with many requiring specific ecological niches and habitats to survive. Some unique lineages are alpine forest dwellers outside of the usual tropical niches, like the kea (*Nestor notabilis*) found in New Zealand. Some species have also adapted to human presences well (e.g., sulphur-crested cockatoos, *Cacatua galerita*, or rainbow lorikeets, *Trichoglossus moluccanus*) and some species have even been successfully introduced into alien urban habitats outside of their natural ranges (e.g., monk parakeets, *Myiopsitta monachus*) as a result of the pet trade (Munn, 2006). Parrot species are often strong flyers and nimble climbers to take advantage of foraging in the trees, however, some do forage on the ground and one species, the Kākāpō, is flightless. Most parrot species utilise secondary nest cavities, often re-used from other species, like woodpeckers, Picidae (Spoon, 2006). Living trees are largely preferred, but many will make use of dead tree cavities as well. Some species use alternate nesting sites, including cliff sides, active termite mounds or burrows in the ground (Munn, 2006). Many species revisit or maintain the same nest site each year, while others may switch sites from year to year.

DOI: 10.1201/9781003208471-15

Table 13.1 Representative species of parrot families and selected ecological information useful for planning husbandry and management in zoos.

Family: Example species	Distribution and distinct features	Social behaviour	Biometric information
Cacatuidae: 21 species Sulphur-crested Cockatoo (*Cacatua galerita*)	Mainly distributed across Australasia from the Philippines and Eastern Indonesia, Solomon Islands, and Australia. Distinguished easily by an independent crest, curved bills, and generally have feather colour that is less bright than other parrot species.	Socially monogamous in pairs, but can live in pairs or large groups/flocks.	Up to 55 cm long. Up to 103 cm wingspan.
Psittacidae: 167 species Scarlet Macaw (*Aro macao*)	Mostly red body with yellow and blue along the wings and tail. Distributed in the wild from Southern Mexico through southern Bolivia and Brazil in Central South America.	Live in pairs, family groups, or medium-sized flocks. Pairs are socially monogamous and remain together year-round.	Up to 89 cm long. Up to 91.5 cm wingspan.
Psittrichasiidae: Three species Pesquet's Parrot (*Psittrichas fulgidus*)	Family consists of two subfamilies distributed in New Guinea, and other members of this group found in Madagascar and other islands in the Indian Ocean. Mostly all black except bright red underbelly. Males have red spots behind their eyes, and both have naked heads, like a vulture.	Behaviour is not well known, but is typically found in pairs or in groups of up to 20. Nests are found in tree cavities.	Up to 46 cm long. Up to 30 cm wingspan.
Psittaculidae: 196 species Rainbow Lorikeet (*Trichoglossus moluccanus*)	Brightly coloured rainbow pattern, small to medium size parrot found across Australia commonly in the Eastern seaboard.	Maintain socially monogamous pairings, but can also be gregarious in large flocks and well adapted to urban settings.	Up to 30 cm long. Up to 17 cm wingspan.
Nestoridae: Three species Kea (*Nestor notabilis*)	Olive-coloured parrots with bright orange under wings/tail. Distributed mostly in the South Island of New Zealand.	Omnivorous and gregarious. Nest sites in natural cavities (e.g., rock crevices, dead trees).	Up to 48 cm long Up to 106 cm wingspan.
Strigopidae: One species Kākāpō (*Strigops habroptila*)	Light brown and olive-green colouring often called the "owl parrot." Previously found on both islands in New Zealand, now only on few islands where no predators are present.	Nocturnal and solitary. Mating system of leks where males gather to attract females. Nests are burrows attended only by females.	Up to 60 cm long. Up to 90 cm wingspan.

The breeding behaviours exhibited by parrots are diverse and can often be more difficult to observe than in other avian species. Most species are sexually monomorphic, making recording courtship or breeding behaviours difficult (Spoon, 2006). Social monogamy is the most common type of breeding system in parrots with a breeding unit consisting of one male and one female. Other systems of breeding seen in parrots include some examples of polygamy or polyandry. Still other species may change mates from year to year rather than maintain a bond throughout their lifetime. Pair bonds in parrots are often vital for breeding success, and for informing a pair's place in social hierarchies. Managing parrots in human care requires careful consideration of these various mating systems and social needs that must be present for breeding to occur and for good welfare to be maintained.

13.2 PARROT ECOLOGY AND NATURAL HISTORY RELEVANT TO THE ZOO

Research on the relevant ecology of wild parrots is a useful starting place to guide management in zoos. Specific topics helpful for guiding management include inter-species interactions and foraging behaviour (e.g., avoiding inter-species aggression and predation in mix-species enclosures), identifying needed resources (proper nesting sites, perching and flying spaces, water and misting needs, height to promote social behaviours, and proper movement) and for determining the proper social structure for zoo exhibits (e.g., small flocks or pairs, seasonal changes in social structures). Parrots are common birds held in many zoos with intense social needs often poorly understood. Using wild information about the social needs and interactions should help produce successful exhibits with the appropriate social environment conducive to successful reproduction, positive opportunities for enrichment, and promotion of positive individual welfare. Parrots forage anywhere from 40 to 75% of their time in the wild, depending on the species (Wilson, 2006; Mellor et al., 2021). The times of day parrots are active are also species-dependent and often also dependent on the type of food being foraged. When mixing multiple species or holding large groups, knowing how parrot species interact can be highly impactful in informing management strategies in zoos. Deeper understanding of the different niches each species inhabits, how they interact with conspecifics and heterospecific, how they forage and what times of day they are active can alleviate aggression or displacement issues in zoo exhibits and allow for more natural behaviours to be shown.

Parrots are specially adapted to be highly intelligent and utilise a variety of complicated environments. Brain sizes in parrots are relatively large and intelligence in most species is high, with some shown to be as intelligent as primate species. They have complex vocalisation behaviours used for a variety of communication behaviours. Some are meant to communicate specifically with a pair, their offspring, and still different vocalisations are used in a flock setting with many conspecifics (Bergman & Reinisch, 2006). Among their other adaptations, vision, touch, and taste all interplay and can hint at different diets and behaviours of birds in the wild. Parrots need to see well to identify and find ripe fruits and foods as well as avoid predators. Physiological adaptations, such as their bill sizes/shapes and unique feet, also allow them to navigate complex habitats and handle foods, and these interplay with intelligence to help parrots survive. Table 13.2 describes some broad examples of physiological and behavioural adaptations that are common or unique to parrot species in more detail. Understanding these adaptations in species-specific management is key to providing good care for parrots in zoos.

Mating systems and reproductive strategies can shape many behaviours of parrots, from social structures, territoriality, and flock dynamics. Differences in hierarchy placement can also impact food availability within flocks or groups. Courtship behaviours can also tell a lot about the nesting behaviours of parrots and can inform care during nesting seasons. "Allopreening" (one parrot preening a conspecific mate) and "Allofeeding" (one parrot feeding a conspecific mate) are both common behaviours observed in parrots in varying configurations. Allopreening, head bobbing by some Amazon parrots and lorikeets and warbling vocalisations by budgerigars *(Melopsittacus undulatus)* are also common courtship displays observed in parrot species (Spoon, 2006). Allofeeding can also be a courtship behaviour signifying imminent nesting behaviours but is more common during nesting. Commonly, males will feed their mates before mating but especially during incubation, as most male parrots do not participate in incubation (e.g., rainbow lorikeets or Puerto Rican Amazon parrot, *Amazona vittata*). When incubation duties are shared between partners, allofeeding may be much less, such as for the cockatiels, *Nymphicus hollandicus* (Spoon, 2006). Care staff can use the knowledge of wild courtship and nesting behaviours to know when breeding is likely to occur and properly prepare with seasonal changes to diets needed or ensuring that nesting sites and appropriate substrates are

Table 13.2 Examples of physical and behavioural adaptations of parrots and how these should be accommodated in zoo management for these species.

General parrot adaptations	Description	Accommodating behaviour in husbandry and housing
Zygodactyl Feet	The feet of parrots are not unique among birds but are a key adaptation that allows them to dynamically explore their complex environments and handle various food types. The term refers to having two toes in the front and two in the back, allowing for high dexterity and gripping.	Provide dynamic spaces with appropriate complexity to allow parrots to use their feet and bill to explore, climb, and feed themselves.
Bill size/shape	Bills of parrots can be very strong and can vary widely in size and shape. Often, these bills inform what kind of diet a bird may utilise (e.g., large bills for large nuts by hyacinth macaws, *Anodorhynchus hyacinthinus*, or smaller more slender bills for excavating fruits and feeding on flower nectar).	Parrot bills grow continuously throughout their lives. Providing suitable substrates, toys, and props that allow birds to chew are helpful. Providing naturalistic food items and complex enrichments will allow them to use their bill and feet to solve puzzles and obtain food rewards.
Vocalisations	Complex vocalisations in the wild and the ability to imitate are unique and valuable adaptations. Unique vocalisations can be used between pairs and their offspring, can differ for different contexts (e.g., feeding, predator avoidance, flock communication).	Ensuring birds are in proper social contexts when housing will allow them the ability to vocalise in the proper manner. Often, screaming and other negative vocal behaviours can be indicative of behavioural challenges.
Socialisation	For most parrots, some part of their life is spent in flocks and this conveys several benefits. Finding food sources, lowering vigilance needed by each individual for predator watches, socialisation for juveniles.	Ensuring that birds are kept in social situations and potentially altering those configurations with the season to mimic how birds may socialise in the wild will benefit the welfare of parrots in zoos and also add variability and intrigue to daily life.
Colour	Colour in parrots serves many purposes including sexual selection, health signalling, and camouflage. Amazon parrots for example can more easily melt into lush green forests to escape predators. Eclectus parrots (*Eclectus roratus*) may use the red plumage of the female to warn other birds from her nest. Bright colours in parrots may also be important for easy identification of flock members or when asserting dominance over nesting sites.	Preventing feather damaging behaviours and ensuring high-quality feather coverage is maintained will ensure that welfare and status of individual parrots is maintained in exhibit settings.

(Continued)

Table 13.2 (Continued) Examples of physical and behavioural adaptations of parrots and how these should be accommodated in zoo management for these species.

General parrot adaptations	Description	Accommodating behaviour in husbandry and housing
Vision	Parrot vision is their most acute sense, much more so than many other birds. It is very complex and needs to be to discern the colours of conspecifics, identify colourful or cryptic food items, etc. Vision of parrots is comparable to humans but birds can see into ultraviolet.	Providing colourful enrichments, new items, and puzzles requiring intense visual scrutiny can stimulate parrots visually. Thinking carefully of neighbouring exhibits and what animals are nearby may also help to stimulate parrots.
Encephalisation (Large Brain Size)	Parrots have relatively large brains and correspondingly high intelligence. This adaptation allows them to be amazing problem solvers and affords them the ability to survive in their environment by finding difficult food sources, avoiding predators, and developing complex social structures.	Parrots should be provided with enrichment in a variety of forms on a regular basis. Many common examples include toys, puzzles, and novel food items along with water features or adding dynamic and variable elements to exhibits. Boredom can lead to stereotypic behaviour, self-harm, challenges to management for care staff and many others.

provided to allow birds to exhibit desired behaviours. If a pair never shows the proper behaviours where breeding is desired, it could be an indication that key elements are missing for the species (e.g., social environments, inappropriate exhibit design, or configuration). Lack of success should warrant an evaluation of current husbandry practices and how they can be altered to encourage success.

Understanding the different reproductive strategies of parrots and their natural histories can allow opportunities for maintaining or increasing overall welfare states. For example, sulphur-crested cockatoos are highly social urban parrots in Australia. Breeding in this species is often done by a pair isolating for nesting, but still associating or in some proximity to other cockatoos. Juveniles are then socialised by the parents and join foraging flocks outside of the breeding season in a fission-fusion system (Aplin, Major, Davis, & Martin, 2021). For zoos wanting to breed a species, knowing the ecological role it plays in their natural habitats will help zoos provide the right social context, good welfare, and allow for successful reproduction.

13.3 ENCLOSURE CONSIDERATIONS FOR PARROTS BASED ON BEHAVIOURAL EVIDENCE

Most species in the various taxonomic families described here live in forested environments, often with thick vegetation and at varying canopy heights, diverse daily activity budgets, and many other considerations. The shape and size of a parrot's bill, general body size, and body shape, and colouration can also convey a great deal of information about how a species interacts with its environment. Understanding the limits of sensory perception of parrots can also guide how to craft exhibits that specifically encourage foraging, social behaviour, and needed physical activity around different points of an exhibit. If housing multiple species together, understanding if one species may bully or dominate another, or specifically with parrots, which species are likely to hybridise or bond with heterospecific is key to avoiding undesired bonding or breeding behaviours. Knowing how to combine multiple species of various niches (i.e., upper canopy species with mid canopy or ground dwelling birds).

Mixing birds from different continents or species commonly known to be more or less noisy may be problematic. For example, grey parrots (*Psittacus erithacus*) tend to be quieter birds, requiring privacy and habitats away from noise when breeding, therefore, they should probably not be exhibited with or near louder parrot species, such as Amazon parrots or cockatoos (Wilson, 2006). Nesting disruption and aggression must also be considered when designing spaces for multiple parrot species. Refugia especially should be provided to allow for animals to avoid each other, have sufficient feeding spaces to reduce competition and height is extremely important for many species to get away from visitors or care staff and add a feeling of safety.

Wild behaviours should be used whenever possible as a resource to guide the shape and size of resources provided within enclosures. Folding natural observations into the design and construction processes will ensure parrots in human care are more likely to exhibit naturalistic behaviours and show healthy activity patterns as documented in free-living birds. One often used method of exhibiting parrots includes free flight aspects, where a few or several birds are let out of an enclosure after extensive training. In this instance, enclosures are given sufficient dimensions, specifically height and width, to allow multiple species groups variety in space. Natural behaviours and displays of communications among birds in these free flight groups are important behaviours to encourage through exhibit design and training where possible.

Before designing new parrot enclosures or programmes, it is important to consider its purpose (e.g., mixed-species exhibit, guest feeding interactions, private breeding pens, breeding on exhibit, etc.). For example, if mate choice is extremely important for breeding success, the size and space provided may be important to consider for successful reproduction. Kea are likely polygynous in the wild and may rely heavily on mate choice, however, traditional management might lead to these birds being managed as individual pairs for exhibit and breeding with little opportunity to choose their own mate in a zoo. In the past several years, management in the North American zoo population of kea has shifted to allow for more mate choice by keeping younger birds in groups in hopes of improving breeding success. Holding larger groups of youngsters together early on allows keepers to evaluate behavioural signs that a pair might be forming a bond. As juveniles mature, formed pairs or small groups are considered for placement in other zoos for exhibits. Managing in this way likely provides valuable socialisation and instruction for young kea on how to be adults. For other existing adult pairs, managers are also considering changes to how they hold the species where breeding is desired, often encouraging group sizes of at least two males to one female. In the wild, multiple females in the same area nesting would not be tolerated and having multiple males for a female to choose may jump start breeding behaviours in a female where previously no or little breeding behaviour was observed. While still early in the implementation, anecdotal evidence suggests that reproduction has been affected already with higher egg production observed annually since the start of this strategy in zoos where multiple birds are held together (Meehan, Klosterman, & Andrews, 2021). Managing in this way allows a larger group, or small flock, to socialise together outside of the breeding season, then pair off and isolate during breeding seasons (Figure 13.1).

Successful reproduction and display of reproductive behaviours (i.e., courtship, mating, parenting) are all key indicators for positive welfare. Parrots not showing appropriate reproductive behaviours may be lacking one or more elements in their environment that could be affecting overall welfare. For example, some macaw species, like the blue-throated macaw (*Ara glaucogularis*), often do not use the same nesting site every year in the wild and may need multiple nesting sites provided in a zoo setting if one that is provided is not preferred. Other species like hyacinth macaws (*A. hyacinthinus*) may need to do some light excavation to illicit the proper motivation to nest. With the prevalence of pet parrots in the world, private pet industries have many nest box products available to fit a wide variety of species shapes and sizes. Sometimes, custom boxes are also useful to provide suitable breeding environments that provide what a species needs. Returning to kea examples, this species breeds underground, often employing a bend in tunnel entrances to allow females more control of the entrance. Figure 13.2 shows an example of a custom box that provides this detail along with some custom doors included for keeper access. Placed on the ground and surrounded by alpine tree enrichment or exhibit features, the scene is set for some successful reproduction. For larger arboreal

Figure 13.1 Top shows a group of adult keas managed in a large group feeding (Photo credit: Cassie Crawford). Allowing birds to socialise in a flock outside of breeding season affects welfare and likely affects reproduction. Bottom photo of juveniles in a mixed flock (Photo credit: Shauna Foster) where juveniles are given access to other young kea early on in life to allow for socialisation and mate choice.

parrots, a wide variety of artificial nest boxes can be provided. Very commonly, large barrels with appropriately sized hols and entrance perches are provided for larger macaws (Figure 13.2 example). Smaller versions, or plastic versions are also used, along with wooden boxes in all shapes and sizes to fit species from budgerigars, cockatiels to Amazon parrots, and hyacinth macaws. Parrot species have a wide variety of nesting cavity requirements, mating systems and habitat preferences, and these must be considered when designing enclosures and providing nest boxes that can ensure reproductive success.

13.4 BEHAVIOURAL BIOLOGY AND PARROT WELFARE

Most species of parrots are naturally highly social animals often with varying and complex diets utilised in the wild. With many species also being highly intelligent, exploring their habitats, making decisions, and solving complex problems can be important factors to consider and encourage when evaluating parrot welfare in human care. Parrot species will provide visible cues to overall welfare states via their activity patterns and body/feather conditions. Understanding the behavioural syndromes (i.e., consistent behaviours expressed by individuals) allows care takers to make plans for when a bird may be experiencing poor welfare and may require management to improve their care. Annual veterinary check-ups of birds are a standard practice to help determine basic measures of welfare like mass. In many animals, changes in body weight are terrific indicators of changes in welfare or health status for individual birds. Regularly reviewing records to spot and identify the likely causes of weight changes, whether a bird is losing mass due to poor social standing in a group, or gaining mass unnecessarily and therefore may need dietary adjustment. For parrots, changing husbandry practices or group management (e.g., providing complex food enrichment, new and varied food items or more and complex spaces to fly and perch) can reduce aggression or ameliorate behavioural issues and move birds towards improved conditions.

Stereotypic behaviours in parrots are particularly challenging, as they are often indicative of welfare concerns. Pet parrots are also frequently surrendered to zoos for a variety of reasons, often including behavioural challenges. A major challenge documented in many birds, but particularly common to parrots involves managing feather damaging behaviours (FDB). Parrots showing this behaviour will pluck, chew, bite and sometimes ingesting their own feathers (Chitty, 2003). In extreme cases, a bird can remove all the feathers from their body except their head, resembling a "naked" parrot. Such self-damaging behaviours can cause issues with thermoregulation, and render birds unable to fly leading to concerns for individual welfare and quality of life (Watters, Krebs, & Pacheco, 2019). Social stature can also be impacted and compound the negative effects of feather loss. In one example from Watters et al. (2019), a group of golden

Figure 13.2 Left, shows a custom nesting box created to fit the natural reproductive behaviour of the kea (Photo credit: Kimberly Klosterman). Note the placement of the box on the ground with an "L" PVC joint used as the entrance to mimic natural burrowing behaviours of kea. Right photo shows a large barrel used for macaw nesting box (Photo credit: Shannon Irmscher). Note that the placement here is supported by a shelf and is horizontal to prevent egg breaking while other placements are possible to fit the biology or size needs for different species.

conures (*Guaruba guarouba*) were exhibiting FDBs and managers set out to find a way to reverse this self-destructive behaviour through dynamic enrichments (see Figure 13.3 for such an example). In this example, several automated feeding devices were incorporated into the birds' enclosures with varying schedules of feeding so that the environment or timing of enrichment available was not predictable. Adding this unpredictable element to their exhibit encouraged exploration, interaction with other parts of their enclosure and even improved interactions with visitors. With the added activity of these birds interacting with their environment, less time was given towards FDBs, pacing, or other self-destructive behaviours, and feathers grew back, improving the bird's welfare and health conditions.

Training parrots to participate in their own care can have many added benefits. Teaching specific behaviours can eliminate the need for stressful and often traumatising physical restraints (e.g., netting, chasing) by veterinary staff and provide a mentally stimulating exercise to mix up the events of day-to-day care. Choice and control are also available to parrots in training for their own care, and animals can express their comfort level from day to day by choosing to participate. A well-rounded training programme allowing for this choice and control and added mental stimulation can vastly improve the welfare of parrot species. The natural histories of different parrots should be considered when training plans are put in place to create the best possible environment to achieve husbandry goals. For example, macaws and many other larger parrots, can be trained for physical handling through various behaviours (i.e., wing opening, clipping of wings, crate training, free flight training).

As previously mentioned, parrots are often highly social. Specialised knowledge of natural history can inform how rigorous training programmes should be in parrots to avoid over-familiarity, preference for trainers, and imprinting by parrot species. For an example, from personal experience, while training a scarlet macaw (*Ara macao*), I encountered a bird with a strong dislike towards male keepers. Conversely, one keeper in particular became a favourite of this animal, creating management challenges. Macaw species often mate for life and create strong bonds with other birds in loose associations or flocks or with people. Managing away from these behaviours and creating a programme that engages participation regardless of trainer proved difficult and arduous, but welfare and care of this animal improved greatly once regular behaviours could be trained by the whole staff. Wings were able to be trimmed by hand without restraint and toe nail trimming was next to be incorporated to allow for handleability and voluntary movement by keepers between enclosures. Less aggression was also directed towards keepers and other conspecifics improving overall welfare and management in this example.

Figure 13.3 Photo of a disassembled time-release feeding device designed to be incorporated into an exhibit and provide unpredictability in zoo animal environments. While not used for parrots, this device is in principle used in the same manner as the example shared in the text by Watters et al. (2019). (Photo credit: Dr. Bethany Krebs).

13.5 SPECIES-SPECIFIC ENRICHMENT FOR PARROTS

Suitable enrichment may perhaps be crucial to providing suitable welfare and high-quality care for parrot species. The types of enrichments can include a broad range of options, including nutritional, occupational, physical, sensory, and social in design and construct. Some options include changing enclosure designs by moving perches around regularly or regularly incorporating new browse items, alterations to feeding schedules, providing new and varied (but appropriate) food items, or providing new or moving existing nesting boxes/structures to illicit breeding behaviours. The type, as well as the timing or amount of enrichment, can be immensely important as well. Zoo settings may often have much less varied schedules of feeding/enrichment, and adding variation and unpredictability may help birds move to more closely match natural activity levels. Knowing the natural history, and some time budget information of the species you are enriching can help to inform a reasonable enrichment schedule that can keep parrots stimulated, healthy, and avoid stereotypic behaviours that could lead to poor welfare and health.

Enrichment for parrots can include:

- Water features such as sprinklers, misting stations, waterfalls, hoses, and novel water interactions.
  - Often providing misting can mimic rain patterns in native habitats for tropical species and improve behaviour patterns and feather quality.

- Changing exhibit structure for non-permanent features. Such as changing perches by adding new or re-arranging existing perches.
- Browse supplementation from approved lists to add fresh greenery and focuses chewing and feeding behaviours away from exhibit features.
- Provision of nesting materials that helps illicit reproductive behaviours in some species.
- Managing social groups to mimic natural age structures or natural fission-fusion systems of association in many species (Aplin et al., 2021).
- Providing minor protein sources (e.g., crickets, carnivore meat) for parrots with an opportunistic feeding strategy that utilises protein (often for breeding).
- Allowing pairs to incubate and rear their own young.
- Automated devices/structures to introduce unpredictability into feeding schedules, which provide added opportunities for exploration and foraging.
- Training programmes that are mentally stimulating and improve husbandry for the parrots and their caregivers.
- Allowing mate choice in breeding programmes.
- Providing proper heights (based on ecological information) for exhibits or holding spaces.
- Allowing for flighted exercises in parrots – often in shows or flock flight behaviours trained for display.
- Dynamic food item provisions to reduce the use of processed diets.

As seen in the previous section, welfare can be greatly improved by provision of welfare, with the reduction of FDB and stereotypic behaviours shown in Figure 13.3 Additionally, enrichment schedules can lead to improvements in breeding output for some parrot species. When presented with various types of enrichment (e.g., food, cognitive, physical), a group of Blue-and-yellow macaws (*Ara ararauna*) housed in pairs near each produced more eggs annually following implementing of more enrichment (Miglioli & Vasconcellos, 2021). This study implemented an enrichment strategy over a longitudinal study and documented significant increases in egg production following enrichment that increased the animal'sphysical behaviour. The sociality of this macaw species plus the vast amount of time spent foraging for complex food items, suggested that a dynamic enrichment strategy that promoted foraging and movement. Sexual behaviour and egg production also increased, though not significantly. Pairing natural history, clear goals towards desired outcome behaviours and accurately measuring those outcomes in zoos are all vital for successful enrichment programmes for parrots.

13.6 USING BEHAVIOURAL BIOLOGY TO ADVANCE PARROT CARE

Managing parrots in populations can often be challenging and many species suffer from a lack of sustainable numbers or low reproductive success. Historically, many parrots, especially larger species, have been housed alone, or in pairs away from any other parrots and may be used in public presentations often. Box 13.1 illustrates major challenges to sustainable population management in parrots and suggest possible solutions for improving long-term sustainability.

BOX 13.1: Considerations and Challenges for population management of parrots in zoos

The order Psittaciformes contains some of the longest living avian species. Some species of parrot are also the most endangered of bird species in the world. These birds are also commonly kept birds in zoos around the globe. Of the 345 or so species of parrots, >40% are threatened or near threatened in the wild. In many cases, captive populations are equal or greater than those in the wild (Mellor et al., 2021). Private owners of parrots collectively hold more birds than zoos and breeding centres combined globally. Despite the breeding success in the pet trade for many species, zoos often encounter challenges to breeding and population sustainability for rarer or more endangered species. The long lifespan of parrots may be lulling many zoos into false senses of security as populations or the number of species seem to remain stable over time, however, the general trend of managed populations, in North America for example, is one of declining numbers (Barkowski, 2020). Some regional associations are challenged by a lack of space, specifically dedicated to reproduction, available to managed parrot species (Barkowski, 2020). Because parrots are so long lived, many individuals that are more successful at breeding are also the original wild birds imported decades earlier, with successive generations less successful.

Wild-caught parrot pairs are generally thought to be more successful and productive breeders than captive hatched birds (Wilson, 2006). Proper socialisation of young parrots is a vital component of reproductive success in wild birds. For many parrot species, young birds need to be near or around conspecifics to learn appropriate vocalisation etiquette and socialising skills needed to be competent adults in the wild. Wild birds would have encountered proper rearing contexts as juveniles to make them more competent breeders. Ideally, birds in the following generations (where the intention is for birds to continue on as future breeders) should be parent raised and then socialised with juveniles of the same species for more successful zoo breeding (Wilson, 2006). In zoos, however, often there is not enough space or enough birds available to provide the proper social situation needed to promote successful breeding. Birds that lack the parent rearing and proper social environment as they grow into adulthood will often not make successful breeders and may develop many other behavioural issues besides (Meehan & Mench, 2006) and be less successful breeders.

Traditional population management in zoos often relies on setting up pairs of animals that are genetically well suited, behaviourally as compatible as possible with an open space at a zoo ready to receive them. Zoos that house various species often only have spaces designated for a single pair or a group that may not be of sufficient size to provide naturalistic social environments, depending on the parrot species. Many parrot species are highly social spending time outside of the breeding season in conspecific flocks and then pairing up during breeding seasons. Additionally, for socially breeding species of parrots like conures (*Aratinga* or *Pyrrhua sp.*) or budgerigars, breeding success may rely on a minimum flock size thus the presence of neighbouring pairs is a benefit to breeding, and success may not be achieved in small numbers. Many zoo regions support cooperative programmes for zoos to combine resources and manage populations for sustainability and assurance populations of many endangered species. The prevalence of pet birds needing re-homing in zoos is making the goal of sustainable population management for parrots more difficult. Previous estimates suggested that 10 million pet parrots were thought to be in private homes (Meehan & Mench, 2006). Not all owners are prepared for the social and environmental needs of a parrot, not to mention the time investment, and many are re-homed, sold on, or surrendered to rescue facilities and, quite commonly, zoos. Often these birds suffer from abnormal behaviours like excessive screaming, stereotypic behaviours, and feather picking. Pet birds are likely to be poor breeders and therefore increases in the number of such birds in zoos may be a barrier to sustainable population management.

Changes to how we plan breeding in zoos may also need improvements and deeper thought to achieve more sustainable populations. Zoos commonly combine resources to share data and exchange animals between facilities to allow for breeding or to fill exhibit spaces. Leveraging this cooperation across facilities to build husbandry knowledge and create more comprehensive plans to manage parrots will lead to more success in breeding. Breeding parrots is intricate and complicated and requires that knowledge of their natural history and behaviour be incorporated in population planning. Before moving a bird(s) and recommending breeding, more in-depth knowledge of spaces used, social settings, and enrichment programmes need to be considered and incorporated into population planning. More dedicated space from zoos for breeding groups is a major first step that is needed for parrot management (Barkowski, 2020). Having dedicated space that can provide the proper for the intricate social needs and fulfil other needs informed by natural behaviour. Engaging in enrichment programmes that specifically target different species can greatly improve breeding success. For example, providing regular and unpredictable types of enrichment can greatly improve both welfare (Watters et al., 2019) and breeding output (Miglioli & Vasconcellos, 2021) by stimulating the use of a parrot's high intelligence.

The examples here suggest that current guidelines around managing may parrot species may not be sufficient to support long-term sustainable populations. While we have kept parrots in human care for many years, the natural history and behaviours of many species remain largely unknown from the wild. Furthermore, zoos often have difficulties providing necessary environments for more endangered or difficult to breed species to be successful. Despite knowledge gaps, zoos can serve an important role by building up husbandry knowledge for species through extensive cooperation. Meticulously documenting husbandry practices and observations can over time build a basis of knowledge for species with little known information for zoo and wild populations to benefit from.

As examples have shown throughout this chapter, providing high-quality care and welfare to parrots is no easy task. Exhibit design must meet the bird's needs and mirror natural history. Social structures should be as close to the wild as possible and mental stimulation through enrichment, enclosure design, diet, and even their exhibit neighbours, is vital. Evidence-based management strategies are needed that are species-specific and dynamic to achieve any long-term sustainability for captive parrot populations. Especially for the species that are threatened in the wild where zoos have the opportunity to manage such a population, providing quality care and high-quality welfare throughout all these areas becomes a conservation imperative.

13.6.1 Further information to inform parrot care

For more resources on welfare assessment in parrots see some example resources below:

The *Manual of parrot behaviour* has a variety of additional resources in more expansive topics of care for parrots. Chapter 16 specifically provides several resources to help diagnose common behavioural challenges for parrots:

https://onlinelibrary.wiley.com/doi/book/10.1002/9780470344651

Welfare of parrots as companion animals: http://www.avianwelfare.org/issues/WelfareAndSuitabilityOfExoticBirds-1.pdf

For more resources on the conservation of wild parrots and accommodating education materials, see some example resources below. Lastly, there is an example of a federal reintroduction programme and a zoo partnership working together to conserve an endangered parrot species.

World Parrot Trust:

http://www.avianwelfare.org/issues/WelfareAndSuitabilityOfExoticBirds-1.pdf

Intertwined conservation corporation https://intertwinedconservation.org/

Conservation of the Puerto Rican Parrot (*Amazona vittata*). Puerto Rican Parrot Recovery Program: https://www.fws.gov/southeast/caribbean/puerto-rican-parrott-recovery-program/

Lincoln Park Zoo's Role in Population Biology expertise: https://www.lpzoo.org/science-project/protecting-the-puerto-rican-parrot/

13.7 CONCLUSION

This chapter has discussed how knowing adaptive traits and behaviour patterns of specific parrot species as well as understanding their social and mental needs is important for zoo management. Parrots require specialised care in the zoo based on the best available information on their natural history, habitat needs, and social and behavioural requirements.

Care for parrots can be improved from the start in the design phase of a new exhibit ensuring the proper space is provided to allow for natural behaviours. Height, flight distances, perch size, nesting options, and structural variability are all facets that should match the natural behaviours of parrots as closely as possible. Then, crafting diets and enrichment schedules that match the food needs and activity of a species in the wild will further benefit parrots. Enrichment, whether through training from keepers or physical enrichment additions, is likely key to maintaining mental health, preventing boredom, and maybe even affecting breeding behaviours. If breeding is desired or part of the goal of exhibiting parrots at your zoo, even more attention should be given to the habitat made and nest structures provided. Who their neighbours are, size and shape and placement of boxes are all important facets to consider and match to the natural history of the parrots you are managing.

On a population level, greater cooperation between zoos in different regions may be needed to meet population goals. Not all zoos can provide the potentially large spaces needed by different species to breed and care for offspring. Traditional breeding schemes that rely on one single pair may not provide success. Thinking more strategically and long-term, applying the principles described here for parrot species can greatly improve the long-term outlook of parrots in zoos.

REFERENCES

Aplin, L., Major, R., Davis, A., & Martin, J. (2021). A citizen science approach reveals long-term social network structure in an urban parrot, *Cacatua galerita. Journal of Animal Ecology*, 90, 222–232.

Barkowski, J. (2020). *Parrot taxon advisory group regional collection plan.* Tulsa Zoo, Tulsa, OK, USA.

Bergman, L., & Reinisch, U. (2006). Parrot Vocalization. In A.U. Luescher (Ed.), *Manual of Parrot Behavior* (pp. 225–231). Blackwell Publishing, Ames, Iowa.

Chitty, J. (2003). Feather plucking in psittacine birds 2. Social, environmental and behavioural considerations. *In Practice*, *25*(9), 550.

Homberger, D. (2006). Classification and status of wild populations of parrots. In A.U. Luescher (Ed.), *Manual of parrot behavior* (pp. 3–11). Blackwell Publishing, Ames, Iowa.

Meehan, C., & Mench, J. (2006). Captive parrot welfare. In A.U. Luescher (Ed.), *Manual of parrot behavior* (pp. 301–318). Blackwell Publishing, Ames, Iowa.

Meehan, J., Klosterman, K., & Andrews, J. (2021). *AZA Species Survival Plan® red program population analysis and breeding and transfer plan for kea (Nestor notabilis).* AZA Population Management Center: Chicago, USA.

Mellor, E.L., McDonald, H., Mendl, M., Cuthill, I., van Zeeland, Y., & Mason, G. (2021). Nature calls: Intelligence and natural foraging style predict poor welfare in captive parrots. *Proceedings of the Royal Society B*, *288*(1960), 1–10.

Miglioli, A., & Vasconcellos, A. (2021). Can behavioural management improve behaviour and reproduction in captive blue-and-yellow macaws (*Ara ararauna*)? *Applied Animal Behaviour Science*, *241*(9), 105386.

Munn, C. (2006). Parrot Conservation, Trade, and Reintroduction. In A.U. Luescher (Ed.), *Manual of parrot behavior* (pp. 27–31). Blackwell Publishing, Ames, Iowa.

Spoon, T. (2006). Parrot reproductive behaviour, or who associates, who mates, and who cares? In A.U. Luescher (Ed.), *Manual of parrot behavior* (pp. 63–77). Blackwell Publishing, Ames, Iowa.

Watters, J., Krebs, B., & Pacheco, E. (2019). Measuring welfare through behavioral observation and adjusting it with dynamic environments. In A. Kaufman, M. Bashaw & T. Maple (Eds.), *Scientific foundations of zoos and aquariums: Their role in conservation and research* (pp. 212–240). Cambridge, University Press, Cambridge.

Wilson, G.H. (2006). Behaviour of captive psittacids in the breeding aviary. In A.U. Luescher (Ed.), *Manual of parrot behavior* (pp. 281–290). Blackwell Publishing, Ames, Iowa.

14

The behavioural biology of hornbills, toucans, and kingfishers

JONATHAN BEILBY
Chester Zoo, Chester, UK

14.1 INTRODUCTION TO THE BEHAVIOURAL BIOLOGY OF HORNBILLS, TOUCANS, AND KINGFISHERS

This chapter examines the hornbills, toucans, and kingfishers – a suite of fairly large, intelligent birds, which in the main are found in the tropics. From a public perspective, there are a lot of similarities – colourful birds with large bills, which are generally active and vocal. The birds covered here are some of the most instantly recognisable zoo birds, not least from their depiction in recent popular media; from "Zazu" the hornbill (a species from the genus *Tockus*) in *The Lion King*, to the toco toucan (*Ramphastos toco*) that was the logo for Guinness for over 40 years! Saying this, all of these species have all been revered culturally for generations. The southern ground hornbill (*Bucorvus leadbeateri*), in many southern African countries, is associated with bad luck – indeed in Zimbabwe a visit to an elderly relative is thought to be the bringer of death! In Polynesia, the sacred kingfisher (*Todiramphus sanctus*) is believed to have power over the seas and waves, whilst in Brazil, emperors would wear cloaks made from the feathers of the channel-billed toucan (*Ramphastos vitellinus*). There are many similarities between the three families, however the next couple of paragraphs will examine each family in more detail, before looking at commonalities in their behaviour biology.

14.1.1 Kingfishers

The 114 species of kingfisher are from the order Coraciiformes, which also includes bee-eaters and rollers. The kingfishers' closest relatives are the todies (Todidae), five diminutive Caribbean species,

DOI: 10.1201/9781003208471-16

and the Neotropical motmots (Kuhl et al., 2021). Kingfishers themselves are found on every continent other than Antarctica, but species diversity is greatest in South East Asia, particularly Papua New Guinea, which is home to 34 species, compared to a mere six species across the entirety of the Americas. The smallest species are the Old World pygmy-kingfishers, of which the African dwarf kingfisher (*Ceyx lecontei*) is the smallest at only 10 cm long, weighing around 11 g. The largest species is the laughing kookaburra (*Dacelo novaeguineae*) from Australia, which measures up to 42 cm long and can weigh as much as 500 g.

Almost all species of kingfisher have a similar appearance, short legs and tail, and a large head supporting a substantial sharp bill that is laterally compressed and triangular in cross-section. Whilst morphology is largely conserved across the family, there are variations in the preferred habitat and ecology:

- the common kingfisher (*Alcedo atthis*) is largely piscivororous, perching on low-hanging branches and diving into small rivers to feed on small fish and the occasional invertebrate;
- the pied kingfisher (*Ceryle rudis*) that has evolved to hover to catch fish, being able to eat them on the wing and is therefore able to live near large, open bodies of water;
- the white-breasted kingfisher (*Halcyon smyrnensis*) feeds on insects, frogs, and lizards in wetland habitat, with composition of diet changing seasonally and with geographic range, with some Indian birds having a more fish-biased diet (Fry & Fry, 2010);
- the laughing kookaburra feeds on a variety of vertebrate and invertebrate prey, largely taken from the ground.

Those kingfisher species that are widely distributed are not threatened by extinction, and some species can thrive in urban environments. Those species endemic to islands are more at risk from human encroachment or natural disasters, with 10% being listed as threatened. The Guam kingfisher (*Todiramphus cinnamominous*) has been Extinct in the Wild since 1986, owing to the introduction of the brown tree snake (*Boiga irregularis*) that predated many adult birds, chicks, and eggs. Thankfully, a captive breeding programme was initiated by the Association of Zoos & Aquariums (AZA) in North America, and the captive population (at the time of writing) numbers around 150 birds. The Javan blue-banded kingfisher (*Alcedo euryzona*) was recently rediscovered in West Java with a population estimated at between 50–249 birds, whilst the Sangihe dwarf kingfisher has not been seen since 1997 (Riley, 1997). Both species are listed as Critically Endangered on the IUCN.

14.1.2 Hornbills

The hornbills were previously placed in the order Coraciiformes, but recent genetic analyses have placed them in the order with the wood hoopoes (Phoeniculidae) and hoopoe (Upupidae) – the Bucerotiformes (Kuhl et al., 2021). There are an estimated 59 species of hornbills, depending on the current taxonomic preference, which occur exclusively in tropical climates of Africa and Asia, with one species (the Papuan hornbill, *Rhyticeros plicatus*) extending to New Guinea and the Solomon Islands. Most species are heavily reliant on large areas of forest, where they forage in the canopy. The smaller African species, and the two species of ground hornbills, are generally found in savannah habitat. The hornbill's most distinctive feature is a very large bill, often complete with a casque (a decorative growth on the top of the bill). In all but one species, this casque is hollow and of a fine honeycomb internal structure. Casques are often brightly coloured, and in many species this colouration originates from pigmented preen oil. The hornbills are the only birds in which the first two vertebrae, the atlas and axis, are fused together, and this is assumed to help support the weight of the bill and casque.

There is a huge amount of morphological diversity within the species of hornbill. From the 100 g black dwarf hornbill (*Horizocerus hartlaubi*) to the 4 kg southern ground hornbill. Whilst closely related to kingfishers, sharing the fusion of digits III and IV at their base, there is elongation of the tarsus bone which allows for much more proficient locomotion – a feature accentuated in the terrestrial ground hornbills. With this range of morphological diversity comes diversity in ecology. The genera of *Tockus* and *Lophocerus* from Africa, rarely feed on fruit, instead relying on insects and small vertebrates. The ground hornbills are carnivores, but most of the other species are frugivorous, with some Asian hornbills having fruit make up 90% of their diet (Kinnaird, O'Brien, & Suryadi, 1999).

14.1.3 Toucans

The toucans (Ramphastidae) sit within the Piciformes, which are found on all continents other than Australia and Antarctica. The Piciformes contain woodpeckers, barbets, toucans, puffbirds, jacamars, and honeyguides. All Piciformes share a zygodactyl foot morphology, where digits II and III face forward, and I and IV face backwards. This allows for excellent manoeuvrability, and easy perching on the trunks of trees. Toucans sit within the New World barbet clade, having evolved from a common ancestor with the prong-billed barbets (*Semnornis*, which has two extant species).

The toucan family evolved very quickly, in evolutionary terms, and as such there is very little in the way of morphological diversity between species. The 36 species are found from Mexico through to South America. All birds have a proportionally large bill and forage for a largely fruit-based diet in the upper stories of the rainforest. There is a variation in size, but proportionally not as great as that seen in either kingfishers or hornbills. The smallest species, the green araçari (*Pteroglossus viridis*) weighing 115 g, whilst the largest, the toco toucan weighing in at around 850 g.

Toucans have nine caudal vertebrae, with the rear three fused together to form a ball and socket joint to the vertebrae in front. This severely limits any sideways movement of the tail but does allow for rapid up and down movement which provides a means of visual communication – the "tail-cock" gesture. This tail movement highlights any patterning, e.g., the chestnut tipped tails of some *Aulacorhynchus* toucanets, and focuses the eye to the bright rump/vent colouration seen in many species (Short & Horne, 2001). This tail morphology also allows toucans to sleep in a unique posture, with the tail raised and put forward to the top of the head, where it is held in position.

14.1.4 Common aspects of behavioural biology

These birds differ in many aspects of feeding and breeding biology but can be seen to share a plethora of important morphological, reproductive, and behavioural traits. When comparing hornbills, kingfishers, and toucans in the text, the term "three groups" will be used. Table 14.1 provides biological information on example species from these groups to highlight both diversity and similarity. All three groups share a proportionally large bill. This has many immediately obvious applications, not least in food acquisition, allowing kingfishers/ground hornbills to attack and kill their prey, and the long bills of toucans and other hornbills can often allow them to reach fruit out of reach of other birds. It has been hypothesised that the bill/casque of hornbills is used to amplify vocalisations throughout a forest habitat, but whilst the casque assists in sound production and resonance, it does not enhance the amplification of dominant call frequencies (Alexander, Houston, & Campbell, 1994). The hornbill's casque does provides structural support for its large bill, not required in the other two groups owing to the relative depth and width of their bills (Kinnaird & O'Brien, 2007). It has been shown that, for toucans at least, the large bill has a thermoregulatory function, providing a huge surface area from which to radiate excess heat. Toucans can modify blood flow to the bill, allowing it to function as a transient heat radiator, and enabling the bird to occupy the forest canopy without overheating.

All three groups are monogamous. As would be expected in birds showing high pair fidelity, there is limited sexual dimorphism. In kingfishers, females are generally larger and there are slight differences in the plumage. For many toucans, there is very little difference other than greater bill length in males. There is mild dimorphism in the green and lettered araçaris (*Pteroglossus inscriptus*), where the male has a black head and the female's is dark brown. In the genus *Selenidera* there is a much greater amount of sexual dimorphism. The birds have an increased intensity and variety of calls with associated tail postures. In hornbills, males are larger than the females, and there are differences in casque colouration/shape between the sexes.

Some hornbills are seen to butt casques in the breeding season, noted in two of the large Asian species – the great (*Buceros bicornis)* and the helmeted (*Rhinoplax vigil*). In the great hornbill, a male bird is documented as flying at other perched males (Raman, 1998), but in the helmeted hornbill, the males fly directly into one another in mid-air, with the resulting impact being enough to push the male backwards. Female helmeted hornbills sometimes accompany the male on this flight, potentially as a means of protecting their territory from invading individuals. This species is particularly reliant on figs (98% of the diet of birds from Sumatran

Table 14.1 Key information on example species of hornbills, toucans, and kingfishers.

Species	Distribution & Red List status (as of December 2021)	Social behaviour	Biometric information & dimorphism
Guam kingfisher (*Todiramphus cinnamominous*)	Formerly the island of Guam, but now restricted to captive populations. Extinct in the Wild	Occur in pairs, with both birds maintaining several nest holes, using one in which both raise the young.	20 cm long, 56–74 g, females average larger than males.
Laughing kookaburra (*Dacelos novaeguineae*)	Mainland Australia, introduced to Tasmania and other offshore islands, and New Zealand. Eucalypt woodland habitat. Least Concern	Birds pair for life. Male feeds the female for around six weeks prior to laying. Cooperative breeding commonly occurs, and female helpers will sometimes lay their eggs in the nest. Helpers incubate for up to $^{1}/_{3}$ of time and play key role in the rearing young.	41–47 cm long, 310–480 g, females average larger than males.
Green araçari (*Pteroglossus viridis)*	Northeast South America. Lowland forest habitat. Least Concern	Travel as pairs or small groups. In the breeding season, male will feed the female, and the pair will excavate the nest together. Both parents rear the young, with the first clutch staying with the parents for at least six months.	30–39 cm long, 110–162 g, female has shorter bill, chestnut cap to the head; male has a black head.
Toco toucan (*Ramphastos toco*)	Guianas, Brazil, Bolivia, Peru, and Argentina. The only toucan that occurs in savannah, open woodlands, and plantations. Least Concern	Feed in small groups, in which all members allopreen. In breeding season, pairs form and establish territories. The pair will take over an old woodpecker nest hole and rear two to four youngsters. Rarely double clutch.	Largest of all toucans. 55–65 cm long, 520–800 g, males average larger than females and with a significantly longer bill.
Southern ground hornbill *(Bucorvus leadbeateri)*	Southern Africa. Found in woodland and savannah habitats. Vulnerable	Reaches maturity after four to six years, occurring in small family groups. Large groups work together to attack and kill large prey. Cooperative breeder, dominant pair assisted by immature helpers. Female does not seal into the nest cavity, partaking in provision of food for the offspring.	90–100 cm long, 2230–6180 g, male has red skin on face and throat patch, female has a blue patch on the red throat skin.
Von der Decken's hornbill *(Tockus deckeni)*	Ethiopia, Somalia, Tanzania, and Kenya. Semi-arid savannah and scattered woodland. Least Concern	Pairs forage mainly on the ground sometimes forming mutualistic relationships with dwarf mongooses *(Helogale parvula)*. Occasionally recorded in small flocks. Female sealed into nest cavity to rear the two to three eggs, with food provided by the male.	35 cm long, 120–212 g, male larger than female with a red bill/casque, which is larger than the female's black bill.

(*Continued*)

Table 14.1 (Continued) Key information on example species of hornbills, toucans, and kingfishers.

Species	Distribution & Red List status (as of December 2021)	Social behaviour	Biometric information & dimorphism
Great hornbill *(Buceros bicornis)*	South and South East Asia, with a disjunct population in South West India. Primary forest habitat but will cross between forest patches. Vulnerable	Travel in pairs or groups in search of food, with groups of up to 200 recorded on fruiting trees in the non-breeding season. When breeding, pairs become territorial and defend nest sites. Female sealed into tree cavity and food provided by the male – up to 185 items daily.	95–105 cm long, 2155–3400 g, males average larger than females. Male has a black rim to the casque, and black skin around a red eye. Female has a smaller casque, with no black, and white eyes surrounded by pink orbital skin which flushes red when breeding.

being composed of figs), so this aggressive behaviour may have evolved as a means of protecting a valuable resource (Kinnaird, Hadiprakarsa, & Thiensongrusamee, 2003). The helmeted hornbill is the only species of hornbill to have a solid casque, and this may have evolved for male-male competition.

14.2 ECOLOGY AND NATURAL HISTORY RELEVANT TO THE ZOO

To fully understand the behavioural biology of an animal it is important to know how it perceives the world. Birds have an additional cone in the eye, for tetrachromatic colour vision, allowing them to perceive colours in the ultraviolet (UV) spectrum (Cuthill et al., 2000). In the rainforest canopy, UV levels are high so the ability to see in this spectrum increases feeding efficiency for toucans and hornbills as ripe fruit is more obvious. Ripe fruit has more sugar and digestible protein – seeing UV confers an adaptive advantage because if these birds are able to see ripe fruit as soon as it is available, it gives them a head start in food collection over mammalian frugivores (that cannot see in UV).

As well as colour, the field of vision is important. Martin and Coetzee (2004) mapped the visual fields of hornbills, showing that they can see 30 degrees above the horizon, meaning the retina is constantly hit by bright light from above. Hornbills have evolved "eyelashes" which provide shade to the bird without creating visual blind spots. Hornbills are also able to see the tip of their beak, meaning they can precisely remove fruit from trees. Owing to this, it has been suggested that hornbills are able to access the nutritious fruit before competitive fruit-eaters. They are also excellent at catching food when tossed up to them, even whilst in flight. This ability to use the beak tip to manipulate food may be particularly useful to hornbills, as they (and the kingfishers) lack the stylohyoideus muscle, which is responsible for retracting the tongue, and influences tongue mobility overall. Kingfishers are seen to whack food against a firm perch before swallowing it whole. Hornbills also have a large gape, allowing them to eat larger fruits which, owing to lack of tongue mobility, are tossed to the back of the mouth to be swallowed. Toucans, on the other hand, have a tongue which is laminated laterally, with laminae that become brush-like at the tip. Toucans can feed on large fruits, using serrations of the bill (called tomia) to rip chunks off. This means that in a zoo setting, whilst toucans can be offered some very soft fruit spiked, hornbills and toucans should be given food in small chunks (about 3 cm cubes). Examples

Figure 14.1 Showing adaptations of toucans and hornbills. Left: a writhed hornbill (*Rhabdotorrhinus leucocephalus*) manipulates a food item. Centre: a rhinoceros hornbill (*Buceros rhinoceros*) shows the "eyelashes" and the ability to see the bill tip. Right: an Ariel toucan (*Ramphastos ariel*) shows the bristle-tipped tongue, and tomia of the bill.

of adaptations possessed by these different species are illustrated in Figure 14.1.

So important are the roles of frugivorous birds that in some Asian forests 50% of trees depend on birds for dispersal of their seeds (Leighton, 1982). Trees have evolved fruit to be eaten, but a diet of just fruit is very high in water and low in protein so hornbills must consume 20–33% of their body weight in fruit each day (Poonswad, Tsuji, Jirawatkavi, & Chimchome, 1998). In the wild, hornbills and toucans eat a variety of fruits, some high in sugars and some high in oils. A diet of sugar-rich fruit will mean more fruit must be eaten in order to achieve a similar energetic intake as an oil-rich diet, however whilst oil-rich fruits provide more calories, the fats must be hydrolysed before digesting, which also takes energy. Some trees (such as figs) provide huge crops of fruit, which have their advantages however the birds must eat a variety of fruits to ensure a balanced diet. Captive hornbills and toucans should be offered a wide variety of different fruits, avoiding citrus and tomato as these fruits can cause iron storage disease (haemochromatosis), to which toucans are especially sensitive. Most fruits available have also been selectively grown for the human palette and are much richer in sucrose than those birds find in the wild. A variety of blueberries, papaya, banana with apple and pear, with some steamed root vegetables (squash, parsnip, sweet potato etc.) works as a general rule for most species. Varying the diet, as well as providing different nutrients, is also a form of enrichment as the birds would not be able to eat the same fruits every day in the wild. There are some low-iron pellets, and others developed especially for hornbills, which provide a complete balance of nutrients, and these should also be offered.

For hornbills, drinking water is very rarely observed in the wild. Owing to the high water content of their food, it is hypothesised that hornbills obtain all the water they need from their diet, an idea supported by the bilobed morphology of their kidney, unique to these birds. Hornbill faeces are also much drier than that produced by toucans, making it an excellent construction material to seal the female in the nest whilst breeding (Kinnaird & O'Brien, 2007). Toucans and kingfisher do drink and bathe regularly. For toucans and hornbills, food and water should be provided off the ground in a rodent-proof feeder as they are highly susceptible to *Yersinia* (a bacterial infection originating from rodent urine).

Kingfishers and ground hornbills are both carnivorous, and their diet is largely composed of whole prey items. Bowls used for kingfishers should be open, allowing them to feed from the wing. The prey items can be placed on the floor, and when kingfisher pairs are kept together there should be two bowls available to reduce competition, particularly when not breeding. In the wild, kingfishers are crepuscular, so a feed first and last thing each day is preferable. Ground hornbills can be fed a variety of prey items, but will enjoy a carcass feed, allowing them to interact socially as a group.

Whilst largely monogamous, important social behaviours are performed. The three groups all engage in courtship feeding, in which males present food to females. Male kingfishers offer nuptial gifts of fish to the female, whereas in hornbills, the pair bond when the male offers high value items to the female rather than consuming them for himself. In both cases, this indicates the male's ability to provide, and can be very important behaviour to observe to determine pair compatibility. The diet of these birds will change during the breeding season, when more protein needs to be offered to initiate breeding behaviours. In the wild, many birds shift themselves to acquiring more prey items during the breeding season, so it is important to recreate this.

Toucans of the genus *Ramphastos* are seen to allopreen one another, taking care to avoid the eyes of their mate. The pairs also duet together, to strengthen pair bonds and defend territories. This is seen in a variety of kingfisher species, with the familiar call of the laughing kookaburra allowing family groups to defend their home range.

Hornbills, toucans, and kingfishers all nest in cavities. Lack (1954) stated that cavity nesting birds can be up to three times more successful than open nesting species in the same habitat, owing to the effects of predation on chick survival. All three groups make use of existing tree cavities for nesting, but some kingfishers also nest in riverbanks, excavating nest tunnels themselves.

Hornbills take this cavity nesting to the next level. The female bird seals herself into the nest cavity using a mixture of regurgitated food and her droppings, leaving just a slit for the male to pass food in to feed her and the chicks. The female will lay the first egg, which owing to the female's continuous presence starts incubating before the rest of the clutch is laid – resulting in hatching asynchrony. In many instances, only the largest chick will survive to fledge. The whole nesting attempt can take up to 140 days in some species. The adaptive function of this nest sealing behaviour is still highly debated. Some studies have been able to show that the female's presence in the nest reduces the effects of sibling aggression, thus increasing reproductive success (Finnie, 2018), whilst others suggest this is the ultimate in predator defence, with the female's stabbing bill acting as excellent protection. Another hypothesis is that by sealing herself in the nest, the female ensures her mate's fidelity as he must provide for her and the young. Indeed, in the most dimorphic of Asian hornbills, it is thought the male's colouration acts as an advertisement for his ability to provide.

In the wild, both hornbills and toucans will flock with others of their kind during the non-breeding season. Records of up to 200 great hornbills are noted as occurring. The young birds gathering here will be able to form bonds with others to form new pairs. This is very hard to replicate in captivity, owing to the space required. Some zoos have tried "dating" programmes for hornbills, with limited success but it has helped try to promote naturalistic behaviours with mate choice. The lack of breeding of many of these species (particularly kingfishers and larger hornbills) is often attributed to pair incompatibility, but there are doubtlessly more contributing factors.

Cooperative breeding has been noted in all groups. In the collared araçari (*Pteroglossus torquatus*) families roost in the nest until the eggs are laid, and only allowed to roost in the nest again after hatching. Bushy-crested hornbills (*Anorrhinus galeritus*) are also seen to frequently have helpers in the nest, but this behaviour is exaggerated in the ground hornbills in which the pair is always accompanied by young from previous years, in order to provide sufficient prey for the developing chick. Two eggs hatch, but the second chick always dies. Ground hornbills have very poor reproductive success with one study showing that pairs were only able to raise one chick every nine years (Wilson & Hockey, 2013). Cooperative breeding in zoos is hard to recreate, with many of the selection pressures that promote this behaviour having been removed due to captive management. The benefits of cooperative breeding are that juvenile birds will gain experience of rearing chicks before raising their own, hopefully resulting in greater reproductive success. Helpers are regularly seen in some wild toucan species, but captive birds can sometimes kill their first clutch of juveniles when going to nest again. To prevent this, keen observational skills are required to identify behavioural changes, and potentially separate the young birds. In ground hornbills, cooperative breeding occurs regularly, but unlike in the wild keepers can intervene to ensure that the second chick does not perish. Wild birds always hatch two chicks, but the second is never raised. Through an accessible nest box, the chicks can both be supplementary fed so that they survive to fledging.

Whilst hornbills and toucans share many morphological and dietary similarities, there are huge differences in their life history. Hornbills, after fledging, develop very slowly and it can take several years before birds are able to reproduce. They do not breed every year, and larger species can live to well over 40 years, with ground hornbills having been reported as living as long as 70 years. Uncommonly for bird species, ground hornbills will generally only breed every third year (triennial breeding). Toucans on the other hand live a fast-paced life. They normally breed around two to three years of age, can often rear two clutches per season, but even the large species rarely live for more than 20 years. In captivity, toucans generally breed earlier than their wild counterparts, potentially because they can form pairs rather than co-operatively breed in their social group.

14.3 ENCLOSURE CONSIDERATIONS BASED ON BEHAVIOURAL EVIDENCE

In the wild, these species all have a specific feeding ecology, and as such will occupy different trophic levels of the same habitat. The white-crowned hornbill (*Berenicornis comatus*) is a very territorial hornbill species, living in lower stories of the forest and taking more animal prey than the sympatric wrinkled hornbill (*Rhabdotorrhinus corrugatus*). Perching for the white-crowned hornbill should be provided lower in the aviary and should be more densely planted above. Canopy-dwelling species prefer to be able to sit on top of the planting, looking down.

Many kingfishers also prefer a good vantage point, such as a pole in the middle of the enclosure. Ground hornbills also readily utilise a log from which they can observe their territory. For other hornbills and toucans, perching should be provided across the aviary, positioned to best allow the birds to be able to fly, allowing them to replicate their long-distance foraging behaviour. Perches must be strong as birds can be heavy, and to allow for courtship and mating behaviours. The perches for all species will be used to clean the bill, and, in kingfishers, to hit prey against.

All birds benefit from UV lighting, but particularly those which live in the canopy. Metabolic bone disease can occur with insufficient UV, but also in birds deficient in vitamin D and calcium. Where possible, access outside allows birds to benefit from natural UV but if this is not possible (in colder climates), indoor UV lighting should be provided. Toucans can be delicate, but provided there is warm housing, many of these species can have outdoor access for some of year.

Tropical habitats do show some degree of seasonality. Sprinkler systems in aviaries can provide a "wet season" for many of these birds. Sprinklers, or hosing of the aviary, allow birds the chance to bathe, and can encourage breeding behaviour, with many zoos turning sprinklers on regularly for the month before the anticipated breeding season. Another cue to breed in captivity has been moving to a new aviary, which mimics natural behavioural biology, as wild birds flock in the non-breeding season, before settling in a territory for the breeding season.

Nest boxes are very important in the enclosure, as all species are cavity nesters, with toucans using the nest to roost in as well. This nest will be a small hollow log for the smaller species but can be a very large oak-barrel for the bigger hornbills. Information from wild birds can inform nest box design and placement. For toucans, to prevent egg breakage, nest boxes can be placed at an angle or have a very tight cavity, so the birds do not drop straight down onto the bottom of the nest. Wild hornbills seal into cavities which naturally have a layer of substrate inside (decaying plant material etc.), but some captive birds remove this when nesting. A concave base to the box prevents eggs rolling from under the female and not being incubated correctly.

Hornbills, kingfishers, and toucans are inquisitive birds that will prey on species much smaller than them. It is best to mix non-breeding birds, as a change in reproductive state can bring on a desire for meat, to the detriment of the resident passerines! In the wild, chestnut-mandibled toucans (*Ramphastos ambiguus*) have been seen hunting as a pair when breeding, taking a variety of prey (Mindell & Black, 1984). A few zoos have managed to successfully mix breeding toco toucans with other large species. Kingfishers are very aggressive to other birds, and several trauma-related deaths have been recorded. Some *Tockus* hornbills have been seen to form mutualistic relationships with dwarf mongoose (*Helogale parvula*) in the wild, and can be mixed together in zoos to create an active enriching exhibit, provided there are areas for both species to retreat into.

As well as being aggressive to different species, hornbills, toucans, and kingfishers can be aggressive to conspecifics. For the Guam kingfisher, some compatible pairs did not breed when able to see or hear other kingfishers. As quite territorial birds, being close to other individuals may cause more stress and reduce reproductive success. For some species of hornbill, the auditory cues of other pairs can increase reproductive success, with *Buceros* hornbills having bred with others of the same genus in close proximity. Perhaps this mimics the duetting between resident pairs in the wild? Visual cues can be a distraction, and it best to limit these for breeding birds. For birds which are imprinted on humans it can also be beneficial to limit contact with the public, where possible, to focus on their conspecifics. When planning an enclosure, the purpose of the enclosure must be carefully considered, as the requirements for a breeding enclosure may be very different from a public-facing naturalistic habitat.

14.4 BEHAVIOURAL BIOLOGY AND WELFARE

As animal keepers, one gets to know the suite of normal behaviours, often referred to as personality, for individual animals under our care. These changes to personality allow us to pick up on any welfare or medical issues. When hornbills, kingfishers, and toucans are unwell, they will change their appetite and willingness to eat, often reduce social behaviours, and sit looking fluffed, often low down. As with many birds, medical issues are often spotted when the condition has progressed too far, and therefore annual health checks can be a very important way to monitor weight and body condition. Weighing can be achieved more regularly by conditioning birds to voluntarily land on scales. This prevents the need to catch, which is stressful, and can cause physical damage to the bird. Training is a valuable technique for getting birds in and out of different parts of the enclosure, but in hornbills the birds can occasionally form a bond with the keeper that may affect the social behaviours performed with members of their own species.

Whilst stereotypical behaviour is not common with these birds, it is possible. Some hornbills can be seen to pluck themselves or their cage mate, whilst others can perform stereotypical movements (repeatedly moving along the same path in the aviary). It has been seen with carnivores that the larger the home range the more likely the species is to stereotype, and so it is to be expected that species with larger home ranges are more likely to stereotype.

For hornbills, resting behaviour, whilst normal at certain times of the day, may be an indicator of boredom. Increased activity is associated with increased interaction with cage mates, and this change in social behaviour may be conducive to increased breeding success. Records can be very useful to monitoring space utilisation and activity budgets that can be compared to wild data for these birds, which, in the main, is fairly well documented.

14.5 SPECIES-SPECIFIC ENRICHMENT

Hornbills, toucans, and kingfishers are very intelligent, inquisitive birds. Hornbills and toucans are known to have huge home ranges, throughout which they forage noisily in the canopy, looking for fruiting trees. The provision of food in a rodent-proof bowl, whilst easy for the keepers and excellent for disease prevention, does not replicate how hornbills forage in the wild – searching for food over the course of an entire day. At Cincinnati Zoo, a puzzle-feeder was used for the rhinoceros hornbill to increase the time the bird took to feed over the course of the day (Salmon, Rinderle, Davidson, & Russell, 2019). At Shanghai Zoo, the behaviour of three great hornbills was monitored before a change to the enclosure was made – in which a series of ladders, ropes, and toys were added, along with novel feeding devices. The hornbills showed a significant increase in foraging, moving, and preening, and reduced time resting (Jun et al., 2014). For ground hornbills, a carcass feed or scatter feed amongst substrate can increase the foraging time.

Owing to susceptibility of both hornbills and toucans to *Yersinia*, it is best to keep birds off the ground. Toucans can also be particularly sensitive to *Capillaria*, a nematode infection, which again mainly originates from substrate. When deciding on enrichment devices, it is best to keep these suspended (where possible) to reduce potential disease risk.

For kingfishers, the provision of live prey is a very effective form of enrichment. As the feeding of live vertebrates is prohibited in many countries, large insects can make an excellent substitute. The most commonly kept species, the laughing kookaburra responds well to hidden food items, in boxes or plastic toys, as well food being scattered across the aviary floor.

Many of the Asian hornbills that originally came into captivity were youngsters taken from wild nests. These birds imprint easily, and this has a significant impact on their behaviour. Some birds will be incredibly aggressive towards their human carers, whereas others will form very close bonds.

For the latter, an excellent form of enrichment is their use in free-flight demonstrations. Hornbills particularly enjoy live food, meaning that these high-value foods can be saved for training. Kookaburras and toucans (most often hand-reared), are also regularly used in free-flight demonstrations. This allows the birds to fly much greater distances than they can in most aviaries and spend longer foraging as they will collect food at irregular intervals from the keepers who are training them.

With all enrichment, it is important to record the behaviour before the device is added to the enclosure as well as afterwards to truly understand the effect of any device. It must also be said that enrichment is sometimes only enriching whilst it is new. For long-lived, intelligent birds, their ability to learn and remember how they have solved previous problems means that for caregivers, the job of crafting and implementing enrichment is as stimulating for us as it is for the bird!

14.6 USING BEHAVIOURAL BIOLOGY TO ADVANCE CARE

Hornbills, toucans, and kingfishers are popular, charismatic animals that engage well with zoo visitors. The greater the level of engagement, the more likely that person will go home being aware of the plight faced by these groups in the wild. Given the rate of deforestation across the tropics, (particularly in South East Asia), their importance as ambassadors cannot be underestimated. The understanding of their complex and fascinating behavioural biology, and its application to captive husbandry, supports the establishment of healthy, sustainable populations so that these birds can continue to be revered by many future generations of humans. Boxes 14.1 and 14.2 provide an example of how the zoo can help wild hornbill conservation and expand on pertinent sources of information for those wishing to know more about these birds.

BOX 14.1: Nest boxes – taking zoo learnings to the wild

The study of wild behavioural biology can enhance captive welfare and reproductive success, but research in a zoo can also benefit wild counterparts. Many Asian hornbills are declining due to habitat destruction, with fewer mature trees in which to nest. Hornbills have always proven to be difficult to breed in zoos. Taking information from wild behavioural biology of these birds advanced nest box design and positioning. The natural cavities that hornbills use can take a long time to form, being started by woodpeckers or a falling limb. A variety of species depend on these cavities for breeding or sleeping throughout the year. The cavity entrance is deceptively small, just enough for a female's head to pass through, but there is a huge range in the volume of the cavity itself, even within a single species' choice of nest-site. Many wild nests do not have a natural perch outside them, indeed in captivity this can sometimes be detrimental with males being able to reach in to destroy eggs. In many zoos, large oak barrels are used, often clad with a mock rock to make them look naturalistic, with a coir/bark substrate added to the bottom, mimicking the natural substrate found in cavities.

There are several excellent examples of nest box provision to wild birds, this example will focus on the HUTAN project in Sabah, Malaysian Borneo.

There are eight species of hornbill native to Borneo, all of which occur around the Kinabatangan River, many of which are in severe decline. Oak barrels are not readily available, so large plastic barrels clad in wood or wooden boxes were put up. The researchers in the HUTAN project have noted that hornbills may show a preference for boxes made with plastic drums rather than wood, and that after erecting these boxes it does take several years before hornbills use boxes, perhaps mimicking their affinity for mature tree cavities.

The key point here is to determine which aspects of hornbill behavioural biology are key to breeding success. The position in the tree is important, and the nest hole must be large enough for a female to enter, but not so big that mudding up becomes too time-consuming. Data loggers in the wild are monitoring temperatures and humidity experienced by wild birds, both in and out of the box, which can be compared to conditions experienced by captive animals. With some species of captive hornbills having poor reproductive success in zoos, these data may prove invaluable in providing the necessary stimuli to promote breeding in the zoo. Awareness of seasonality and potential cues from rain or periods of relative dry may also help. The fact that hornbills take several years to use the boxes indicates a need for patience from our perspective – unlike many other birds, they will breed only when they are ready (Figure 14.2).

Figure 14.2 Left: The box used for captive rhinoceros hornbills. Right: photos of the box under construction in Sabah, before being placed in-situ in the forest (photo: W. McLeod).

BOX 14.2: Additional reading on hornbills, toucans, and kingfishers

IUCN Hornbill Specialist group: https://iucnhornbills.org/

For detailed account of individual species, and conservation efforts: http://datazone.birdlife.org/home

The South East Asian Action Partnership focuses on conservation of endangered and evolutionarily distinct animals, including several hornbills and kingfishers: https://www.speciesonthebrink.org/

Hornbills in the City – a conservation approach to hornbill study in Singapore – Marc Cremades and Soon Chye Ng. An excellent book describing a long-term study focusing on the learnings of captive birds and application to in-situ conservation.

A good chapter that examines the behaviours associated with imprinting of these species: Witman, P. & LaGreco, N. (2020). Hornbills, kingfishers, hoopoes, and bee-eaters. *Hand-Rearing Birds*, 549–565.

14.7 CONCLUSION

Hornbills, toucans, and kingfishers provide zoologists with excellent subjects for behavioural studies. They are long-lived, intelligent birds which, in the main, have a wealth of wild ecological and behavioural data available. This chapter shows how knowledge of their wild behaviour can (and should) be applied to many aspects of captive management, e.g., nutrition, daily husbandry, and enclosure design, which ultimately results in improved welfare and breeding success. There is still a long way to go in order to totally understand the captive requirements of these birds, with some significant outstanding problems (e.g., the very poor breeding success of *Buceros* hornbills) a challenge; but with continued research in the field and in the zoo, the future looks positive for these charismatic birds where they occur in ex-situ populations.

ACKNOWLEDGEMENTS

I would like to thank Mr Wayne McLeod for his training and guidance, and Mr Mark Vercoe for his assistance with this chapter.

REFERENCES

Alexander, G.D., Houston, D.C., & Campbell, M. (1994). A possible acoustic function for the casque structure in hornbills (Aves: Bucerotidae). *Journal of Zoology*, *233*(1), 57–67.

Cuthill, I.C., Partridge, J.C., Bennett, A.T., Church, S.C., Hart, N.S., & Hunt, S. (2000). Ultraviolet vision in birds. In *Advances in the study of behavior*, pp. 159–214. Academic Press.

Finnie, M. (2018). *Conflict & communication: Consequences of female nest confinement in yellow-billed hornbills*. PhD thesis, University of Cambridge, Cambridge, UK.

Fry, H.C., & Fry, K. (2010). *Kingfishers, Bee-eaters and Rollers*. A&C Black, London, UK.

Jun, M., Jianqing, Z., Kangning, H., Jianyi F., Hongyun, S., Ji, X., & Qunxiu, L. (2014). Effect of environmental enrichment on the behavior of captive great hornbills. *Chinese Journal of Wildlife*. http://en.cnki.com.cn/Article_en/CJFDTOTAL-YSDW201401017.htm

Kinnaird, M.F., Hadiprakarsa, Y.Y., & Thiensongrusamee, P. (2003). Aerial jousting by Helmeted Hornbills Rhinoplax vigil: Observations from Indonesia and Thailand. *Ibis*, *145*(3), 506–508.

Kinnaird, M.F., & O'Brien, T.G. (2007). *The ecology and conservation of asian hornbills: Farmers of the forest*. University of Chicago Press, Chicago, IL.

Kinnaird, M.F., O'Brien, T.G., & Suryadi, S. (1999). The importance of figs to Sulawesi's imperiled wildlife. *Tropical Biodiversity*, *6*(1&2), 5–18.

Kuhl, H., Frankl-Vilches, C., Bakker, A., Mayr, G., Nikolaus, G., Boerno, S.T., Klages, S., Timmermann, B., & Gahr, M. (2021). An unbiased molecular approach using 3'-UTRs resolves the Avian Family-level Tree of Life. *Molecular Biology and Evolution*, *38*(1), 108–127.

Lack, D. (1954). *The natural regulation of animal numbers*. Oxford University Press, Oxford, UK.

Leighton, M. (1982). *Fruit resources and patterns of feeding, spacing and grouping among sympatric Bornean hornbills (Bucerotidae)*. PhD thesis, University of California, Davis, USA.

Martin, G.R., & Coetzee, H.C. (2004). Visual fields in hornbills: Precision-grasping and sunshades. *Ibis*, *146*(1),18–26.

Mindell, D.P., & Black, H.L. (1984). Combined-Effort Hunting by a Pair of Chestnut-Mandibled Toucans. *The Wilson Bulletin*, 96(2), 319–321.

Poonswad, P.A.N.V.P., Tsuji, A., Jirawatkavi, N., & Chimchome, V. (1998). Some aspects of food and feeding ecology of sympatric hornbill species in Khao Yai National Park, Thailand. *The Asian hornbills: Ecology and conservation, Thai Studies in Biodiversity*, *2*, 137–157.

Raman, T.R.S. (1998). Aerial casque-butting in the Great Hornbill *Buceros bicornis*. *Forktail*, *13*, 123–124.

Riley, J. (1997) *Biological surveys and conservation priorities on the Sangihe and Talaud islands, Indonesia*. CSB Conservation Publications, Cambridge, UK.

Salmon, H., Rinderle, J., Davidson, B., & Russell, F. (2019). Rhinoceros Hornbill Puzzle Feeder Enrichment. In *Undergraduate Scholarly Showcase*. University of Cincinnati, Cincinnati, USA. https://journals.uc.edu/index.php/Undergradshowcase/article/view/1223

Short, L., & Horne, J.F. (2001). *Toucans, barbets, and honeyguides: Ramphastidae, Capitonidae and Indicatoridae (Bird Families of the World)*. Oxford University Press, Oxford, UK

Wilson, G., & Hockey, P.A. (2013). Causes of variable reproductive performance by Southern Ground-hornbill *Bucorvus leadbeateri* and implications for management. *Ibis*, *155*(3), 476–484.

15

The behavioural biology of passerines

PHILLIP J. GREENWELL
Lieu dit Salce, Saint Georges, France

JONATHAN BEILBY
Chester Zoo, Chester, UK

15.1 INTRODUCTION TO PASSERINE BEHAVIOURAL BIOLOGY

Representing more than 60% of all known bird species, Passeriformes is the largest order of birds, (Dorrestein, 2009) with a broad ecological and biological diversity. They represent some of the most numerous of species, with population estimates of over a billion individuals (red-billed quelea, *Quelea quelea*, American robin, *Turdus migratorius*, red-winged blackbirds, *Agelaius phoeniceus*, and common starlings, *Sturnus vulgaris*), illustrating their ability to exploit changing or modified environments. Passeriformes also contains some of the rarest avian species, those most vulnerable to anthropogenic pressures, such as the Hawaiian crow, *Corvus hawaiiensis* (Extinct in the Wild), Cebu flowerpecker, *Dicaeum quadricolor* (Critically Endangered, population 60–70) and Stresemann's bristlefront, *Merulaxis stresemanni* (Critically Endangered, population 1–49). For some species, captivity is their last refuge from extinction.

Passerines are found on all continents except Antarctica, and across a range of habitats. Some have narrow ecological and dietary niches (Hawaiian honeycreepers, Carduelinae) while others can occur across several habitat types, particularly when migrating (e.g., Eurasian robin, *Erithacus rubecula* and common myna, *Acridotheres tristis*). As landscapes move from natural to urban, many passerines are successfully exploiting these newly created niches. Habitats naturally influence feeding systems, and these can vary greatly in passerines. Some are predominantly aerial hunters (swallows and martins, Hirundinidae), many are generalists,

DOI: 10.1201/9781003208471-17

and one family is aquatic (dippers, Cinclidae). Nectar, fruit, small land vertebrates, and seeds are all dietary items of specialised feeders within the passerine order. There are approximately eight different types of morphological or functional adaptations to passerine bills, for example, to successfully exploit these resources (see Table 15.1). It is important to note that despite the specialised bill structure, many species will have broad or varied feeding habits; for example, many predominantly seed-eaters will take also consume food and vegetable matter (particularly during breeding periods).

Table 15.1: Examples of feeding adaptations of passerines commonly kept in captivity and how this can be accommodated in zoo management for these species (adapted from Austin, Clench, & Frank, 2021)

Two species of wheatears, the Northern (*Oenanthe oenanthe*) and pied (*O. pleschanka*), have some of the longest migrations of all birds – 30,000 km and 18,000 km, respectively. To the other extreme, some species have reduced flight, becoming essentially flightless. New Zealand wrens (*Acanthisittidae*) are well-known examples, and this limited flight has quickly resulted in extinctions caused by introduced land predators. Some members of the South and Central American tapaculos (*Scytalopus*) are also considered essentially flightless and have restricted or localised range, with populations easily becoming easily isolated due to this flightlessness (Hermes, Jansen, & Schafer, 2018).

Passerines are distinguished from similar bird orders by their foot type; four well developed toes (three that face forwards and one to the rear to enable perching) in a conformation known as anisodactyly. Whether a trunk-climbing species such as a nuthatch (Sittidae), a terrestrial species of lark (Alaudidae) or the predominantly aerial swallows, all have the same basic anatomical foot structure, though with adaptations suited to specific ecological niches.

Social and reproductive behaviours are wide-ranging across the order. Smith's longspurs (*Calcarius pictus*) and Alpine accentors (*Prunella collaris*) are polygyandrous (males and females have multiple mating partners during a breeding season), pied flychatchers (*Ficedula hypoleuca*) and American redstarts (*Setophaga ruticilla*) are polygynous (one male mates with multiple females but each female only mates with a single male). Polyandry (one female mates with several males in a breeding season) is seen in dunnocks (*Prunella modularis*) but this species shows great plasticity in its reproductive behaviour, utilising monogamous, polygynous, polygynandrous, and polyandrous reproductive behaviours, depending on several factors). Across all bird species, monogamy is the most common strategy; in passerines, zebra finches (*Taeniopygia guttata*) and Northern cardinals (*Cardinalis cardinalis*) show strong mate fidelity for example. However extra-pair copulations and brood parasitism occur even in strongly monogamous species, with higher incidences observed in captive populations of some species than in wild populations (Griffith, Holleley, Mariette, Pryke, & Svedin, 2010).

Some passerine species employ lekking (where an aggregation of males come together, displaying against one another to attract visiting females) as a mating system. This behaviour has evolved independently five times within the passerine family, in the cotingas (Cotingidae), manakins (Pipridae), birds of paradise (Paradisaeidae), weavers (Ploceida) and a single species of bulbul (the Yellow-whiskered Greenbul, Eurillas latirostris). In these birds, increased sexual selection has led to the evolution of increased ornamentation in the males, with much stronger female mate choice. In most of these species, the male plays no part in rearing the young, and so the female has very drab plumage to camouflage her while incubating and rearing the young. Knowledge of the mating system of the individual passerines can have profound impacts on aviary design, stocking densities, management of non-breeding birds, and species management across the course of a breeding season.

Information on breeding behaviour is crucial to conservation action. Many passerines are managed in captivity and those species of high conservation importance are currently safeguarded in zoological collections. Wild populations have been supplemented through ex situ conservation breeding (and reintroduction) activities. In addition to being the most numerous order of birds, passerines also account for nearly half of extinct, threatened, and near-threatened avian species. Given their taxonomic weight and level of threat, this order is also the one most commonly translocated as part of conservation practices (Skikne, Borker, Terrill, & Zavaleta, 2020)- an activity where captive husbandry skills and knowledge are essential in success.

Table 15.1 Adaptations of passerines that should be considered in captive housing and husbandry.

Adaptation	Promoting feeding behaviour in husbandry and housing
Conical bill:	
Primarily seed eater. Bill used to crack the tougher outer coating of seed. With some exceptions, many seed specialists consume grass seeds which are either taken from the grass stem, or from fallen seeds on the ground as they are released from the plant. Examples include Gouldian finches (*Chloebia gouldiae*), Zebra finches (*Taeniopygia castanotis*), Java sparrows (*Lonchura oryzivora*) and canaries (*Serinus canaria*).	Smaller seed-eaters, encompassing many finch species, may be suitable for planted exhibits. Planting can include a wide range of suitable ornamental grasses which provide naturalistic feeding options, give additional shelter/nesting provisions, and may also attract beneficial insects. Where appropriate seeds can be scatter-fed across a range of substrate types (e.g., gravel/grass) to increase foraging time, while germinating seeds have a high nutritional value for breeding birds. Fresh or dried seed heads (e.g., millet) can be hung to encourage alternate feeding options and increase activity levels. Ground-feeding birds may feel more secure when provided with denser vegetation close by to shelter in, or with sentinel species in the same enclosure.
Frugivorous feeders: Generally have a wider gape for swallowing whole fruits, or more heavy bill for tearing apart fruits Examples include tanagars from the *Tangara* genus, bird-of-paradise, golden-fronted leafbirds (*Chloropsis aurifrons*) and the Asian fairy bluebird (*Irena puella*).	Fruiting shrubs, vines and other plants can be integrated into the aviary to encourage natural foraging behaviour, provide nesting sites, and attract insects. Plantings to support feeding variety/activity must be of appropriate size or texture to enable whole-ingestion (e.g., berries such as *Pyracantha* or currants) or soft enough to tear (e.g., papaya or pear). Larger fruits can be spiked to stabilise to enable tearing of the flesh.
Probing bill:	
Generally slim bill that curves downwards. Slender in species that search flowers for nectar or insects (e.g., some sunbirds; Hawaiian honeycreepers). Examples include purple honeycreepers (*Cyanerpes caeruleus*) and beautiful sunbirds (*Cinnyris pulchellus*).	Nectivorous species can be kept in well planted enclosures rich in flowering shrubs, perennials, and annuals. Thin stems of new growth encourage foot dexterity and perching instability, promoting natural feeding behaviours. Insect-attracting flowers also encourage natural hunting behaviour and supplement artificial diets.

(*Continued*)

Table 15.1 (Continued) Adaptations of passerines that should be considered in captive housing and husbandry.

Adaptation	Promoting feeding behaviour in husbandry and housing
Toothed bill: Tip tends to be hooked, notched along upper mandible sides; ability to tear up larger prey or fruit. Examples include the shrikes (Laniidae), Australian bell-magpies (Cracticidae) and some tanager species.	Though rarely kept in captivity, shrikes present interesting feeding behaviours that translate into unique enclosure design considerations. Well-known for impaling prey on thorns or pointed material, such opportunities should be present in aviaries. Impaling appears innate, and motivation is high to perform the behaviour. Food caching influences sexual selection in some species of shrike, therefore provision of suitable larder sites may be important in breeding success. Aviary cohabitants, if any, should be chosen with care for these predatory birds. Aviary height has been deemed important in captive-breeding for release situations, with heights of approximately 3 m considered suitable to accommodate the hunting/feeding behaviour of such species.
Insectivorous feeders: No specific bill type as varies with main prey items taken., swallows have shorter finer bills with a wider gape. Bills are generally lightly built, though dependent on principal prey items. Examples include rainbow pittas (*Pitta iris*), verditer flycatchers (*Eumyias thalassina*) and dusky wood swallows (*Artamus cyanopterus*).	Aerial insectivorous, such as wood swallows, should be offered feed from an elevated platform, as prey is rarely taken from the ground. Sufficient space is required for flight and aerial manoeuvring, lightly planted enclosures with insect-attracting plants, particularly taller shrubs or climbers, will increase natural feeding behaviours as well as dietary variation. Knowledge of species-specific hunting strategy is key in food presentation, whether prey is taken from foliage, ground or airborne. For ground-foraging species (e.g., pittas), a deep bed of leaf-litter can encourage foraging behaviour when live food is hidden amongst it. Most insectivores will transition to frozen or freshly killed live foods after some time in captivity, though hunting for prey, as well as a variable or unpredictable feeding times, will support welfare needs, and increase foraging time. Active fruit fly cultures can be kept in enclosures for example, a species favoured by many small insectivores.
Omnivorous feeders: Unspecialized in shape and function but usually strongly built. Examples include Javan green magpies (*Cissa thalassina*), Hawaiian crow (*Corvus hawaiiensis*) and many jay species.	Rich, complex enclosures are recommended for members of the corvid family. Naturally intelligent and curious, puzzle feeders can be used successfully for many species. Similarly hiding/caching food in suitable material can also stimulate foraging behaviour. Whole carcass feeding can engage birds for longer periods of time in a naturalistic manner, relative to natural prey type and within hygiene parameters. Live foods which encourage hunting behaviours can be offered on unpredictable feeding routines, using enrichment devices to deliver prey items.

Zoos, and the knowledge derived from captive populations, have been instrumental in safeguarding wild populations. For critically endangered populations where knowledge may be lacking in basic biology and behaviour, and where the establishment of an ex-situ population may be high risk, close congeners of a less threatened status have been kept in captivity for aviculturists to gain valuable, transferable management skills (Kerr, 2021). Notable examples of this practice include the critically endangered Puaiohi (*Myadestes palmeri*) of Hawaii and the vulnerable ʻŌmaʻo, *Myadestes obscurus* (Tweed et al., 2006), and the chestnut-capped laughingthrush (*Pterorhinus mitratus*) acting as a "model" for the Sumatran laughingthrush (*Garrulax bicolor*) (Owen, Wilkinson, & Sozer, 2014). The latter is particularly noteworthy, with the European Association of Zoos & Aquaria (EAZA) launching "Silent Forest," a campaign to tackle the Asian songbird crisis, where some endemic species have suffered population declines due to the wildlife trade.

Passerines are generally well-represented in most zoological collections, though diversity across the order may be limited given its breadth. Unlike many larger mammal species, birds are generally easier to provide adequate space for, with aviaries being a common feature of most collections. Generally, Passeriformes have small body masses and are therefore more easily housed when compared to larger avian species that may require more substantial housing. It is within the constraints of captivity that the understanding of the behavioural biology of the occupants often means the difference between successful husbandry and failure. Species, despite belonging to the same order, might have dramatically different requirements. Aviaries, as static environments, might not be suitably designed (or planted) to accommodate the needs of one species if it is to replace that of another. For example, a grassland species may not feel at ease in a heavily planted aviary, and similarly so for a canopy species in a sparse and open enclosure. Biologically relevant aviary furnishings will provide for, and therefore induce, more species-appropriate behaviours, particularly when used in conjunction with correct food presentation, social grouping, sites for roosting and nesting and, if applicable, suitable cohabitants.

15.2 PASSERINE ECOLOGY AND NATURAL HISTORY RELEVANT TO THE ZOO

Zookeepers should have a sound understanding of the natural history of the species under their care as it is only through the application of this knowledge that welfare and husbandry can be maintained and improved. Advances are being made constantly to our understanding of avian behaviour and biology. Members of the *Cissa* genus, including the critically endangered Javan green magpie (*Cissa thalassina*), frequently fade from vivid green to dull turquoise in captivity. Plumage colour is known to be a sexual signal in many bird species, an indicator of mate health or infers a degree of protection from predators. Supplementation of lutein (a yellow carotenoid) via gut-loading of insects at the breeding facility for the Javan green magpie resulted in a return to the vibrant green plumage that is seen in wild populations. Such information can be disseminated to collections housing *Cissa* species to support dietary (and husbandry) review.

Carotenoid pigments are also a visible indicator of fitness in some species, e.g., blackbird (*Turdus merula*) where the more brightly coloured the bill, the lower the parasite load. As carotenoids are used in the immune response, a male bird with a brighter bill suggests that he is fit enough to partition dietary carotenoids away from immunocompetence to sexual signalling, thereby increasing his potential reproductive success. This may apply to the Javan green magpies, and while turquoise birds have bred in captivity, the breeding success is much higher for birds with green plumage. In the green magpie, plumage pigmentation is considered unstable and may fade in bright sunlight; the EEP (European endangered species breeding programme) for the Javan green magpies recommends that aviaries are covered with shade cloth to prevent this from occurring. As a species from the forest floor of evergreen forests, light penetration is low and therefore light-excluding material imitates this effect in captivity (Owen, 2019) and further replicates the natural habitat of the species (Figure 15.1).

Observations on free-living communities can suggest potential improvements in captive situations. Aviaries are often themed on a geographical basis. In the wild, Gouldian finches (*Chloebia gouldiae*) readily associate with another Australian endemic passerine, the long-tailed finch

Figure 15.1 Enabling behavioural biology and diversity through aviary design. The aviaries above represent suitable species-specific housing for the Javan green magpie (Left) and Gouldian finch (Right). The green magpie enclosure supports natural foraging through the use of deep leaf-litter as a substrate and areas of denser and thinner vegetation. Sturdy and secured perches are placed to enable and encourage flight, with varying diameters and angles. The roof of the enclosure filters direct sunlight, representing similar conditions to forest floor ecosystems and helps retain the carotenoid pigmentation which can fade under bright sunlight. Screens are mounted between aviaries to reduce conflict with neighbouring pairs. Natural nestbuilding can occur in the vegetation, with free choice of location. Food can be spiked onto branches to enable the birds to tear off smaller sections as would be seen with wild counterparts (Photo: A. Owen). The Gouldian finch aviary offers a greater level of light, and therefore heat, to enter, with a dappled shade to permit the occupants to find microclimates for thermoregulation. UV light is important in sexual selection. Gouldian finches are renowned as heat-loving birds and sheltered suntraps permit basking opportunities, important for plumage maintenance. Live plants provide natural foraging opportunities, while flushes of seed production replicate elements of the natural environment and add additional nutrients throughout the stages of maturity. Leaves also offer nest building material and plants provide cover from dominant individuals as well as shaded/sheltered areas. Natural perches are placed at various heights and in areas of different luminosity. The spacing of perches provides ample flight space throughout the enclosure and between resources. Food and water resources may be wall-mounted or floor-based, permitting birds to move between feeding areas, and distribution of food ensures that dominant individuals cannot monopolise a resource (photo: P Rose). Both aviaries are constructed to suit local climate conditions and are likely to be in the natural temperature range of the species housed. In colder climates, heated and sheltered accommodation is provided, possibly with supplementary lighting to extend feeding hours in northern hemisphere locations.

(*Poephila acuticauda*) and with black-faced wood swallows (*Artamus cinereus*) that act as sentinels for predator presence. Gouldian finches will refrain from descending to the ground to drink until the long-tailed finches decide to land and drink (O'Reilly, Hofmann, & Mettke-Hofmann, 2019). Despite this association, long-tailed finches are an aggressive nest site competitor (Pearce, Pryke, & Griffith, 2011). In a large enough aviary with multiple nesting sites to choose from, long-tailed finches may be a suitable mixed-species cohabitant, as their presence may make the Gouldian finches feel more secure when feeding on the ground. However, within the confines of an aviary, any harmonious mixing may depend on individual personalities as well as species-specific requirements. Australia has many small passerines that may be more ideally suited for a Gouldian finch mixed-species enclosure, if accurate biogeographical representation is the basis for such an exhibit (Figure 15.1).

Many mixed-species support compatible communities until the breeding season begins. Competition for resources can be particularly intense in a closed environment. Considering breeding strategies of the occupants are important; communal/colony breeders, pairs that isolate and cooperative breeders all require different approaches to aviary set-up and inhabitants. Cooperatively breeding species, where one male is dominant over subordinate males such as the white-browed sparrow weaver (*Plocepasser mahali*), may see increased aggression if the lead male weakens or dies, though this may only be a temporary phase as a new dominant male becomes apparent (Wingfield, Hegner, & Lewis, 1992), therefore the continual removal of the "aggressor" to manage the group will continue to create antagonistic interactions. Not all cooperative breeders exhibit aggression upon the loss or removal of a flock member; the white-browed babbler (*Pomatostomus superciliosus*) shows significant non-confrontational social restructuring on the removal of individuals from a captive population (Oppenheimer, 2005). Alloparental care, where an individual feeds dependent young that are not its own, appears to be a strong indicator in mate or collaborator selection for this species, and allopreening and allofeeding were also noted at high frequency.

For many cooperatively breeding species, in captivity the first clutch of juveniles can often be left in the aviary while the second clutch is being raised, provided the aviary is large enough. Careful observation is required, but in some species such as laughingthrushes (Leiothrichidae), juveniles can gain valuable experience and participate in the rearing of young, forming closely related groups as would be seen in the wild. These birds are often able to be kept together outside of the breeding season, but changes in behaviour will be noticed when the breeding pair attempt to nest again, and in most cases it is advisable to remove the previous year's juveniles at this point.

If maintaining genetic diversity is the aim of a breeding programme, then knowing the potential breeding lifespan of the species can prevent genetic over-representation of certain individuals. African village weavers (*Ploceus cucullatus*) can have a reproductive lifespan of approximately 15 years in captivity (Collias, Collias, Jacobs, Cox, & McAlary, 1986). If one male supresses or dominates other males, successfully out-competing other males in attracting females, decreased genetic diversity of the resulting population can be seen. Some songbird species can live and reproduce into their twenties. Senescence (the processes of ageing) in captive populations is poorly studied for passerines, even though many songbirds have a life span of approximately 15 years (Dorrestein, 2009); understanding reproductive output and behavioural change associated with ageing can help support end of life care as well as well as more peaceful species continuity.

15.3 BEHAVIOURAL BIOLOGY AND PASSERINE WELFARE

Disputes and conflicts in captive flocks or between individuals can be reduced by spacing resources around an enclosure and offering refuge areas for shier birds, to avoid harassment. Behaviours also fluctuate across seasons, with placid birds becoming highly defensive during breeding – defending nest sites vigorously or chasing, harassing, or attacking potential mates (or competitors) excessively. In mixed-species aviaries, some species might be tolerated during this time, others may not, and similarly, some species that congregate outside of breeding might need larger territories during nesting, resulting in aggression if this extra space is not provided. Enclosure and resource use can also change, resulting in unrest or dispute within and between species. For example, a pair of Raggiana bird-of-paradise (*Paradisaea raggiana*) can be mixed in a free-flight exhibit until the breeding season commences when the female will hunt and kill birds smaller than herself.

Birds are renowned for masking illnesses, often only showing physical signs once the problem is far advanced. Aviculturists with a keen understanding of the normal behavioural pattern of their charges will be able to detect any slight changes in behaviour and ideally intervene to treat the problem rapidly. Annual health checks, body condition scores, precise and documented health protocols and treatments can all assist in keeping passerine populations in good health.

Behavioural changes can be seen throughout extreme weather events. For example, in a mixed-species aviary in Adelaide, Australia, behavioural changes were noted across a species, with some birds spending more time in shaded areas or on the substrate during heatwaves (Xie, Turrell, & McWhorter, 2017). Microsites in enclosures, i.e., areas which offer small changes in environmental conditions such as temperature, luminosity, or humidity, allow for behavioural diversity. Allowing multiple microsites (where small environmental gradients such as heat or humidity can be achieved) to be present (i.e., via running and still water, use of sprinklers or misters, varying shade density, differing air currents, and different substrates) offers individual birds options in how to best achieve thermoregulatory needs. Basal metabolic rate and body temperature are usually higher in passerines than non-passerine birds (Dorrestein, 2009); taking this into consideration across all husbandry scenarios, such as weather events or reproductive cycles, permits caregivers to ensure that adequate resources are available. To ensure metabolic requirements are met across seasons, multiple heated areas in cold weather, or additional water and food resources in times of heat or cold stress, should be presented.

Larger birds, such as corvids, can be trained to step onto scales for weighing. Alternatively, perches can be placed within the enclosure and attached to weighing devices so that for smaller birds, mass can be recorded discretely. Birds can be lured to land for weighing by having a favoured food item placed next to the scales.

Members of the corvid family are intelligent and curious and may interact with artificial devices such as those made for parrots or even for human children. Corvids can be offered problem solving devices or tools to access food rewards, thereby increasing foraging time and providing opportunities for enhanced welfare states. However, corvids are also known to be neophobic and birds bred for conservation purposes should be managed to retain neophobic responses so that they are better suited for release. In a dynamic setting, such as a busy zoological park, prolonged neophobia can be detrimental to individual welfare states and most species on display quickly habituate to novel objects in or around their enclosures. Some behaviours may be counterproductive in captivity (i.e., neophobia, heightened aggression, strong flight impulse), inducing chronic stress and thereby causing welfare concerns. Zookeepers must strive to find the balance between maintaining behavioural diversity across multiple generations against that of individual bird welfare. Ultimately the destination of the species, or individual, (e.g., to supplement captive populations, or to be part of an in-situ release programme) will influence how this balance is decided upon.

The Hawaiian crow, or "alala," (*Corvus hawaiiensis*) demonstrates an unusual neophobic response whereby neophobia is higher in younger individuals. This goes against the norm of object neophobia in most other species, where young animals begin to explore their surroundings and are generally more neophilic to support this exploratory behaviour. This species has a heightened neophobic response, with latency to feed in the presence of new stimuli exceeding that of all known responses from other passerines and other avian or mammalian object neophobia studies (Greggor, Masuda, Flanagan, & Swaisgood, 2020). Knowing how species respond to potentially threatening stimuli can be a key factor in release and reintroduction success, however it must be tempered with the need to explore novel environments and potential food items. Training programmes can be instigated to expose birds to particular stimuli as either a deterrent (i.e., predator) or attractant (i.e., novel food resources) to improve and shape behavioural diversity that is important to conservation outcomes.

15.4 SPECIES-SPECIFIC ENRICHMENT FOR PASSERINES

Aviaries designed with the natural habitat of the species in mind will give a sound foundation for appropriate enrichment options. To promote the widest range of behaviours possible enclosures should offer the most feasibly varied landscape; areas of shade and light, denser planting, and free-flight space all need to be considered. Perches should be varied – some stable, others flexible, and with multiple diameters and textures. Perches at varying heights and distances from each other promote movement and activity, as does placing feeding resources throughout the enclosure. Substrate, particularly for ground-feeding or ground-dwelling birds, should also offer similar diversity (Figure 15.2). Gravel areas, for example, are ideal for scattering seed for finches to forage, bark areas can be turned over by forest floor feeders and may attract small invertebrates, as would areas of grass or low-growing ornamental plantings. Supplementary heat may be needed depending on climate and species, and many smaller passerines will bask under heat lamps and enjoy free movement throughout temperature gradients.

Hygiene must always be a key consideration of avian care. Dietary needs and therefore excrement production will influence enclosure design. Frugivorous or nectivorous birds may need a principal substrate that is easily cleaned, due to their copious and/or liquid droppings. An enclosure for such species also needs to offer areas of mixed substrates to encourage feeding behaviours. Substrates can harbour parasites and consideration must be given to of the trade-off between enhancing behavioural diversity and supporting optimal health.

Figure 15.2 Aviary design and perches are influenced by the needs of the occupants and how they move through their habitat in the wild. A Javan green magpie (top left) perching vertically, a behaviour seen in this species and likely influenced by forest floor vegetation which grows tall and narrow as it reaches towards the light. This species tends to flit rapidly from perch to perch in search of prey, and as such branches and other perches must be stable enough to take this impact. Top right illustrates a red-headed Gouldian finch feeding on seeding grasses. Flexible stems like this encourage dexterity and activity as the balance is continually challenged. Presenting food items in this manner increases foraging and feeding activities and can be easily replicated in a captive environment. Seeding grasses and flowering plants are also likely to attract insects, a supplementary feed source for many passerines, as well as provide nesting material (Photo: G. Postle). African pied crows (Corvus albus) (bottom left) require sturdy branches that can take the weight of the birds, though flexible and smaller branches also support foot health. In most situations, natural branches, which offer varied diameters and bark textures are preferable to smooth man-made plastic or wood equivalents, and while possibly less easy to clean, are easily replaced as they become soiled or brittle. Fresh branches must always be cleaned with a suitable disinfectant prior to use to reduce the risk of contamination from wild birds. A Javan banded pitta (Hydrornis guajanus) on a soil and sand-litter (bottom right). Ground-dwelling birds need careful management and maintenance in aviaries, with a balance of humidity, drainage, and airflow to prevent stagnant and overly wet ground. However foraging activities are easily managed with the addition of deciduous leaves piled into the enclosure. Logs, stones, and smaller branches can be placed on the floor to provide additional surfaces for these species. Soil and leaf-litter enclosures are ideal for natural plantings, with the humus continually improved through the addition of mulch, natural leaf fall, and bird faeces.

Areas under perches or feeding zones should be easily cleaned for example, while other zones of substrate may be managed under roofed/covered sections to avoid contamination from wild bird faeces or to prevent damage from weather.

Most passerines are social. Keeping appropriate pairings or together as a flock will ensure that individuals can interact with other members of the species and a social environment will allow for personality types to have their own niche within the group. Where free-mate choice is permitted breeding results are likely to improve. Mate selection, nest site selection, nest building, and raising a brood all promote behavioural diversity.

Planted enclosures, or dense plantings around aviaries, attract insects which many species will naturally predate, offering additional nutritional benefits at the same time. Live foods can be offered, particularly for omnivores and insectivores, and whole carcass feeding (e.g., "pinkies," small vertebrate species) is suitable for some omnivores. Critically endangered blue-crowned laughingthrushes (*Garrulax courtoisi*) show increased foraging activities when food was presented in a more naturalistic style, i.e., out of the food pot and presented on the ground, amongst foliage or suspended from branches. Intake of certain food items increased when they were presented in a different manner (Daoqiang et al., 2017). This species, a cooperative breeder, has increased foraging frequency when housed in larger groups compared to smaller ones, suggesting that flock size as well as social structure promotes more naturalistic feeding behaviours.

Care must be taken with some species when housed in planted aviaries that contain plant species that produce fruits or berries. Certain species of birds, notably mynas and starlings (*Sturnidae*) and bird-of-paradise and other frugivorous species, are sensitive to iron storage disease (ISD), also referred to as haemochromatosis. ISD causes iron accumulation in the liver, generally resulting in the death of the bird unless identified and treated early. Fruits high in vitamin C, such as kiwis, blueberries, mango, tomatoes, and sweet peppers, may influence iron absorption and should be avoided when planning aviaries and designing diets for iron-sensitive species (Cork, 2000).

15.5 USING BEHAVIOURAL BIOLOGY TO ADVANCE PASSERINE CARE

Understanding behavioural biology in wild populations enables the transfer of knowledge into managing situation to give greater insight into the suitability of care. Behavioural research in non-zoo populations (private aviculturists/research facilities/conservation centres, etc.) yields new information that can be applied to zoo populations. To understand these ideas, recent research into Gouldian finch behaviour and ecology and the application of these findings to husbandry and management routines will be evaluated as a case study.

15.5.1 Case study on Gouldian finch behaviour and husbandry

One of the most distinctive and colourful of all birds, the Gouldian finch has, unsurprisingly, been a popular zoo and avicultural standard for decades. Endemic to northern Australia this finch is of conservation concern (listed as Near Threatened on IUCN Red List as of 2022) and is subject to in-situ conservation action plans. The species experienced a population decline in the 1970s but it is unclear what the contributing factors of this were. Australia banned the export of its native species in 1960, therefore the current population outside of the native range is derived from multiple generations of captive-bred birds. Commercial trapping for the domestic market ended in the 1980s, and the species is well established as an aviary bird in its native country and internationally. Multiple colour mutations have been bred by hobbyists, and our understanding of this species in aviculture (i.e., husbandry, reproduction, genetics) is therefore well established as a result.

Gouldian finches inhabit open tropical woodland in northern Australia, and are mainly granivorous, though invertebrates are occasionally consumed. These finches are highly mobile, possibly nomadic, in their search for ripe seed availability. Seeds are taken from the ground, or the finches balance on grass stems to access unripe/unfallen seeds. Populations are impacted upon and influenced by fire-burning regimes and grazing densities of livestock. Fire burning alters the quantity and quality of grasses, and therefore the Gouldian finch population as a result. Fires also create mosaic habitats, which offer grasses ripening at different times, and as such fires are managed within the range of this species as part of their ongoing recovery plan. These areas are favoured for breeding sites as Gouldian finches mostly feed on open burnt ground, which exposes seeds of annual grasses. Habitat and feeding resources change between dry and wet seasons, furthering complications for in-situ monitoring as the birds move in search of food.

Gouldian finches travel in flocks, however, these are often composed of juveniles as the natural lifespan is relatively short – approximately 5 years. Wild Gouldian finches are polymorphic, with three head-colour types being described: black (the most common, estimated around 70–80%), red (around 20–30%) and yellow being a rarity (approximately 0.03%), though these percentages change slightly between male and female birds. Populations as a whole have a male bias, and there is a genetic incompatibility between the three existing polymorphs. The Gouldian is the only Australian finch that is an obligate cavity nester, constructing a crude nest in a hollow of a limited number of native tree species. Nest cavities are a naturally rare resource, and competition for them is high, particularly with the long-tailed finch.

Outcomes from field studies, such as those detailed above, can be added to evidence gained from captive populations. For example, research conducted on captive Gouldian finches suggests that, after multiple generations, captivity has had a negative consequence on the genetic diversity and fitness of the birds, with 32–48% reductions in genetic variation noted when compared to the wild population (Bolton & Griffith, 2021). As behavioural richness decreases with duration in captivity, it is important to note that some behaviours may have been reduced or deselected over the previous decades of captive breeding. Carrete and Tella (2015) suggest that captivity induces a rapid decline in anti-predatory behaviours in some passerine species and this is important information for those wishing to use captively-bred birds for wild population supplementation programmes.

To maintain behavioural and genetic diversity in captivity, flocks that represent natural populations should be held, with a larger number of males than females (to support free-mate choice) and an appropriate ratio of head-colour morphs. It might not be immediately clear to see the relevance of head-colour selection on zoo management strategies, or indeed any behavioural biology implications. However, research shows that head colour is linked to personality and fitness traits. Red-headed colour morphs tend to be more aggressive and vigilant compared to black-headed morphs. Black-headed morphs are bolder and more neophilic. In the wild, individuals typically assort with mates of the same head-colour type (Williams, King, & Mettke-Hofmann, 2012). Maintaining this social system in captivity, with frequent ingress of new individuals to bolster genetic diversity, can help counter against inbreeding depression, and also provide a naturalistic display that presents an opportunity for education about, and research on, the role of polymorphism and genetic diversity in managing population-level behaviours.

Free-mate choice is important in breeding Gouldian finches. Females paired with non-preferred males showed stress responses substantially higher than females who chose their own mate. Females choose a male based on sexually dimorphic traits (greater UV contrast of the blue collar around the head mask and longer pin-tail feathers) whereas males select a female of the same head colour as themselves. Breeding between colour morphs results in reduced hatching and survival of young (Pryke & Griffith, 2009).

Gouldian finch aviaries should be large enough to hold a colony of several pairs, plus surplus males. If kept inside, suitable lighting should be available to enable the female to select a preferred male based on key sexually selected traits. Lighting for birds is specifically designed to provide the correct balance of UVA, used in identifying mates and ripe foods, and UVB, used to produce vitamin D. Nest boxes should be greater than the number of potential pairs to support free nest site selection, and to reduce competition and offer nesting opportunities for other aviary occupants (if present). Ideally, nesting sites will be based on the characteristics of cavities found in the wild. Researchers in Australia have replicated these characteristics to create supplementary nest boxes to bolster wild populations (Brazill-Boast, Pryke, & Griffith, 2013). Hollows in trees permit very little light to enter the interior – the hypothesis for the iridescent gape markings seen in chicks. The design of the box should also ensure that chicks are unable to fall out by accident and can only fledge at the correct moment of maturity when they are mobile enough to access the elevated opening.

Aviaries should be appropriately designed in respect to the natural habitat where possible (see Figure 15.3); scattered trees with a grassy understorey, including grasses like the endemic sorghum (*Sarga* spp.) to demonstrate natural feeding strategies and provide nutrient-rich fresh forage. In general *Eucalyptus* spp. do not cast heavy shade due to their open branch structure and narrow leaves, therefore heavy shade should be limited, which in turn would facilitate grass growth as an underplanting. If planting is not possible, then cut branches offer a suitable alternative, placed inside holders to emulate the natural habitat structure. Mosaic-type habitat structures can be replicated in aviaries with varying substrates to encourage foraging behaviours. Additional feed or part of the daily ration can be scattered across open gravel/sandy areas to increase time spent on natural behaviours. Grassed areas can have staggered maintenance, to allow grasses to seed naturally or to create a short sward to encourage foraging. Offering different seed mixes in rotation and in addition to the standard diet, as well as providing fresh or dried ripening grasses (either strewn across substrate or held upright in clamps), can help replicate the seeding events that are important to the daily rhythms of wild finches. Suitable planting also offers opportunities for individuals to evade others if needed. Larger areas, clear of perching, encourage flight – particularly if key resources are distributed throughout the enclosure.

Explanation of Figure 15.3:

Eucalyptus trees provide height in the aviary, allowing for vigilance behaviour from a vantage point; the trees provide a light dappled shade which enables thermoregulatory behaviours and also permits plants to grow under the canopy. The growth of eucalyptus is generally fast and regular cutting of branches for maintenance encourages slender new branches, which provide multiple diameters of perches, improving foot health through regular changes in position. The trees are dispersed throughout the aviary to encourage flight and multiple trees enable birds to assort within their preferred groupings.

Several grass species are planted through the aviary, with different heights and flowering periods. A key aspect of the wild behaviour is feeding from seeding grasses, and mobility is encouraged as birds perch on slender stems to feed, as well as providing

Figure 15.3 Illustration of an ecologically relevant Gouldian finch aviary, incorporating key behavioural biology information of this species into the design.

additional nutritional benefits. A shorter sward is also maintained into which feed seed can be lightly scattered to encourage foraging behaviours; this can also be done in areas of the sand and gravel substrate which surrounds the plantings or on the access path. A sandy gravel mix enables birds to take small particles into the gizzard to help breakdown seeds.

Larger shrubs, ideally flowering native Australian species, provide further perching opportunities as well as denser canopies which may be used to create microsites for thermoregulation as well as refuge against aggressors in the aviary. Flowering plants may also attract insects and have pollen, which may also be consumed.

A shallow water source with pebble embankments and margins permits natural bathing opportunities, important for overall plumage health. Other aspects of the enclosure such as rocky outcrops and termite mounds help embed the concept of an Australian aviary and provide habitats areas for other suitable aviary inhabitants.

Maintenance must also be considered; the ease of cleaning the water source for example, as well as horticultural practices must all be part of the overall design ethos. Principle feeding and watering areas, along with suitable shelter and nesting sites, are not illustrated here. However, nest boxes can be placed throughout the enclosure, either by fixing to aviary walls, on posts, or secured to larger tree trunks.

BOX 15.1: Future research questions to further support evidence-based husbandry using behavioural information

Do mixed aviaries with county-appropriate species (wood swallows, long-tailed finches) alter daily behaviour patterns in Gouldian finches, such as vigilance and feeding behaviours? What other naturally occurring species associations or interactions can be drawn from for accurate representation in captivity and possible welfare enhancement for other species?

How are microsites, where small degrees of temperature differences exist, used by aviary inhabitants on hotter or cooler days? Knowledge of how captive birds respond to increasing temperatures may be pertinent to appropriately managing climate change predictions in captivity.

Which species present behaviours that can contribute to better care? Many species will take live food from the hand – can this be modelled onto a weighing scale so that weight monitoring can be part of the care strategy?

What species selection works best in multi-species aviaries? Can species be selected to occupy specific niches within the aviary to capitalise on limited space yet provide a harmonious collection of birds of conservation or educational merit?

How does flock size influence behaviour and reproduction? Are larger groups less neophobic than smaller ones? How can we determine the best flock size for each species kept?

How do we manage flock fluctuation to separate birds in breeding seasons and then form creches for new young to develop important socialising skills and for natural mate selection?

BOX 15.2: Further information to inform passerine care

Detailed information and recommended breeding practices for many songbirds threatened by the trade in Indonesia, with transferrable information to similar, less threatened species can be found at: https://www.silentforest.eu/

Excellent range of videos and field reports from finch conservation projects in Australia, with useful images of habitat and other species, at: https://www.finchesqueensland.org/

A range of live camera feeds across multiple habitats and species, the Cornell University bird cams feature a host of videos and species and discussions on interactions captured by the cameras. See: https://www.allaboutbirds.org/cams/

15.6 CONCLUSION

In the wild, passerine species must make multiple choices every day; daily choices may be curtailed in captivity, but behavioural needs will not be. Ensuring that aviaries are richly designed, meeting the behavioural and ecological needs of the occupants, supports good avian welfare and improves breeding success. Natural social groupings permit free-mate choice; a range of microhabitats allows individuals to regulate body temperature; while larger components of the aviary, such as shrubs and denser plantings, allow birds to avoid aggressors or stressful interactions. Distribution of food or nesting resources can reduce conflict and encourage movement through the enclosure to maximise the useful space available. This chapter has highlighted how the care that is provided to captive passerines is continually evolving as we learn more about their behavioural biology, alongside advances in genetic and nutritional studies. Zookeepers and other care providers are gently reminded to continually learn about their charges, to engage in research, and to contribute to the furthering of our knowledge of captive passerine care.

REFERENCES

Austin, O.L., Clench, M.H., & Frank, G. (2021). Passeriform. *Encyclopaedia Britannica*. https://www.britannica.com/animal/passeriform

Bolton, P.E., & Griffith, S.C. (2021). Evolution in aviculture: Loss of genetic diversity and head-colour morph frequency divergence in the domesticated Gouldian Finch (*Erythrura gouldiae*), *Emu – Austral Ornithology, 121*(1–2), 55–67.

Brazill-Boast, J., Pryke, S.R., & Griffith, S.C. (2013). Provisioning habitat with custom-designed nest-boxes increases reproductive success in an endangered finch. *Austral Ecology, 38*, 405–412.

Carrete, M., & Tella, J. (2015). Rapid loss of anti-predatory behaviour in captive-bred birds is linked to current avian invasions. *Scientific Reports, 5*, 18274.

Collias, N.E., Collias, E.C., Jacobs, C.H., Cox, C.R., & McAlary, F.A. (1986). Old age and breeding behaviour in a tropical passerine bird *Ploceus cucullatus* under controlled conditions. *The Auk, 103*, 408–419.

Cork, S.C. (2000). Iron storage diseases in birds. *Avian Pathology, 29*(1), 7–12

Daoqiang, L., Zhiyong, W., Hailing, H., Dongtao, L., Xiaohong, W., & Zoo, N. (2017). The effects of hung up fruit feeds on feeding behaviors of blue-crowned laughingthrushes (*Garrulax courtoisi*). *Chinese Journal of Wildlife*, 1.

Dorrestein, G.M. (2009). Passerines. In T.N. Tully, G.M. Dorrestein, A.K. Jones, & J.E. Cooper (Eds.), *Handbook of avian medicine* (2nd edition). Saunders Ltd., Missouri, USA, 169–208.

Greggor, A.L., Masuda, B., Flanagan, A.M., & Swaisgood, R.R. (2020). Age-related patterns of neophobia in an endangered island crow: Implications for conservation and natural history. *Animal Behaviour, 160*, 61–68

Griffith, S.C., Holleley, C.E., Mariette, M.M., Pryke, S.R., & Svedin, N. (2010). Low level of extra-pair parentage in wild zebra finches. *Animal Behaviour, 70*(2), 261–264.

Hermes, C., Jansen, J., & Schafer, M. (2018). Habitat requirements and population estimate of the endangered Ecuadorian Tapaculo *Scytalopus robbinsi*. *Bird Conservation International, 28*(2), 302–318.

Kerr, K.C.R. (2021). Zoo animals as "proxy species" for threatened sister taxa: Defining a novel form of species surrogacy. *Zoo Biology, 40*(1), 65–75.

O'Reilly, A.O., Hofmann, G., & Mettke-Hofmann, C. (2019). Gouldian finches are followers with black-headed females taking the lead. *PLoS One, 14*(4), e0214531.

Oppenheimer, S. (2005). *Endocrine correlates of social and reproductive behaviours in a group-living Australian passerine, the white-browed babbler*. Doctoral dissertation, University of Wollongong, Australia.

Owen, A. (2019) *EAZA Best practice guidelines for the Javan Green Magpie Cissa thalassina*. 1st ed.

Owen, A., Wilkinson, R., & Sozer, R. (2014). *In situ* conservation breeding and the role of zoological institutions and private breeders in the recovery of highly endangered Indonesian passerine birds. *International Zoo Yearbook*, 48, 199–211.

Pearce, D., Pryke, S.R., & Griffith, S.C. (2011). Interspecific aggression for nest sites: Model experiments with long-tailed finches (*Poephila acuticauda*) and endangered gouldian finches (*Erythrura gouldiae*), *The Auk, 128*(3), 497–505.

Pryke, S.R., & Griffith, S.C. (2009). Postzygotic genetic incompatibility between sympatric colour morphs. *Evolution*, *63*, 793–798.

Skikne, S.A., Borker, S.L., Terrill, R.S., & Zavaleta, E. (2020). Predictors of past avian translocation outcomes inform feasibility of future efforts under climate change. *Biological Conservation*, *247*, 108597.

Tweed, E.J., Foster, J.T., Woodworth, B.L., Monahan, W.B., Kellerman, J.L., & Lieberman, A. (2006). Breeding Biology and Success of a Reintroduced Population of the Critically Endangered Puaiohi (*Myadestes Palmeri*), *The Auk*, *123*(3), 753–763.

Williams, L.J., King, J.A., & Mettke-Hofmann, C. (2012). Colourful characters: Head colour reflects personality in a social bird, the Gouldian finch, *Erythrura gouldiae*. *Animal Behaviour*, *84*(1) 159–165.

Wingfield, J.C., Hegner, R., & Lewis, D.M. (1992). Hormonal responses to removal of a breeding male in the cooperatively breeding white-browed sparrow weaver, *Plocepasser mahali*. *Hormones and Behavior*, *26*(2), 145–155.

Xie, S., Turrell, E.J., & McWhorter, T.J. (2017). Behavioural responses to heat in captive native Australian birds, *Emu – Austral Ornithology*, *117*(1) 51–67.

16

The behavioural biology of captive reptiles

STEVE NASH
Wild Planet Trust, Paignton, UK

16.1 INTRODUCTION

Reptilia represents a hugely diverse group and is, by some margin, the most speciose terrestrial vertebrate class (Uetz, Freed, Aguilar, & Hošek, 2021). This chapter follows the traditional Linnaean view of reptile taxonomy, which distinguishes these species from birds on the basis of physical features and metabolism. Phylogenetically, of course, the relationship between reptiles and birds must be recognised and it is certainly worth remembering when considering husbandry and welfare as the similarities between these two groups are often greater than either aviculturists or herpetoculturists realise. As a group, these animals occupy a huge range of habitats within almost all terrestrial (and many aquatic) biomes, and this diversity has resulted in a bewildering array of life cycles, behaviours, and adaptations. Many hundreds of species are maintained in captive settings and this can pose substantial challenges to a reptile keeper who may have daily responsibility for the members of an entire taxonomic class whose needs may vary substantially.

Perhaps more so than with any other taxa, the behaviour of captive reptiles is a topic that many zoo visitors have a predetermined expectation on. Anyone who spends time in a reptile house or similar exhibit will have heard the refrain "it's just sitting there," a comment usually followed a couple of seconds later by the sound of frantic tapping on the glass front of an exhibit, as the viewer attempts to waken the animal within and encourage it to "do something." What may be surprising to hear is that the apparently restricted behavioural range in captive reptiles is not a result of a reduced capacity for behavioural diversity on the part of the animal; rather, it is a consequence of husbandry parameters that do not meet the species-specific needs of a remarkably diverse and behaviourally complex group.

DOI: 10.1201/9781003208471-18

16.2 REPTILE ECOLOGY AND ENCLOSURE CONSIDERATIONS

Given the huge diversity of reptile species kept in captivity, there is an inherent risk in taking a generalised approach to enrichment, as even closely related species can exhibit markedly different behaviours and life histories. There are however several overarching aspects which can be considered irrespective of species. When it comes to promoting behavioural diversity in captive environments, the key considerations must be the provision of complexity, choice, and control, underpinned by an awareness of the environmental parameters that the species in question has evolved within. Giving the animal the necessary agency to utilise an appropriately diverse space in accordance with its individual need should be seen as fundamental, rather than forcing it to survive in an overly simplified, uniform environment based on a human perception of need. The addition of enrichment items, as is common with captive mammals and birds, should certainly be considered (for a summary, see Eagan, 2019). It is however reasonable to suggest that the most profound changes in a captive reptile's behaviour are likely to be realised through the provision of a thermally complex and ecologically relevant enclosure that takes the animal's specific social needs into account, and allows the expression of innate behaviours that may otherwise be hindered by circumstance.

16.2.1 Historic misconceptions

We can identify a variety of basic needs for most species, things such as warmth, shelter, food, etc. Many species have proven to be remarkably resilient and will survive in captivity so long as these basic requirements are met. This should not be taken as evidence that all of the animal's needs are being met however, and it should be the aim of every person involved in captive reptile management to ensure that they provide optimal husbandry conditions that extend beyond simple survival of their charges. Our understanding of an animal's needs will continue to increase with further study; however, the behavioural needs of many commonly kept species have been significantly underestimated and oversimplified in the past. There is a growing awareness of both the prevalence of abnormal behaviours in captive reptiles, the limitations of many accepted practices, and also (encouragingly) an increased interest in correcting past assumptions and promoting an evidence-based approach to husbandry and welfare.

Reptile welfare can be difficult to quantify and measure. Although the body of research is increasing, there remains a lack of data pertaining to captive welfare and this is undoubtedly a factor in the historic absence of enriching environments for such species (Burghardt, 2013). In addition, the phylogenetic distance of these taxa from humans results in a lowered intuition regarding the recognition of behavioural abnormalities as well as the appropriateness of the captive environment as a whole. The innate nature of reptile behaviour requires the provision of specific environmental conditions analogous to those in which the species has evolved, and consequently, their ability to thrive, or even survive, in novel environments may be reduced. The impact of this "innateness" cannot be overstated; deficient, or depauperate environments that lack critical features, or which deviate markedly from wild parameters, will result in an animal that is unable to display its full behavioural repertoire.

The level of ecological knowledge for many commonly kept species of reptile is often restricted (Arbuckle, 2013). The speciose nature of these taxa exacerbates this issue as an ever-growing body of research highlights the extent, not only of inter, but also intra-specific variation. The frequently encountered limitations of captive environments mean that studies of captive animals cannot always be relied upon to inform our understanding of natural behaviours, and indeed behavioural divergence of captive animals from their wild progenitors is well documented. Although this divergence typically manifests as an absence of normal behaviour rather than the expression of abnormal or deviant ones (Michaels, Gini, & Preziosi, 2014), it is certainly true that there are several frequently witnessed behaviours that fall into the latter category; for example, repetitive interaction with transparent boundaries, hypoactivity, and prolonged periods of basking that may often include close proximity to heat sources, leading to burns.

Reptiles have gained a reputation as "stoic automatons," unremitting stimulus-response creatures (Burghardt, 2013) with a tendency towards behavioural inertia in captivity. In stark contrast to their captive counterparts, wild reptiles frequently demonstrate a behavioural diversity that more than rivals that of mammals and birds. It is of note that certain health issues such as thermal burns and rostral abrasions are prevalent in captivity, non-existent in wild individuals, and indicative of the lack of understanding regarding specific requirements of many taxa. Captive environments are invariably depauperate in comparison to nature (controlled deprivation as outlined by Burghardt, 1996) and it is the role of the keeper to identify and provide the most salient features. This seemingly simple task is often hindered by a lack of species-specific natural history data, and further compromised by a tendency to resort to folkloric husbandry techniques (as outlined by Arbuckle, 2013). Such techniques are prevalent within herpetoculture and are often the causal factor behind the perpetuation of several persistent myths; notably, those relating to enclosure size and to the provision of overly simplistic or generalised housing. There remains a tendency to maintain captive herpetofauna under comparatively simple, size-restricted conditions. This is particularly the case for animals housed off exhibit in zoos but can often be encountered in display exhibits as well. Exhibit enclosures may convey the appearance of natural environments, but this aesthetic can mask a functionally deficient space for the inhabitant.

16.2.2 The thermal environment

The ectothermic metabolism of reptiles dictates that all activity is determined by the thermal parameters that the animal is exposed to, with behavioural diversity linked inexorably to thermal heterogeneity. A captive environment that fails to provide the thermal and structural complexity required by a given species will consequently fail to provide the conditions necessary for a full, and natural, behavioural repertoire to be displayed. Although some specialised species may require their environment to be cooled, and some species may do well under local/ambient conditions, the majority of species that we are considering will need some form of supplemental heating. It should therefore be self-evident that a carer should understand how these animals obtain their heat in the wild, and that this understanding will determine the suitability of their captive provision.

Ultimately, the heat that is used by wild ectothermic animals will have originated from the sun as solar radiation and will be utilised in different ways depending on the thermal needs of the animal. Some species may try to avoid direct sun, others will actively seek it, and many will do both depending upon circumstance and need. At a basic level, we can identify two key strategies for acquiring heat:

1. **Heliothermy:** A heliothermic animal is one that obtains its heat by basking in overhead, radiant heat.
2. **Thigmothermy:** A thigmothermic animal obtains heat by putting itself in contact with a warm object, for example, a rock that has been exposed to the sun.

Although presented here as a binary choice, many species will utilise both of the above heat acquisition methods to varying degrees, as well as being influenced by other methods of heat transfer such as convection and evaporation. Active, diurnal species tend to be heliothermic whereas secretive, crepuscular, and nocturnal species have a tendency to be thigmothermic. Just because an animal has a tendency to do something, does not mean that that should be the only thing they are able to do for their lifetime in captivity. A crepuscular snake may tend to emerge at dusk and warm up on a sun-warmed rock, but they may also choose to discretely bask in the sun on another day. Viewing heat provision in overly simplistic terms is almost certainly a key factor in the reduced behavioural repertoire of captive reptiles. It is not simply a matter of providing an animal with heat, the heat must be provided in a manner that the animal is able to use effectively, and it is essential that the carer understand that this manner will change according to need. In addition, one must also consider the changes that occur during the course of a 24-hour period, as well as seasonally, and these fluctuations must be replicated in captivity if we are to provide our animals with a suitable environment. Due to the inescapable link between thermal environment and behaviour, it must also be recognised that behaviour will also change according to time of day and season.

When we think of a basking reptile, we often imagine a diurnally active animal, sitting exposed on a rock or log in full sun. Many species do not do this however and instead display a behaviour known as cryptic basking (for example, Bertoia, Monks, Knox, & Cree, 2021). Many reptiles are prey items for other animals, so sitting in an exposed position actually puts them at increased risk of predation. Consequently, many reptiles will partially conceal themselves while at the same time, warming themselves in the sun's heat. This should be borne in mind when arranging a vivarium – a reptile should not be forced to choose between warmth and security – however, traditional arrangements featuring single heat sources and single refugia often force animals to do exactly that.

Most reptile species are thermoregulators, actively utilising behavioural and physiological methods to regulate their body temperature. To gain heat and raise their body temperature above the ambient, they may bask, lie on a warm rock, turn a different colour, change their body posture, etc. To avoid overheating they may seek shade, burrow, turn a paler colour, etc. Reptiles can maintain their body temperature within very particular limits so long as they can behave naturally. Successful thermoregulation requires access to a range of temperatures or, more specifically, the provision of a thermal gradient. The phrase gradient implies an orderly, linear transition, whereas one should instead view this as a matrix, a multidimensional abiotic component of an animal's living space that should also be extended beyond heat to include other parameters such as light and humidity. Needless to say, such complex matrices are easier to achieve in larger enclosures.

It should be clear that these animals operate within a range of temperatures with a preferred optimum temperature (POT) which represents the temperature at which the reptile metabolism functions most effectively. Most reptiles will spend their waking hours trying to achieve this POT to ensure metabolic efficiency. If the environment does not allow the achievement of such a temperature, their behavioural repertoire will be compromised, forcing the animal to spend extended periods engaged in core survival behaviours rather than anything else.

The provision of a basking area that corresponds to the size of the animal is key, enabling the animal to warm its entire body at once as it would do if basking in the sun (see Figure 16.1). This does not mean that a big animal needs a big heat lamp, however. For many larger species, the best arrangement would be to have two or three smaller lamps in a cluster to provide a larger area of even heat. The occurrence of thermal burns in captive reptiles is a phenomenon that one would not expect in an animal so intimately in tune with its thermal environment. Although caged heat sources may prevent the incidence of burning, they do not address the underlying reason behind their occurrence. This can invariably be traced to a deficient heating system that fails to allow the animal (often a larger species) to warm its entire body at once. If an animal can achieve its desired temperature at a distance, what need would it have to put itself in contact with the heat source itself?

16.2.3 Light

The world of light and light intensity can be confusing, however, understanding the importance of correct lighting and its impact on behaviour is vital. It must be remembered that in nature, light and heat are not separate entities, representing a continuum of wavelengths along the electromagnetic spectrum (EMS), and both must be considered when looking at the health and behaviour of captive reptiles. Ultimately, the aim should be to best replicate the parameters of natural sunlight that are experienced in the animal's home range. The light intensity can vary in an animal's environment and throughout the day and year. As with temperature, the aim should be to provide light gradients – areas of bright light, dappled shade, and seclusion. Choice and control should be key! There are three main "types" of light, occurring in the middle of the EMS. In order of increasing wavelength, these are ultraviolet, visible light, and infrared, although the latter is generally referred to as heat. Proper care, and successful provision of the conditions needed for a full repertoire of behaviours to occur, require the inclusion of all three.

There exists substantial evidence that all reptiles should be given access to a source of UVB light. Until recently, the prevailing view amongst many reptile keepers was that nocturnal species did not require or benefit from the provision of UV (or indeed any) lighting and could be successfully maintained using only indirect or room lighting. It is now increasingly recognised that this view is flawed and the provision of lighting for nocturnal species should be considered equally as important as for diurnal species. Rather than being truly nocturnal, many species are actually crepuscular (active at dawn and dusk) or cathemeral (active at intervals during day and night), meaning that such species have ample opportunity to expose

Figure 16.1 Provision of heat and light should be linked, with basking areas that enable an animal to illuminate and warm its body in its entirety. For some species, this may require very sizeable basking spots. Seen here for Aldabran giant tortoise, *Aldabrachelys gigantean*.

themselves to natural sunlight. It is also clear that many "nocturnal" species will bask under UV lights in captivity if provided, although basking duration may be brief or may involve exposing only parts of their body (i.e., a limb). In addition, natural UV is reflected into the entrances of burrows and other hiding places, meaning exposure is possible even if the animal is hiding, while many species that are active at night will still sit out during the day. Leaf-tailed geckos (*Uroplatus* sp.) for example will sit motionless on tree trunks during the day, so exposure to UV is unavoidable. It is important to stress that the provision of adequate hiding places is essential. These animals need to be able to escape from exposure to UV or high-intensity lighting as sustained exposure can cause ocular damage (as it would do in humans).

It is important to remember that UVB is not the only light requirement of captive reptiles. UVA wavelengths have a huge impact on behaviour, appetite, and colour perception, and fall within the visual range of reptiles. Visible light intensity is also frequently overlooked, and it is not uncommon to see enclosures that provide UVB lighting but no other. The ideal way to illuminate an animal would be to expose them to natural, unfiltered sunlight but this may not be possible or practical. Artificial lighting will therefore be necessary, with T5 fluorescent tubes generally recognised as the best available means of providing artificial UVB at the time of writing. Metal halide floodlights are an excellent way to provide high-intensity visible light and UVA, as well as valuable short-wave infrared (IR-A). UVB is generally measured using the UV Index (UVI) and can be readily measured using a handheld meter (for example, Solarmeter 6.5/6.5R). Such an item should be viewed as essential equipment for reptile keepers, enabling accurate mapping of UVB gradients within an enclosure. It has been suggested that the need for UVB can be overcome by making use of a dietary D3 supplement. Dietary D3 is not suitable for all species and may be extremely dangerous if overdoses are administered. In nature, behaviour often provides the "off-switch" for D3 synthesis via exposure to the sun; in captivity, with oral administration and confusing signals from artificial lighting provision, no such option exists.

16.3 BEHAVIOURAL BIOLOGY AND REPTILE WELFARE

There is a risk that human perception of habitat can play a detrimental role in husbandry, with consequent impacts on behaviour. The lack of species-specific natural history data requires the keeper to provide an environment based on their interpretation of what an environment may include. This means that microhabitat features can easily be overlooked, or that assumptions of what is needed may fail to recognise specific niches with associated evolved behaviours. The leopard gecko (*Eublepharis macularius*) is incredibly common in captivity, with many keepers believing it to be a desert dwelling species. The reality of its natural distribution is very different, however (see Khan, 2009), and the species inhabits a far more diverse range of habitats than is commonly assumed. Its generalist nature has allowed it to survive in captive environments that may not fully represent its habitat preference, an achievement that more specialist species may be unable to replicate in a constrained environment that does not offer diversity and choice.

16.3.1 Enclosure size

The issue of enclosure size is one that raises numerous viewpoints, and it is regretfully still the case that the view that many reptile species can be kept successfully in small enclosures still persists. Snakes in particular are often considered to be inactive, sedentary animals. It is often stated that if placed in a large enclosure they will not use the space, will feel insecure, and may not feed, and for this reason, the use of small enclosures is promoted. There are several false assumptions with this view, and it should be the aim of all reptile keepers to promote the best possible method of husbandry for the animals in their care.

A captive animal should have sufficient space to stretch out, move freely, and carry out normal behaviours, and there are clear benefits to its doing this (Hoehfurtner, Wilkinson, Walker, & Burman, 2021). Many commonly kept species display activity patterns that do not mirror those of their carer's work hours. Consequently, the behaviour of nocturnal and crepuscular species often goes unstudied, or unappreciated. The use of cameras to reveal what takes place when carers are not present should be encouraged, particularly for animals that may have a favoured refuge or perch to which they tend to return to rest. It would be easy to assume that in such cases, the animal rarely moves, when in reality it may be moving a great deal.

When considering enclosure size, mention should also be made of the concept of "flight distance" – the minimum distance that an animal will allow between itself and a potential predator before it attempts to escape. Captive animals may view the keeper or the viewing public as a potential threat and try to escape from their advances or retreat from their presence. An enclosure should be big enough that you can look in and view its occupant without crossing this invisible barrier. The COVID-19 pandemic and associated lockdown closures provided interesting opportunities to quantify the impact of the viewing public on captive reptile behaviour, with Carter et al. (2021) suggesting that this impact, where observed, is species-specific.

In nature, sedentary behaviour is typically transient, following, for example, the consumption of a meal and should not be used as a basis for determining spatial needs (Warwick, Arena, Lindley, Jessop, & Steedman, 2013). Perhaps conversely, it is often larger specimens (notably large pythons, Pythonidae and boas, Boidae) that are housed in the smallest enclosures relative to their size. Such an approach serves to further minimise the opportunities for the animal to display a natural range of behaviours while also exacerbating a range of health issues because of enforced immobility.

16.3.2 Complexity and choice

Where complex, enriched environments are provided, this may often be done for aesthetic reasons of the display rather than for a belief that the animals require or benefit from such enhanced stimulation (Burghardt, 2013). Sterile housing conditions differ dramatically from the structurally complex wild environments and are likely to result in elevated stress levels for the inhabitant (Mendyk,

2015). Typical stress behaviours include, but are not limited to, behavioural inhibition, heightened vigilance, hiding, aggression, decreased exploratory behaviour, and reduced behavioural complexity (Morgan & Tromborg, 2007) as well as hyper- and/or hypoactivity (Warwick et al., 2013) and it is clear that many captive reptiles display such behaviours.

When it comes to the prevalence of abnormal behaviours, perimeter pacing and interactions with transparent boundaries (ITB) are frequently seen in many reptile taxa and both can be explained by inadequacies in the captive conditions. Solid, transparent boundaries have no parallel in nature and it is perhaps unsurprising that a captive reptile might interact with them, with repetitive scratching and head rubbing commonly seen. A common approach to preventing this behaviour is to cover the glass in the hope that an opaque boundary will deter the interaction. While this may reduce the frequency, one has to question the cause of the behaviour and ask why the animal was pacing and interacting with the barrier in the first place. If such behaviours are viewed as exploratory or escape behaviours, this suggests that it may be the conditions within the enclosure that the animal is having to endure that have triggered their expression. Rose, Evans, Coffin, Miller, and Nash (2014) describe the impact of enclosure furnishings on the prevalence of ITB in chuckwalla (*Sauromalus ater*). In this instance, the creation of more ecologically relevant features within the vivarium led to a reduction in ITB, supporting the view that such behaviours can be "designed out" by ethologically informed approaches to enclosure design and husbandry. Similarly, Case, Lewbart, and Doerr (2005) suggest that the provision of more complex environments reduced the occurrence of escape behaviours in captive eastern box turtles (*Terrapene carolina*), as well as measurable physiological differences when compared to counterparts in barren environments.

Consideration should be given to how individual species seek refuge and security. For a terrestrial species, refuge may be taken in a narrow crevice between rocks. For fossorial species, substrate composition and depth must be considered (Spain, Fuller, & Allard, 2020). Arboreal species may rely on dense vegetation or by resting, motionless against a background they blend into. The colouration and patterns on an animal's integument can provide clear clues to keepers here – green-coloured lizards are not green by chance! Although technically true that providing an arboreal species with a hide-box at floor level is providing a refuge, it does not come close to providing a biologically meaningful environment. Refuge selection by free-living individuals also relies on an animal's assessment of a range of temporally variable parameters (such as thermal and hydric properties), as well as structural features such as refuge size and shape (Croak, Pike, Webb, & Shine, 2008).

16.4 SPECIES-SPECIFIC REPTILE ENRICHMENT

Exhibits housing reptiles are often set up with a high degree of naturalness in terms of their appearance. Theming (for example artificial rockwork), natural substrates, branches, lianas, and live plants are commonly seen, with attempts made to replicate desert, rainforest, and riparian biomes amongst others. While this approach has much to commend and can serve a valuable function in engaging zoo visitors with wider messages regarding habitats and adaptations, there must also be a note of caution in terms of functionality for the inhabitant. Aesthetics should not be allowed to trump the needs of the animal. Sections of thick bamboo are often seen in enclosures for tropical species and can effectively convey the appearance of such environments to visitors. They are however of limited value for many arboreal species, lacking, for example, the textured bark found on typical tree branches that provides the necessary traction for claws to grip. The furnishings within an enclosure will make a huge difference to the amount of useable space. Many enclosures tend to contain a lot of unused space; considered use of furnishings can easily create more complex and enriching environments that can dramatically increase the opportunities for movement to the animal contained within.

An animal's physical appearance can provide valuable clues to its lifestyle, environment, and subsequent behaviour. Long toes and sharp claws suggest an ability to climb. Green colouration suggests a life spent amongst vegetation. An animal's appearance may appear quite dramatic against a sparse or artificial background, but in its native environment, seemingly gaudy colours become cryptic and provide the perfect foil to allow the animal to blend in. Against a background of aspen bedding, the well-known corn snake (*Pantherophis guttatus*) does not blend in at all, but if the same snake is set among soil and leaf litter, the distinctive saddle markings allow

Figure 16.2 An enclosure's furnishings must facilitate the behaviour of the species it houses. This helmeted iguana (*Corytophanes cristatus*) requires vertical trunks on which to perch and scan the surroundings for food and threats.

it to disappear. Different species occupy different niches and utilise their environments in different ways. The helmeted iguana (*Corytophanes cristatus*) in Figure 16.2 for example rests in an upright position on vertical perches, whereas other rainforest lizards may require limbs of different thicknesses or orientation.

Elevated stocking densities and spatial restriction have been implicated as causative factors leading to elevated stress levels and aggression towards conspecifics. Overcrowding manifests as overt and covert (Warwick et al., 2013) with the former relating to the physical number of animals in a given space, and the latter the ability of all animals to access all features of said space. Covert overcrowding can therefore be an issue in any enclosure housing more than one animal, and keepers must be aware that the features that individual animals compete for may not all be visible. While it is easy to identify a problem where multiple animals are trying to access a single hide-box or branch, it is not as straightforward to see multiple animals vying for access to a specific area of UV intensity or humidity. The provision of multiple basking opportunities and UV sources within each enclosure provides a logical solution, however this approach may go against the common orthodoxy seen in many (often small) enclosures, where a single heat source is provided at one end in order to maximise a linear heat gradient.

For many insectivorous species, the provision of live feeder insects provides opportunities to promote natural hunting and feeding behaviours. These behaviours can be encouraged by the simple act of introducing multiple food insects into an enclosure and allowing the animals to hunt them, but this can be further enhanced by using other methods. Hanging perforated containers of insects above an enclosure provides a means of randomising the appearance of food items, as insects crawl through the holes and drop into the enclosure below. Varying the type of food insect is also useful, not only because it increases the nutritional profile of the animal's diet, but because it encourages different methods of prey handling and capture; a mealworm requires a very different approach to a locust, a moth, or a phasmid for example. Puzzle feeders, commonly used with captive mammals, can also be used (Bryant & Kother, 2014). Extractive foraging has been frequently reported in small Varanid lizards for example (Mendyk & Horn, 2011) and adoption of such methods can lead to increased foraging times in captivity. For species that do not feed on insects, food-linked behaviours can still be encouraged. Prey items such as pre-killed rodents can be used to create scent trails for snakes and larger lizards. Herbivores can be provided with whole food items rather than chopped and diced, and food items can also be hung off the floor to promote browsing behaviours.

The tendency to maintain reptiles under conditions of close captivity (i.e., within self-contained enclosures where all aspects of the animal's life are controlled by a carer) is a situation that can easily lead to heightened deprivation. Where local conditions allow, outdoor maintenance for at least some of the year offers many benefits. This approach is frequently seen in Chelonia, but less often with other reptile orders. In addition to the unarguable benefits of exposure to unfiltered natural sunlight, such open enclosures have a greater likelihood of being dynamic habitats, exposed as they are to climatic variation and sensory stimuli from visitors and other factors. For species that have evolved to contend with the complexities of surviving in complex wild habitats, the lack of variation and challenge encountered in a predictable, closed vivarium, should be viewed as a husbandry challenge to be overcome. Such features can readily be incorporated into the design of reptile enclosures and can lead to much more stimulating and enriching environments for their inhabitants. Similar consideration should be given to the inclusion in mixed-species environments, although of course these should be based upon a well-considered choice of species.

Novelty is a feature that has been shown to promote exploratory behaviour in certain taxa. Olfactory novelty is particularly suitable for consideration, either through the provision of prey-linked scent trails, or the addition of non-prey-linked items (i.e., substrate from another enclosure, seasonal leaf litter) although in such instances, care should of course be taken to avoid any biosecurity risk. Londoño et al. (2018) demonstrated further benefits from chemosensory enrichment using the scent of rival male Catalan wall lizards (*Podarcis liolepis*) with resultant decreases in abnormal and escape behaviours.

The use of scent from conspecifics requires keepers to consider communication between such animals. Clearly, olfactory communication is important in the example given previously, but other methods of intra and indeed inter-specific communication also demand our attention. Many species may be solitary in nature and may not benefit from the continuous presence of other individuals; however, this does not mean that they should live their entire existence alone. Although the permanent presence of other animals may be detrimental, controlled, transient interactions may in fact be beneficial, particularly for taxa such as chameleons that utilise colour change to reveal and communicate mood, or for species where territoriality and social hierarchy have been observed.

Further research on reptile enrichment will reveal multiple avenues for carers to investigate. Play behaviour has been documented in many species, from turtles (Burghardt, Ward, & Rosscoe, 1996) to crocodilians (Murphy, Evans, Augustine, & Miller, 2016), and there is growing awareness of more advanced cognitive functions such as problem-solving and counting abilities in monitor lizards (Manrod, Hartdegen, & Burghardt, 2008). Many species respond positively to training, and as well as the behavioural benefits of this approach, such endeavours have the potential to transform husbandry routines and healthcare interventions. Tactile stimuli are a largely unexplored field, although the inclusion of stiff broom heads as shell scratchers for tortoises has been shown to be particularly beneficial.

16.5 USING BEHAVIOURAL BIOLOGY TO ADVANCE REPTILE CARE

So what does informed husbandry look like, and how can behavioural diversity, and positive welfare states, be promoted? Enclosure size is a key factor, not only because it allows for free movement and the expression of exploratory and active behaviours, but because it provides the diversity of environmental parameters necessary to allow for complexity and choice. It overcomes the limitations of an anthropocentric view of reptile need and provides the animal the necessary agency to demonstrate the full scope of its innateness. Scepticism over accepted practice should be encouraged.

Awareness of the environment from a reptile's perspective must also be encouraged. Current technology allows accurate measurement and mapping of key parameters, with tools like infrared cameras transforming our understanding of useable space and thermal complexity within an enclosure. Visualising heat in this way (as in Figure 16.3)

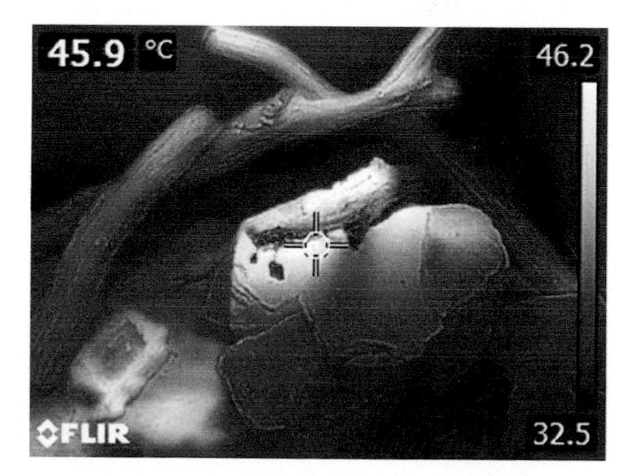

Figure 16.3 Thermal imaging equipment allows keepers to visualize the animal's environment in a way that is biologically meaningful to a reptile. Different heaters provide different thermal environments and can help explain deviations in behaviour and activity budgets when compared to wild counterparts.

enables keepers to recognise deficiencies that may previously have been overlooked.

While there are clear challenges in providing appropriate environments, this provision is by no means unachievable. It should be within the scope of all dedicated carers and institutions to design and provide enriched and complex environments for the reptiles in their care. Many conservation programmes recognise the growing need for a captive component to species recovery, and it is beyond doubt that there are significant potential benefits to be had from engaging zoo visitors with this incredible group of animals. Examples of innovative and forward-thinking husbandry abound, and reptile keepers should strive to implement the findings of new research and adapt past practices to ensure that reptile welfare continues to advance.

To further evidence excellent reptile husbandry and management in the zoo, Box 16.1 suggests research questions that could/should be answered to further our knowledge of the needs of this diverse and evolutionary complex group of animals. Similarly, Box 16.2 provides further pieces of literature and suitable published information for those wishing to know more about specific aspects of reptile behavioural biology and how this underpins captive care.

BOX 16.1: Future research questions pertaining to reptile behaviour

There is huge scope for further research that will substantially improve the behavioural husbandry and welfare of captive reptiles. Two particularly pertinent avenues are detailed below:

1. Enclosure use studies: Several tools exist to enable carers to better understand how the animals they look after utilise their enclosure space. Research that utilises the Modified Spread of Participation Index is perhaps the most easily conducted and offer great insights for carers to evaluate enclosure suitability (Plowman, 2003). When used with mammals, such studies enable an understanding of how an animal uses the physical features of its environment. With reptiles (and indeed, other ectotherms) these studies become even more impactful if the environmental parameters are incorporated as well. The availability of thermal cameras and UVI/Lux meters means that it is now possible to see how enclosure use is affected by these factors and can go some way towards overcoming the issues associated with the phylogenetic distance mentioned previously. In enclosures housing multiple animals, they can also reveal potential areas of conflict caused by covert overcrowding, where social factors affect access to essential resources (i.e., basking spots).
2. Documenting the impact of different IR sources: Infrared radiation comes in different forms and they are not created equally. There are three types of infrared; A, B, and C. Natural solar radiation is referred to as Infrared A (or near-infrared, or short-wave infrared) and provides effective deep tissue penetration, allowing the animal to warm itself "all the way through." This is the type of infrared that reptiles need to bask effectively. Infrared C on the other hand does not do this and instead provides localised surface heating, but very little in the way of deep tissue warmth. Infrared C is the warmth that is emitted from an object that has previously warmed up from exposure to the sun. This difference has a profound effect on how a reptile can warm its body and retain heat, with subsequent "knock-on effects" for health and behaviour.

BOX 16.2: Suitable directed reading on reptile husbandry and housing

In 2010, a paper was published that investigated the basking behaviour and UV exposure of certain snakes and lizards in different habitats. This paper categorised reptiles into different groups based on how they basked and the level of UVB that they were exposed to in the wild. These groups, or zones, are now widely referred to as "Ferguson Zones" after the lead author of the paper.

Ferguson, G.W., Brinker, A.M., Gehrmann, W.H., Bucklin, S.E., Baines, F.M., & Mackin, S.J. (2010). Voluntary exposure of some western-hemisphere snake and lizard species to ultraviolet-B radiation in the field: how much ultraviolet-B should a lizard or snake receive in captivity? *Zoo Biology, 29*(3), 317–334.

In 2016, members of the BIAZA Reptile & Amphibian Working Group (RAWG) published a landmark paper in the *Journal of Zoo and Aquarium Research (JZAR)*, which detailed the suggested UVB needs of over 250 species. This paper should be considered essential reading for any keeper involved in the husbandry of captive reptiles.

Baines, F.M., Chattell, J., Dale, J., Garrick, D., Gill, I., Goetz, M., Skelton, T., & Swatman, M. (2016). How much UVB does my reptile need? The UV-Tool, a guide to the selection of UV lighting for reptiles and amphibians in captivity. *Journal of Zoo and Aquarium Research*, *4*(1), 42–63.

16.6 CONCLUSION

The need for evidence-led, ethologically informed husbandry of captive reptiles should be clear. As a group, reptiles have endured a captive history where their complex needs have been largely misunderstood and misinterpreted, resulting in a tendency to maintain them under depauperate conditions that impede their behavioural repertoires. Those involved in reptile husbandry should strive to ensure that current thinking and advances in understanding are actively employed and applied, while simultaneously questioning the often-simplistic methodologies of the past. An awareness of the interplay between animal and environment is essential, with recognition of reptilian cognition and behavioural potential being equally necessary. Complexity, choice, and control are key in enclosure design and approaches to daily husbandry.

REFERENCES

Arbuckle, K. (2013). Folklore husbandry and a philosophical model for the design of captive management regimes. *Herpetological Review, 44*(3), 448–452.

Bertoia, A., Monks, J., Knox, C., & Cree, A. (2021). A nocturnally foraging gecko of the high-latitude alpine zone: Extreme tolerance of cold nights, with cryptic basking by day. *Journal of Thermal Biology, 99*, 102957.

Bryant, Z., & Kother, G. (2014). Environmental enrichment with simple puzzle feeders increases feeding time in fly river turtles (*Carettochelys insculpta*). *Herpetological Bulletin, 130*, 3–5.

Burghardt, G.M. (1996). Environmental enrichment or controlled deprivation? In Burghardt, G.M., Bielitski, G.M., Boyce, J.T., & Schaefer, J.R.D.O. (Eds.), *The wellbeing of animals in zoo and aquarium sponsored research*. Scientists Center for Animal Welfare, Greenbelt, USA, pp. 91–101.

Burghardt, G.M. (2013). Environmental enrichment and cognitive complexity in reptiles and amphibians: Concepts, review, and implications for captive populations. *Applied Animal Behaviour Science, 147*, 286–298.

Burghardt, G.M., Ward, B., & Rosscoe, R. (1996). Problem of reptile play: Environmental enrichment and play behavior in a captive Nile soft-shelled turtle, *Trionyx triunguis*. *Zoo Biology, 15*(3), 223–238.

Carter, K.C., Keane, I.A., Clifforde, L.M., Rowden, L.J., Fieschi-Méric, L., & Michaels, C.J. (2021). The effect of visitors on zoo reptile behaviour during the COVID-19 pandemic. *Journal of Zoological and Botanical Gardens*, 2(4), 664–676.

Case, B.C., Lewbart, G.A., & Doerr, P.D. (2005). The physiological and behavioural impacts of, and preference for, an enriched environment in the eastern box turtle *Terrapene arolina arolina*. *Applied Animal Behaviour Science, 92*(4), 353–365.

Croak, B.M., Pike, D.A., Webb, J.K., & Shine, R. (2008). Three-dimensional crevice structure affects retreat site selection by reptiles. *Animal Behaviour, 76*(6), 1875–1884.

Eagan, T. (2019). Evaluation of enrichment for reptiles in zoos. *Journal of Applied Animal Welfare Science, 22*(1), 69–77.

Hoehfurtner, T., Wilkinson, A., Walker, M., & Burman, O.H. (2021). Does enclosure size influence the behaviour & welfare of captive snakes (*Pantherophis guttatus*)? *Applied Animal Behaviour Science, 243*, 105435.

Khan, M.S. (2009). *Natural history and biology of hobbyist choice leopard gecko Eublepharis macularius*. Talim ul Islam College, Rabwah, Pakistan.

Londoño, C., Bartolomé, A., Carazo, P., & Font, E. (2018). Chemosensory enrichment as a simple and effective way to improve the welfare of captive lizards. *Ethology*, *124*(9), 674–683.

Manrod, J.D., Hartdegen, R., & Burghardt, G.M. (2008). Rapid solving of a problem apparatus by juvenile black-throated monitor lizards (*Varanus albigularis albigularis*). *Animal Cognition*, *11*(2), 267–273.

Mendyk, R.W. (2015). Life expectancy and longevity of Varanid lizards in North American zoos. *Zoo Biology*, *34*, 139–152.

Mendyk, R.W., & Horn, H.G. (2011). Skilled forelimb movements and extractive foraging in the arboreal monitor lizard *Varanus beccarii* (Doria, 1874). *Herpetological Review*, *42*(3), 343–349.

Michaels, C.J., Gini, B.F., & Preziosi, R.F. (2014). The importance of natural history and species-specific approaches in amphibian ex situ conservation. *The Herpetological Journal*, *24*(3), 135–145.

Morgan, K.N., & Tromborg, C.T. (2007). Sources of stress in captivity. *Applied Animal Behaviour Science*, *102*(3–4), 262–302.

Murphy, J.B., Evans, M., Augustine, L., & Miller, K. (2016). Behaviors in the Cuban Crocodile (*Crocodylus rhombifer*). *Herpetological Review*, 47(2) 235–240.

Plowman, A.B. (2003). A note on a modification of the spread of participation index allowing for unequal zones. *Applied Animal Behaviour Science*, 83(4), 331–336.

Rose, P.E., Evans, C., Coffin, R., Miller, R., & Nash, S. (2014). Using student-centred research to evidence-base exhibition of reptiles and amphibians: Three species-specific case studies. *Journal of Zoo and Aquarium Research*, *2*(1), 25–32.

Spain, M.S., Fuller, G., & Allard, S.M. (2020). Effects of habitat modifications on behavioral indicators of welfare for Madagascar giant hognose snakes (*Leioheterodon madagascariensis*). *Animal Behavior & Cognition*, *7*(1), 70–81.

Uetz, P., Freed, P, Aguilar, R., & Hošek, J. (Eds.) (2021) *The reptile database*. http://www.reptile-database.org

Warwick, C., Arena, P., Lindley, S., Jessop, M., & Steedman, C. (2013). Assessing reptile welfare using behavioural criteria. *In Practice*, *35*, 123–131.

17

The behavioural biology of amphibians

JACK BOULTWOOD
Wild Planet Trust, Paignton, UK

17.1 INTRODUCTION TO AMPHIBIAN BEHAVIOURAL BIOLOGY

There are over 8000 extant species of amphibian with more being discovered or reclassified every year (Frost, 2021). Just a fraction of this diverse group of animals can be found in captive collections globally. Amphibians are split into three groups: Anura, the tailless amphibians (frogs and toads); Urodela, the tailed amphibians (salamanders and newts); and Apoda, the limbless amphibians (caecilians). These three groups can be found in a range of habitats which include but are not limited to the tops of tree canopies and mountains, the bottom of cave water systems or deep underground, as well as water systems such as lakes and rivers including the surrounding areas. The diversity of amphibian habitats, niches and behaviours leaves many captive species' daily care a true challenge; creating a husbandry protocol based on scientific evidence, which is then applied to the species' wild behavioural ecology, provides an opportunity to create a captive environment based around the captive species' biological, behavioural, and ecological needs. Using an evidenced base approach should lead to an increase in positive welfare, a reduction in negative welfare, and a further understanding of species specifics and concepts that may be applicable across multiple species. Many amphibian species are incredibly sensitive to sudden change due to their specialised evolutionary traits; these evolutionary traits make them highly efficient within their natural habitat. Although many species are extremely sensitive to environmental change, a few amphibian species are more robust and are considered invasive due to their ability to aggressively adapt, outcompete, and fill the niche of other species; for example, the cane toad (*Rhinella marina*) in Australia. As ectotherms, external environmental parameters govern the lives of amphibians by triggering their complex and diverse breeding behaviours and strategies, dictating their activity budget, regulating the accessibility and type of food available, and directing their

DOI: 10.1201/9781003208471-19

natural biological processes and behaviours. These external parameters are often not detectable by humans without appropriate recording equipment, pushing captive amphibian husbandry further into a science.

Providing an appropriate evidence-based captive environment for amphibians increases the chance of discovering the complex, diverse, and often secretive breeding behaviours and strategies performed by these animals. Triggering these breeding events by replicating the complex conditions they need, that will lead to the production of viable offspring, can often be just the start of the puzzle. Next comes the care of constantly developing larvae or tadpoles and consideration of metamorphosis and the development of adult life stages. Another challenge when keeping amphibians in captivity is that the full-grown adult's habitat can be vastly different to the conditions needed to rear and raise tadpoles or larvae, with the continuous growth periods for these juveniles being incredibly sensitive and dangerous. All of these elements need to be taken into consideration and validated with evidence to develop successful and effective captive management plans that allow for successful reproduction but also high animal welfare standards.

17.2 AMPHIBIAN ECOLOGY AND NATURAL HISTORY RELEVANT TO THE ZOO

Over the years of captive amphibian care, many sources of literature have been produced that compare aspects of wild amphibian life history strategies to captive amphibians. However, comparatively, amphibians are vastly understudied against other taxa but there are still published examples available that can help inform their captive management. There is little information on the effects of nutritional quality on captive amphibian health, behaviour, and wellbeing but what has been investigated shows that even minor changes can have a lasting effect. Often, when the colour of a captive amphibian is compared to a wild counterpart, the captive animal lacks the vibrancy of its colours. Ogilvy, Preziosi, and Fidgett (2012) showed that the introduction of dietary carotenoids, a pigment amphibians use for colouration in their skin, at a certain time during tadpole development in red-eyed tree frogs (*Agalychnis callidryas*) increased the vibrancy of adult colours. This introduction of carotenoids into the captive diet improved growth rates in female tree frogs, improved colouration in adult tree frogs and improved reproductive success later on in life. Although vibrancy was not measured, similar results were seen when carotenoids were introduced as a supplement to adult captive strawberry poison dart frog (*Oophaga pumilio*) diets, improving reproduction by increasing the quality of fertilisation and the number of trophic eggs (unfertilised eggs provided as nutrition for tadpoles) that could be produced by the mother of the developing tadpoles (Yeager, Dugas, & Richards-Zawacki, 2013).

Providing nutritional content to amphibians usually must be introduced to the feeder items, which are then given to the amphibian. Wild amphibians eat a vast array of live foods ranging from invertebrates, small mammals, birds, and even other amphibians. The availability of live food diversity in captivity is reduced in comparison to the availability in the wild, this restricts the nutritional value on offer to those live food species that are readily available. This restriction in captive nutrition has caused complications in geriatric amphibians, something that is unlikely to occur in the wild due to the unlikelihood of amphibians reaching captive ages and the diversity of the nutrition. The high fat content of captive live food has caused instances of lipid deposition (lipidosis) on the eyes of certain species of amphibian (Lock, 2017). This unique and unusual condition displays another instance where captive nutrition needs to be monitored and adapted throughout amphibian life stages. These examples highlight how amphibians are highly dependent on the nutritional quality provided by resources in their native habitat, without an attempt at replicating this biologically relevant natural nutritional range gaps occur in improving and sustaining positive captive welfare or even allowing for basic biological processes.

Amphibians demonstrate varied feeding ecologies, based on their evolutionary traits or stage of their life. Currently, zoological institutions are often able to offer captive amphibians something that is not available to other taxa, live prey items – most amphibians are insectivores as adults and insects can be offered as a live food source, creating wild-type opportunities that promote natural hunting and foraging behaviours in amphibians. Amphibians in early life stages, tadpoles, and larvae, can have species specifics feeding strategies and specially adapted mouth parts that are lost after metamorphosis. Examples of specific strategies or morphologies can be seen in Malaysian leaf frog (*Megophrys nasuta*) tadpoles which have a specially adapted funnel-shaped mouth that only allows them to forage and filter food and debris from the surface of the water (Wildenhues et al., 2012). These strategies and species-specific morphologies can be seen in the differing dentition of the juvenile Taita African caecilian (*Boulengerula taitana*) when compared to adults of the same species, leading to the mother of the offspring to develop a nutritious and modified thick skin layer after egg-laying that the juveniles eat (Kupfer et al., 2006). Looking for these species-specific traits and behaviours is key to successfully feeding juvenile amphibians in captivity, reducing the amount of trial and error, and increasing welfare for young amphibians. The provision of captive feed items for adult amphibians should also be based on the specie's specifics traits and behaviours as although most adult amphibians are insectivores the suitability of the captive feed item can be limited. Bumblebee toads (*Melanophryniscus klappenbachi*) remain a microphagy (i.e., feeding on items much smaller than the animal itself) from metamorphosis to adulthood, with hardly any change from being able to eat items the size of springtails the moment they make their way onto land, this is not the same for other small amphibians such as *Dendrobates* sp. who can be a similar size to bumblebee toads, as their food item size increases as they grow from juvenile to adult. Seasonality also needs to be considered for the availability and suitability of live food items for amphibians, especially when an amphibian is in its aquatic stage, provision of terrestrial food items while the amphibian might be spending the vast majority of their time in a water source means that food source is no longer readily available, providing a live food source in the water such as invertebrate larvae or aquatic worms offers the same advantages to amphibians as when they are offered terrestrial food items on land. As these live food sources often thrive in the same environments that amphibians do, there is an opportunity to have living enclosure custodians that can help keep the captive environment healthy and also be a continuous or supplementary food source; this increases enclosure usage and space use, as amphibians hunt for live food items within the enclosure.

As amphibians cover such vast and diverse habitats, they have developed specialised morphologies and behaviours, governing the use of space by the species specialism, the four main habitat styles being arboreal, terrestrial, fossorial, and aquatic with most amphibians using a mixture of the four. Without a biologically relevant captive habitats welfare can be compromised, natural behaviours are not achieved, whilst mating and reproductive systems are not accommodated. Considerations on an amphibian's evolutionary traits and natural behaviours can highlight how an amphibian uses space and what needs to be considered in captive management, careful planning and strategic management can create a suitable space for multiple amphibian species, within one captive environment, with little to no compromise to their welfare. Life stages also need to be considered when determining an amphibian's use of space and habitat specialisation, as this can drastically change over the course of each life stage with the possibility of changing from fully aquatic (when larval) to being terrestrial or arboreal once full metamorphosis has been completed. Some amphibians, such as newts, are considered to be semiaquatic once adults with a preference to be in the water during the times they are most active; these semiaquatic species often have a period of an early temporary terrestrial life stage after metamorphosis – the transition periods between these stages can prove to be a dangerous period for these animals so noting how space is used and understanding how habitat specialisation can change over an amphibian's life is important for the maintenance of a successful and sustainable captive population.

As amphibians are ectotherms, their lives are governed by a need to thermoregulate, and they have behaviours and life strategies to meet this basic biological process. Thermoregulation and access to a biologically relevant heat source define an amphibian's ability to naturally use space and affects the suitability of their habitat (and hence an enclosure). If the opportunity to achieve an optimal body temperature for a specific season is not achieved within a zoo enclosure, all other considerations around space use and species-specific specialisms are compromised and a full behavioural repertoire is unlikely to be achieved.

Reproductive strategies and mating systems in amphibians are extensive and complex, with many species being cryptic and secretive. Captive breeding success can be dependent on the quality of captive care and seasonal triggers. Most amphibian fertilisation occurs outside of the body through a sequence of behaviours and strategies, known as amplexus, that lead to fertilised eggs. An exception to this can be seen with most caecilians and salamanders, which use different methods of internal fertilisation; caecilians use an external sex organ, known as a phallodeum, which is inserted into the female for insemination, salamanders use sperm packets which are collected and internalised by a female salamander's cloaca. Before external or internal fertilisation occurs, amphibians will undergo a series of courtship behaviours, these behaviours often whittle down the competition to a breeding pair, which are often the result of a reproductive strategy that increases biological fitness in the offspring. These courtship behaviours can range from vocalisation, wrestling, discharge of pheromones, or visual displays such as crest size or vibrancy of colour, which allow for competition between males, an example of this can be seen in the aggressive battling seen between male African bullfrogs (*Pyxicephalus adspersus*), which often leads to the strongest and largest male becoming the successful mate after a long territorial battle with other hopeful males.

Courtship behaviours are often triggered by seasonal changes, such as a rainy season or the transition from winter into spring. These triggers are important to replicate or reproduce in captivity to encourage the initial stages of breeding events and stimulate seasonal biological changes to begin. Once a successful breeding event occurs, a series of strategies are used to create appropriate nests or egg-laying sites, some species even use a variety of non-traditional internal incubation methods. Studying the wild preference for egg-laying sites can increase the likelihood of achieving successful and healthy captive offspring, offering a variety of biologically relevant egg-laying sites to red-eyed tree frogs and Morelet's tree frog (*Aglychnis moreletii*) has proven to have a significant positive effect on oviposition, such consideration for oviposition sites should include the presence of predators, humidity, laying substrate, water temperature, and water quality (Sanchez-Ochoa, Perez-Mendoza, & Charruau, 2020). Using evidence from the wild and applying it to captive attempts at stimulating breeding (e.g., encouraging seasonal biological changes and promoting appropriate oviposition site selection) could reduce stress caused by inaccurate breeding attempts and any prolonged replication of seasonal breeding triggers.

Population structures for amphibians are usually seasonally controlled, with most amphibians leading a solitary life until the triggers for a breeding season start to occur. Some exceptions to this rule occur and looking at the behavioural patterns of a life history strategy of species can help inform an appropriate stocking density and level for captive amphibians. The mimic poison frog (*Ranitomeya imitator*) forms monogamous pairs that stay together throughout their lives; providing a population structure that allows for pair bonding early on in life is key for a natural social structure for this species in captivity (Brown, Morales, & Summers, 2010). The provision of inter- and intraspecies interactions in captivity can be of benefit for captive amphibians as such interactions occur in wild populations. However, the ability to retreat and remove themselves from any interaction is also important. Territorial opportunities in captivity can improve breeding outcomes and boost an individual's health status if done correctly in a way relevant for each species being housed (Anderson et al., 2014).

17.3 ENCLOSURE CONSIDERATIONS FOR AMPHIBIANS BASED ON BEHAVIOURAL EVIDENCE

Amphibians inhabit a multitude of different environments across the world. Consequently, they have evolved a vast array of behaviours, anatomical forms and physiological functions that help them to fill their chosen ecological niche. Observing amphibian behaviour in their natural habitat, and how they interact with their environment, can help us to provide captive amphibians with a biologically relevant captive enclosure that exploits their physiological adaptations. There is some opportunity to provide captive amphibians with a wild-like captive environment that is not practical or achievable for many other taxonomic groups; an almost like-for-like wild environment can be created for many amphibian species and quite often is a necessity for a healthy and successful captive population. Categories that should be considered when designing and building an enclosure for an amphibian are many, but with the overarching theme that, as ectotherms, their behavioural biological processes are governed by their ability to thermoregulate appropriately for their species' requirements. The strategy used to provide these requirements changes depending on the species' ecological niche and if their evolutionary adaptations lead them to lives of being arboreal, fossorial, terrestrial, aquatic, or a combination of the four.

Thermoregulation is based on the behaviours performed around the presence of the sun and the interactions it has with the environment. The behaviour of thermoregulation is linked to the process of achieving optimal body temperature that allows basics biological processes such as digestion, passively other benefits come from thermoregulation such as the exposure of UVB (Ultraviolet B) that allows for the synthesis of vitamin D3, which controls appropriate calcium distributions throughout bones and organs within the amphibian. Thermoregulation can be directly involved with basking in the patches of sunlight or interacting with an object or body of water that is affected by the sun. Other types of amphibian thermoregulation include thermoconformers; these are animals that control their body temperature through the shuttling between environmental temperatures that are not in direct sunlight and maintain a body temperature within an optimal range, whereas other amphibians use the method of basking in direct or defused sunlight. Recognising the type of thermoregulation strategy used by the amphibian will allow for accurate and proper control of their body temperature, ensuring they are able to achieve and complete basic biological processes. Evaluating temperature ranges within the habitats available to wild counterparts can inform the replication of appropriate climatic conditions and temperature ranges for captive amphibians. Such conditions may need to change with season as amphibians can be strongly connected to their environment and seasonal fluctuations. Without seasonal fluctuations amphibians are unlikely to be able to go through their full behaviour repertoire across each life stage, such as hibernation, courtship, and breeding events.

Provision of a light source is also needed to stimulate and encourage natural behaviours. The length of daylight hours (photoperiod) is believed to work in tandem with the triggering of seasonal behaviours, and it is important to replicate and match similar or identical daylight hours that the amphibian experiences in its natural range. Light intensity (LUX) also encourages an increased activity level and is also believed to have an influence on thermoregulation choice and improve thermoregulation quality, these light levels are important to match with the light levels achieved in their wild habitat to promote a wild-like activity budget and behavioural repertoire. Matching temperatures, UVB output, LUX levels and day length with the environmental parameters found the home range and changing them with the season is the basis for creating an appropriate enclosure for the amphibian being housed. Using the UV-Tool (highlighted in Section 17.6) to identify suitable biomes, heat, UVB, light levels, and photoperiod whilst including biologically relevant enclosure furnishings, will create an overall appropriate environment and microclimates that enable accurate thermoregulation.

Creating microclimates can be produced by including planting and furnishings within a captive environment. Provision of plants and enclosure furniture should be based on what is biologically relevant to the amphibian being considered and how it will promote natural behaviours. Provision of planting that either is the exact plant found in the natural habitat or replicates the function of that plant can allow for an increase in natural behaviours, providing tall grasses for bumblebee toads creates an opportunity for them to climb which they will readily do (Figure 17.1), this changes an enclosure for a species that is typically considered to be terrestrial to being adept climbers at low heights. Viewing the life history of dart frogs, such as the mimic poison frog, highlights how strongly amphibians can be linked to their environment and this can even include the species of plants around them. Bromeliads (Bromeliaceae) are used by mimic poison frogs as an established territory within which their young are laid, hatched, transported, and reared. Without these observations on how amphibians use specific plant species, it would be hard to replicate such conditions in captivity. Once such links between amphibians and their habitat are understood, a naturalistic alternative (or even artificial options are known to be successful) can be offered to fulfil this behavioural function. Many South American tree frogs are nocturnal, using large leaves to create a sealed humid and protected area that allows for sustained hydration via a drinking patch that is used during the heat of the day. As rain falls during the day, the tree frog can be observed pulling water into this sealed area, which increases its ability to stay hydrated before being able to go about their lives during the cooler

Figure 17.1 Bumblebee toad (*Melanophryniscus klappenbachi*) climbing long grasses in enclosure (photo: F. Boor/WWT Slimbridge).

Figure 17.2 Black-webbed treefrog (*Rhacophorus kio*) climbing along a branch.

and more humid night-time conditions. Viewing an amphibian's relationship and interactions with plants helps a captive enclosure in promoting relevant use of microclimates, natural behaviours, breeding events, territory establishment, parental care, and maintenance of body condition.

Substrates and surfaces have a significant impact on captive amphibian behaviour, affecting fossorial burrowing behaviours, hiding between fallen or dead leaves, movement speeds in arboreal species, and foraging suitability for aquatic species. The intrinsic characteristics of substrate types become more complicated as we learn more about the lives of amphibians; there is a clear preference for substrate types that provide advantages (e.g., during foraging) and allow for natural behaviours to be performed. For example, research that tested soil preferences of a captive population of Gaboon caecilians (*Geotrypetes seraphini*) showed that when two different soil types were provided and the position of the caecilians recorded, a preference for the substrate that allowed for better burrow retention was clear (Tapley et al., 2014). Identifying something as simple as substrate preference for a fossorial species is critical to how they spend perform an activity budget. Providing a substrate that allows for better burrow construction reduces energy expenditure and potentially reduces stress as the animal does not need to constantly reconstruct burrows after their collapse. Differing surfaces are not only found at ground levels- arboreal amphibians also come into contact with surfaces such as branches and vines; monitoring how animals move across these surfaces can inform keepers on their suitability for particular species. For example, narrow branches can be harder to use relative to the species being observed, and can causes an amphibian to change its grip type to a less effective style when compared to use of a wider branch (Herrel et al., 2013), Figure 17.2.

It is also important to look at the biological relevance of a substrate or surface for a captive amphibian and how it uses space. Some arboreal amphibians spend most of their lives in the tree canopy and very rarely come into contact with the ground except on rare occasions. In captivity we often want to provide everything that we view as "their environment" into an enclosure a fraction of the size of their wild environment. Therefore, some aspects (such as soil) are almost biologically irrelevant and it could be argued a cause of stress if the amphibian comes into contact with that substrate. The inclusion of an organic, ground-based substrate may be detrimental to the welfare of some arboreal amphibians, and other opportunities such as a clean hydration area may be more beneficial to these species. Such methods have been incredibly successful when used for fully arboreal species (Gray, 2020).

The provision of relevant substrates for aquatic amphibians is also extremely important – not only can it affect water quality in the enclosure but it also offers potential egg-laying sites (both in the water and around the water's edge). Without these substrate types, many species would not reproduce in captivity. The need for a substrate that is suitable for the transition between terrestrial to submerged-under-water is crucial for species such as marbled salamanders (*Ambystoma opacum*) that lay their eggs on the shore of seasonally fluctuating water bodies. These eggs, which are in stasis and partially developed, are waiting for the stimulation of flood waters before they hatch. Aquatic substrates, such as small gravel, silt, and larger pebbles allow for escape routes for aquatic live food, which can then establish within that environment and provide a continuous supply of food for larval and juvenile amphibians. Establishment of live food populations within an amphibian exhibit needs to be continuously monitored to ensure the overall health and cleanliness of the enclosure can be managed optimally.

Other factors of how amphibians interact with their wild environment should also be considered when recreating an appropriate captive enclosure, and such considerations can be how amphibians interact with other organisms that live alongside them. Custodians is a term used when discussing small organisms, such as springtails (Collembola) and woodlice (Oniscidea), that live in the same environment as our amphibians. In captivity, their job is to clear the enclosure of waste produced by the amphibians in a confined space, and if done correctly, custodians can help keep an environment healthy by reducing the build-up of a bio-load and enclosure toxicity. The interactions that amphibians have with these custodians in the wild are mainly as a food resource; often a fallen or rotten piece of fruit or animal waste will be the food source for these custodians and this creates a concentrated population of these small organisms that becomes an excellent opportunity for a feast for insectivorous amphibians.

Amphibians have a multitude of behaviours that are centred around their use of water and how it interacts with their everyday life and with different life stages – all amphibian species have an extreme dependency on water in their early development stages. Understanding the provision of water is key to having healthy and successful captive amphibians that perform a wide range of their diverse behavioural repertoires. Some aquatic life that is based around water, fish for example, have benefitted from years of research to further understand their biological relationships with water. Such considerations have only emerged in the captive amphibian world relatively recently and the relationship between amphibians and water is still being fully discovered and understood (Odum & Zippel, 2008).

Amphibian skin is highly permeable, and this allows for the absorption of water. Often, amphibians are used as indicator species to see how well an environment is doing, but incorrectly disposed water from captive amphibians can also create a direct and potentially devastating route for pollutants and contaminants in natural watercourses. Providing managed and safe water for amphibians is key for captive population success, using many aspects of water management from fish keeping strategies in combination with behavioural observations of water usage by captive amphibians to create biologically relevant water provision and allow for the display of natural behaviours. Using rainforest species as an example, misters and rain bars help to simulate a rainy season – providing hydrating wet leaves for red-eyed tree frogs or vital water changes for bromeliads (to prevent them drying out) that are home to dart frog tadpoles. However, these provisions would not be appropriate for a fully aquatic species such as the Lake Titicaca frog (*Telmatobius culeus*) that requires chilled water sustained at a relatively high pH to match the high-altitude waterbody of Lake Titicaca. Other montane species, such as the emperor crocodile newt (*Tylototrition shanjin*), breed in small, slow-moving water bodies such as ponds or ditches, and these slow-moving water bodies allow for courtship displays, the release of pheromones and the safe provision and collection of sperm packets (Ziegler, Hartmann,

Van der Straeten, Karbe, & Bohme, 2008), all of which could be disrupted by an artificially introduced fast water flow.

Enclosure sizes are changing as the amphibian industry learns more about the natural history and behaviour of amphibians as they are observed in the wild and subsequently kept in captivity. What would be considered as an acceptable enclosure size of a standard "hobbyist" vivarium may no longer be suitable to fulfil the needs of a captive amphibian. At a minimum, there needs to be space for an amphibian to experience a heterogeneous thermal environment that allows for optimal thermoregulation. Territories and aggressive interactions between tank mates need to be understood when selecting enclosure size – an enclosure needs to be large enough to fit an individual's territory, allowing for the option of other tank mates to enter and leave territorial disputes without the constant pressure of one individual dominating the only available area. The biological relevance of enclosure size and structure is key. Providing glass walls for arboreal amphibians seems to increase the useable surface area of their enclosure (Gray, 2020). Comparatively, providing an arboreal option for a fully fossorial species, such as a tiger salamander (*Ambystoma tigrinum*), would hold little value to the captive species and using this space for a deeper substrate level or the provision for a seasonally relevant water body would be a better use of space.

Overall providing an amphibian with choice and control over its environment is a key enclosure consideration- one that is essential to animal welfare. The options that a captive amphibian has choice and control over need to be well thought-out, have biological functionality, be safe to use, seasonally relevant, and promote natural behaviours.

17.4 BEHAVIOURAL BIOLOGY AND AMPHIBIAN WELFARE

Multiple species of amphibians can have their welfare assessed by their overall body condition and physical health, but this does require extensive knowledge of a standard physical condition of an amphibian. These physical welfare assessments do not cover all types of welfare impact, the results of which are often only the aftermath of an ongoing welfare problem. Using regular weighing and body condition scores can help record and compare responses to care (and hence welfare state). As amphibians can drastically lose condition, often hiding issues that cause a reduction in mass or body condition, regular monthly weighing should be undertaken (unless there is a requirement for more regular intervention) – this, in combination of daily visual checks that do not involve a disturbance or cause further stress, can give a good judge on the condition of the amphibian and identify when an animal may be experiencing poorer welfare. Managing groups of captive amphibians creates a competitive social environment where dominant individuals can outcompete subordinate individuals or quieter, less bold species. Watching for any behaviours indicative of (excessive) social dominance and (continued) negative responses to it, as well as recording enclosure usage, noting territories, visually assessing posture, and having visual body condition checks can help inform group management and if more resources need to be included. This can also inform decisions regarding whether individuals need to be removed or introduced to a different group dynamic; this creates a more holistic management style and reduces the cost of continuous stressors to the individual.

Although the idea of a "personality" in amphibians has not been well studied keepers can tell that individuals under their care have different personality traits that suit different captive situations. The expression of such personalities can increase and decrease as seasons fluctuate, allowing for a prediction on how an amphibian might behaviour during a breeding season or within a territorial dispute. Understanding the traits of our individual amphibians creates opportunities for better individual welfare assessment, by understanding if an individual is not behaving as it would normally and helps inform on group dynamics.

Understanding that amphibians use species-specific differences in visibility, whereby differences are related to the specific camouflage or anti-predation tactics (e.g., aposematism- being colourful and visible to advertise a warning message) offer an opportunity to understand the expected behaviour of a species. If an aposematic species is hidden from view for an extended period, we can assume that there are other factors that are stop it from using its "confidence" to sit out in the open, advertising its toxicity to predators. In the opposite situation, a camouflaged species that has become highly visible could be distressed, trying to move away from a factor that is causing the stress. This information can help to best select a species for public viewing and visitor engagement programmes, using life history strategies and natural behaviours to inform us on what species find stressful and therefore influencing species choice when creating an enclosure that allows for a good visitor experience plus a safe and useable enclosure for the amphibian (Boultwood, O'Brien, & Rose, 2021).

Recording captive environmental parameters that are matched to biologically relevant ranges can help to ensure that the correct environmental provisions are met. Daily accurate enclosure temperatures and air humidity should be taken from multiple points in an enclosure. Measuring UVB output and adjusting daylight intensity and photoperiods should also be recorded and changed dependent on wild seasonal data. This ensures that enclosure parameters are as biologically relevant as possible for the captive species and lets them perform behaviours dependent on seasonal requirements.

17.5 SPECIES-SPECIFIC ENRICHMENT FOR AMPHIBIANS

Enrichment for captive amphibians can encompass variation in environmental parameters, nutrition, and consideration of feeding ecology, seasonal fluctuations, heterogenous enclosure furnishings, social choices, and specific environmental changes based on life stage. All enrichment should enhance or encourage natural behaviour or stimulate a biological trigger that brings on a seasonally appropriate behavioural response. Appropriate enrichment for captive amphibians is shown to reduce mortality and injury rates, improve growth rates, and maintain good body condition. Many captive amphibian diseases are caused by incorrect husbandry that does not meet the biological or behavioural requirements of the amphibian. Small changes to environmental conditions can bring about huge benefits to animal behaviour and welfare. For example, an increase in appropriate basking behaviours reduces metabolic bone disease and encourages foraging behaviours by enabling custodians (also within the enclosure) to increase their activity levels (Michaels, Downie, & Campbell-Palmer, 2014).

Suitable ideas for amphibian enrichment include:

- Shelter provision – natural and non-natural shelters reduce panic behaviours and increase activity, also allowing for thermoregulation and controlled exposure to UVB.
- Increased waterbody surface area maximises oxygenation and therefore reduces surface swimming or breaking-the-water's-surface actions of some aquatic species.
- Deeper water can increase growth rates for species naturally occurring in larger water bodies.
- Visual barriers between tank mates help to establish territories and reduce overly aggressive interactions.
- Complex environmental furnishings (levels within a vivarium, perches, hollows, caves, and planting) improve body condition (increases exercise) and reduce stress caused by unwanted social interaction.
- Devices to slow feeding (e.g., coconut shells with a small hole in to let out fruit flies, *Drosophila*), or populations of enclosure custodians, or concentrated feed areas increase foraging duration, reducing rapid feeding behaviours and increasing the time between hunting events.
- Gradients in environmental parameters and mimicking seasonal fluctuations (by using misters, foggers, and rain bars) stimulates physiological responses and variation in natural behaviours and seasonal activity budgets.
- Introduction and removal of mate competition during breeding events can improve breeding strategies and improve overall mate choice.
- Allowing the performance of parental care (when applicable to the species and population management plans).
- Heterogeneous substrate types across the enclosure creates differing microhabitats.
- Complex planting to match the form and function of plants used by wild counterparts.

17.6 USING BEHAVIOURAL BIOLOGY TO ADVANCE AMPHIBIAN CARE

This chapter has examined the importance of enclosure design, husbandry routines, and population management techniques for captive amphibians being based on species' biological traits and natural behaviours. Amphibians are a diverse animal group, and their captive requirements are complex and specialised. All captive management decisions need to be based on evidence rooted in the species' habitat preference, life-history strategy, and ecology to ensure that any zoo enclosure is biologically relevant.

The main reasons for zoos existing are to create opportunities for education, research, conservation, and entertainment, and from this standpoint the animals chosen to be represented within the zoo should fall under one or more of these categories. Amphibians can have one of the lowest research focuses in the zoo and can have some of the fewest number of species held in zoological collections, and can be considered less "charismatic" species for the public when compared to mammals (Rose, Brereton, Rowden, Lemos de Figueiredo, & Riley, 2019). These factors become a greater issue when you look at the predicted extinction crisis that amphibians are currently facing, and that they are experiencing the greatest recorded loss of biodiversity that is attributed to a specific disease, chytridiomycosis (Scheele et al., 2019).

A cohesive approach to expanding captive amphibian research output would help improve our understanding of their husbandry, looking at simple yet fundamental questions such as soil type preferences has a significant impact on the lives of many captive amphibians. More critically endangered amphibian species should be considered as serious options for ex-situ zoo conservation outputs, with some species taking up little of the finite and valuable resources zoos have – expenditure of relatively little resource would have a significant impact on the conservation outcome for these critically endangered species.

A focused and collaborative approach between institutions breeding these species would also be useful to the overall captive amphibian population. Often large numbers of individuals of one species are held within a single institution, these individuals are often genetically valuable or from a single captive population and are valuable to the specific zoo. But sharing information and offering opportunities to become familiar with such a species (by increasing the number of zoos holding it) only benefits the potential successes of ex-situ populations in conservation, research, educational programmes, and visitor engagement.

Possibilities of future research options that could have a significant impact on individuals and populations are listed in Box 17.1.

To further inform our understanding of amphibian care and how to build in capacity for more zoos to be successful with amphibian species, Box 17.2 describes some useful sources of further information.

17.7 CONCLUSION

Amphibians' form and function, as well as the environments they inhabit, are incredibly diverse. Observing how they interact with their environment informs best practices for captive animals. Providing biologically relevant, evidence-based enclosures for captive amphibians promotes natural behaviours, breeding success, and our ability to maintain healthy captive populations. This provision of biologically relevant parameters gives choice and control over their use of the environment. Basing these parameters on information gathered from the wild means that we can meet safe working practices and improve welfare, even for parameters that we cannot easily perceive (such as UVB or thermoregulatory needs). For many amphibian species, we know very little about how they behave in the wild and this lack of knowledge can be caused

BOX 17.1: Future research options for captive amphibians

- Develop evidence-based dietary supplementation for commonly kept (overlooked) amphibians.
- Create body condition scoring standards for commonly kept species.
- Identify how captive amphibians use infrared wavelengths for thermoregulation.
- Identify how captive amphibians interact with UVB.
- Identify the effects of seasonal fluctuations in light intensity and photoperiod on amphibian behaviour.
- Evaluate whether the usual sizes of captive enclosures are large enough for specific tree frog species, for example, the giant waxy monkey frog (*Phyllomedusa bicolor*), to thrive and breed within.

BOX 17.2: Further reading for captive amphibian management:

- The UV-Tool, guidance for appropriate use and selection of UVB for reptiles and amphibians: Baines et al. (2016). How much UVB does my reptile need? The UV-Tool, a guide to the selection of UV lighting for reptiles and amphibians in captivity. *Journal of Zoo and Aquarium Research*, *4*(1), 42–63.
- An example that follows wild ecology and behaviours and applies those to captive breeding attempts: Bland, A. (2015). A record of captive reproduction in the red bellied toad *Melanophryniscus klappenbachi* with notes on the use of a short-term brumation period. *Herpetological Bulletin*, *131*, 15–18.
- Database that holds species-specific information and maps of species distributions: https://amphibiaweb.org/
- Climate graphs that can help inform annual seasonality for captive amphibians: https://en.climate-data.org/

by species' inaccessibility to researchers. Breaking down each part of captive care and evaluating its relevance to that species and its reaction to it is paramount for reducing the number of overlooked areas of captive husbandry and housing. The use of ecological information can turn a generic enclosure into a fossorial habitat or a tree top canopy, suitable for a specific species. Increased observations of amphibian behaviour (both in captivity and in the wild) and sharing this information more widely will only improve our ability to enhance captive environments, in turn improving amphibian welfare and increasing the opportunity to protect this declining group of animals.

REFERENCES

Anderson, T., Hocking, D., Conner, C., Earl, J., Harper, E., Osbourn, M., Peterman, W., Rittenhouse, T., & Semlitsch, R. (2014). Abundance and phenology patterns of two pond-breeding salamanders determine species interactions in natural populations. *Ocelogia*, *177*, 761–773.

Boultwood, J., O'Brien, M., & Rose, P. (2021). Bold frogs or shy toads? How did the COVID-19 closure of zoological organisations affect amphibian activity? *Animals*, *11*(7).

Brown, J., Morales, V., & Summers, K. (2010). A key ecological trait drove the evolution of biparental care and monogamy in an amphibian. *The American Naturalist*, *175*(4), 436–446.

Frost, D. (2021). Amphibian species of the world 6.1. Retrieved from American Museum of Natural History: https://amphibiansoftheworld.amnh.org

Gray, A. (2020). Cruziohyla captive husbandry guidelines. Retrieved from FrogBlogManchester: https://frogblogmanchester.com/about/costa-rican-frogs-2/cruziohyla-captive-husbandry-guidelines/

Herrel, A., Perrenoud, M., Decamps, T., Abdala, V., Manzano, A., & Pouydebat, E. (2013). The effect of substrate diameter and incline on locomotion in an arboreal frog. *Journal of Experimental Biology*, *216*(19), 3599–3605

Kupfer, A., Muller, H., Antoniazzi, M., Jared, C., Greven, H., Nussbaun, R., & Wilkinson, M. (2006). Parental investment by skin feeding in a caecilian amphibian. *Nature*, *440*, 926–929.

Lock, B. (2017). Corneal Lipidosis/Xanthomatosis in Amphibians. Retrieved from Veterinary Partner: https://veterinarypartner.vin.com/default.aspx?pid=19239&catId=102919&id=7996856

Michaels, C., Downie, R., & Campbell-Palmer, R. (2014). The importance of enrichment for advancing amphibian welfare and conservation goals: A review of a neglected topic. *Amphibian & Reptile Conservation*, *8*(1), 7–23.

Odum, R., & Zippel, K. (2008). Amphibian water quality: Approaches to an essential environmental parameter. *International Zoo Yearbook*, *42*(1), 40–52.

Ogilvy, V., Preziosi, R.F., & Fidgett, A.L. (2012). A brighter future for frogs? The influence of carotenoids on the health, development and reproductive success of the red-eye tree frog. *Animal Conservation*, *15*(5), 480–488.

Rose, P., Brereton, J., Rowden, L., Lemos de Figueiredo, R., & Riley, L. (2019). What's new from the zoo? An analysis of ten years of zoo-themed research output. *Palgrave Communications*, *5*(1), 1–10.

Sanchez-Ochoa, D., Perez-Mendoza, H., & Charruau, P. (2020). Oviposition site selection and conservation insights of two tree frogs (*Agalychnis moreletti* and *A. callidryas*). *South American Journal of Herpetology, 17*(1), 17–28.

Scheele, B.C., Pasmans, F., Skerratt, L.F., Berger, L., Martel, A., Beukema, W., Acevedo, A.A., Burrowes, P.A., Carvalho, T., Catenazzi, A., De la Riva, I., Fisher, M.C., Flechas, S.V., Foster, C.N., Frías-Álvarez, P., Garner, T.W.J., Gratwicke, B., Guayasamin, J.M., Hirschfeld, M., Kolby, J.E., Kosch, T.A., La Marca, E., Lindenmayer, D.B., Lips, K.R., Longo, A.V., Maneyro, R., McDonald, C.A., Mendelson, J., Palacios-Rodriguez, P., Parra-Olea, G., Richards-Zawacki, C.L., Rödel, M.-O., Rovito, S.M., Soto-Azat, C., Toledo, L.F., Voyles, J., Weldon, C., Whitfield, S.M., Wilkinson, M., Zamudio, K.R., & Canessa, S. (2019). Amphibian fungal panzootic causes catastrophic and ongoing loss of biodiversity. *Science, 363*(6434), 1459–1463.

Tapley, B., Bryant, Z., Grant, S. Kother, G., Feltrer, Y., Masters, N., Strike, T, Gill, I., Wilkinson, M., & Gower, D.J. (2014). Towards evidence-based husbandry for caecilian amphibians: Substrate preference in *Geotrypetes seraphini. The Herpetological Bulletin, 129*, 15–18.

Wildenhues, M., Rauhaus, A., Bach, R., Karbe, D., Straeten, K., Hertwig, S., & Ziegler, T. (2012). Husbandry, captive breeding, larval development and stages. *Amphibian and Reptile Conservation, 5*(3), 15–28.

Yeager, M.B., Dugas, J., & Richards-Zawacki, C.L. (2013). Carotenoid supplementation enhances reproductive success in captive strawberry poison frogs (*Oophaga pumilio*). *Zoo Biology, 32*(6), 655–658.

Ziegler, T., Hartmann, T., Van der Straeten, K., Karbe, D., & Bohme, W. (2008). Captive breeding and larval morphology of *Tylototriton shanjing* Nussbaum, Brodie & Yang, 1995, with an updated key of the genus. *Der Zoologische Garten, 77*, 246–260.

18

The behavioural biology of freshwater fishes

*CHLOE STEVENS
RSPCA, Wilberforce Way, Horsham, UK

MATTHEW FIDDES
CJ Hall Veterinary Surgery, London, UK

PAUL ROSE
University of Exeter, Exeter, UK
Slimbridge Wetland Centre, Slimbridge, UK

* The views expressed in this chapter do not necessarily reflect those of the RSPCA, its RSPCA Council, staff, members, or associates.

DOI: 10.1201/9781003208471-20

18.1 INTRODUCTION TO FRESHWATER FISH BEHAVIOURAL BIOLOGY

Public aquaria are common multi-taxa attractions that provide unique insights into an unfamiliar environment. As well as specific, purpose-built attractions, many zoos and wildlife parks also contain aquaria that house diverse ranges of species and habitats from wetland ecosystems. Freshwater fishes are being given more space in the exhibits at large, accredited, or membership institutions due to the conservation needs of many species. Zoological organisations have been instrumental in the conservation and extinction prevention of several species that have been extirpated from the wild, such as the butterfly goodeid (*Ameca splendens*) and the Potosi Hole pupfish (*Cyprinodon alvarezi*), listed as Critically Endangered (CR) and Extinct in the Wild (EW) on the IUNC Red List, respectively. Intensive management and expert care, based on sound knowledge of fish ecology, helps build and maintain sustainable populations within ex-situ facilities. Within BIAZA (British & Irish Association of Zoos and Aquaria) institutions, several populations of EW fishes are managed as part of global conservation action, including La Palma pupfish (*C. longidorsalis*), a close relative of the Potosi Hole pupfish, which shares a similar fate in its native habitat. Approximately half of fish species are found in freshwater and they make up a quarter of all known vertebrate species, yet over a third of these species are threatened with extinction (IUCN, 2021). The need for the aquarist to not only be an animal keeper but also a conservation biologist, with an understanding of the behavioural biology of the species being managed, is more important than ever.

Despite covering a relatively small proportion of the Earth's surface, the habitats of freshwater fishes can vary widely in water chemistry, temperature, structure, seasonality, and food availability. As a result, the fishes that live in these habitats exhibit an extraordinary amount of morphological, ecological, and behavioural diversity – which has important implications for their captive care. One of the largest of all freshwater fish, the arapaima (*Arapaima gigas*) from the basin of the Brazilian Amazon, can reach three metres in length, while some fish species are among the smallest of all living vertebrates. The smallest freshwater fish, the Indonesian superdwarf fish (*Paedocypris progenetica*), which is a mere 1 cm long, will have very different aquarium husbandry needs compared to the arapaima! This chapter will therefore aim to highlight some of this behavioural diversity and discuss how this relates to husbandry. It aims to focus on representative species and examples that are commonly seen in public aquaria (some examples that help illustrate the potential species diversity of a public aquarium are provided in Table 18.1), but will also draw examples from species more closely associated with aquaculture or scientific research. The most important take-home message for aquarists is the need to have detailed knowledge of the natural environments and behavioural ecology of each species (such as their movement patterns, seasonal influences on behaviour, social, reproductive, and foraging activities) to provide the correct "environment in miniature" in the aquarium.

18.2 FRESHWATER FISH ECOLOGY AND NATURAL HISTORY RELEVANT TO THE AQUARIUM

18.2.1 Ecology

Freshwater fish species are found from the Arctic to the tropics, in water bodies including streams, lakes, and rivers. The features of these water bodies can be drastically different – from the small, ephemeral pools inhabited by the African turquoise killifish, *Nothobranchius furzeri* (Valdesalici & Cellerino, 2003), which lives for approximately three months, to the huge and highly speciose environments of the African Great Lakes. Some water bodies are highly vegetated, and fishes may use these plants as food, shelter, or breeding substrate, while others are more barren; for example, the blind Mexican cavefish, *Astyanax mexicanus*, lives in darkness with little food or oxygen in isolated caves (Elliott, 2018).

The conditions of the water itself are a key aspect of the environment. Fishes are sensitive to the physico-chemical properties of water, including temperature, pH, dissolved oxygen, salinity, hardness, conductivity, turbidity, flow rate, and concentration of nitrogenous compounds, so providing them with conditions that replicate those found in their natural environment is one of the most important aspects of captive care. Changes in water conditions may also be important – for example, they can

Table 18.1 Example species of commonly exhibited freshwater fishes in public aquaria.

Order	Species	Geographic range	Preferred water conditions (wild)
Characiformes	Red-bellied piranha (*Pygocentrus nattereri*)	Whitewater wetlands in South America, e.g., River Amazon.	Neutral to mildly acidic pH. Wide temperature range of 15–30°C.
Cichliformes	Discus (*Symphysodon* sp.)	Floodplains, lakes, and rivers of the Amazon Basin, South America. Some species are confined to blackwater rivers and flooded forests.	Dependent on species, neutral to strongly acidic pH. Temperature range of 28–31°C.
Anabantiformes	Giant gourami (*Osphronemus goramy*)	Wide range of rivers, lakes, and swamps in Southeast Asia, e.g., lower Mekong in Cambodia and Vietnam.	Water pH around neutral. Temperature range of 20–30°C.
Siluriformes	Redtail catfish (*Phractocephalus hemioliopterus*)	Tropical South America, Amazon, Orinoco, and Essequibo river basins.	Acidic to neutral Ph range, normally around pH 6.5. Temperature range of 20–26°C.
Characiformes	Congo tetra (*Phenacogrammus interruptus*)	Tropical Africa, central basin of the River Congo.	Acidic to neutral Ph, optimal around Ph 6.5. Temperature range of 22–27°C.
Cypriniformes	Koi carp (*Cyprinus rubrofuscus* "koi")	Japan, domesticated form of the Amur carp.	Neutral Ph. Avoid extreme temperature ranges; 15–25°C is optimal.
Atheriniformes	Boeseman's rainbowfish (*Melanotaenia boesemani*)	Endemic to Western New Guinea, lakes, and tributaries of Bird's Head Peninsula.	Neutral to alkaline pH (up to pH 8). Temperature range of 25–27°C.
Cyprinodontiformes	Guppy (*Poecilia reticulata*)	The Caribbean (e.g., Trinidad & Tobago) and South America (e.g., Venezuela).	Tolerant of many water conditions but ideally pH neutral to slightly alkaline. Temperature range of 22–28°C.
Characiformes	Cardinal tetra (*Paracheirodon axelrodi*)	Upper reaches of the Orinoco River and Rio Negro in the southern Amazon.	Very acidic pH (around pH 4 to 5.5) with zero hardness. Temperature range of 23–26°C.
Cypriniformes	Clown loach (*Chromobotia macracanthus*)	Streams and flood plains of inland Borneo and Sumatra.	Shows highly seasonal fluctuation in habitat choice. Variable pH range of pH 5–8. Temperature range of 25–30°C.
Osteoglossiformes	Silver arowana (*Osteoglossum bicirrhosum*)	Flooded forests, white- and blackwater habitats of the Branco River and basins of the Rivers Amazon, Essequibo, and Oyapock.	Acidic pH (5.5 to 6) is best but will withstand neutral conditions. Temperature range of 24–30°C.

be a breeding cue for tropical species that spawn when the rainy season begins.

18.2.2 Diet and feeding behaviour

Freshwater fishes may be carnivores, omnivores, herbivores, or detritivores, and even within each of these broad groups, feeding and foraging strategies may vary. For example, among carnivores, some predators actively hunt prey while others employ "lie-in-wait" strategies; some use teeth to capture prey while others use suction to pull prey into their mouth (Helfman, Collette, Facey, & Bowen, 2009). Some species have generalist diets, while others are more specialist, such as *Perissodus microlepis*, a scale-eating cichlid from Lake Tanganyika (Takeuchi, Hori, & Oda, 2012). In an aquarium, diets need to be chosen carefully with the natural tendencies of the species and life stage in mind: for some generalist herbivores and omnivores, such as guppies and tetras, this may be as simple as providing flake or pellet food along with plant material and dead invertebrates. Others prefer live prey – for example, archerfish (*Toxotes* spp.) are known for their interesting strategy of shooting jets of water at insects or other arthropods to knock them into the water (Simon & Mazlan, 2010).

18.2.3 Social behaviour

Sociality is common in freshwater fishes, with many species forming aggregations that range from small, temporary groups to larger, more organised schools or shoals at some point in their lives (Wright, Ward, Croft, & Krause, 2006). There are also examples of solitary or territorial species, and some fishes that are solitary as adults but form shoals as larvae, e.g., jewel cichlid, *Hemichromis bimaculatus* (Chen, Coss, & Goldthwaite, 1983). Many species are capable of differentiating between kin and non-kin, e.g., striped kribensis, *Pelvicachromis taeniatus* (Hesse, Bakker, Baldauf, & Thünken, 2012), or familiar and unfamiliar individuals, e.g., rainbowfish (Brown, 2002) and will use this information to make decisions about which groups to join. Other factors such as group size are relevant to these decisions: many species will choose larger shoals over smaller shoals when given a choice in laboratory conditions, e.g., swordtails, *Xiphophorus helleri* (Wong, Rosenthal, & Buckingham, 2007). Cues from different sensory modalities are used for recognition and communication among individuals and within groups. For example, red-bellied piranhas produce sound using jaw movements and muscles around the swim bladder, and different sounds are associated with different behaviours, including display, aggressive, and social behaviours (Millot, Vandewalle, & Parmentier, 2011).

Aggressive behaviours may be used to maintain dominance hierarchies within groups, or to maintain ranges among territorial species. In social species, dominance is often based on body size, although sex, prior residency, experience, or age may also play a role (Helfman et al., 2009). In the aquarium, housing social fishes in too small a group, or altering group size or composition may result in an unstable dominance hierarchy and lead to increased aggression. For territorial species, providing fishes with enough space to form territories without encroachment is key for minimising aggression and promoting good welfare.

18.2.4 Reproductive behaviour

Given the importance of public aquaria to the conservation of many freshwater fish species, sound knowledge of the breeding of freshwater fishes is crucial. Reproductive strategies vary from parthenogenesis to monogamy to polygamy to promiscuity. Strict monogamy seems to be relatively uncommon among freshwater fishes, as some fishes like the arapaima are socially monogamous but engage in extra-pair mating (Farias et al., 2015). Both polygyny and polyandry are seen in freshwater fishes – in polygynous species, this may result in the formation of harems, where one male defends multiple females, and the males of some fish species display lekking behaviour to attract females. Although most freshwater fishes lay eggs that are externally fertilised, a number are internal fertilisers and bear live young (e.g., Poeciliidae).

Parental care occurs in many freshwater fish species. Some species build nests from substrate, plant matter, or bubbles in which spawning can take place and where eggs can be guarded. Others spawn on surfaces such as rocks or on plant leaves. Other forms of parental care might include fanning water over eggs to supply them with oxygen or guarding and provisioning larvae. A number of species, most famously among Cichlidae, engage in mouthbrooding of eggs and larvae (Kuwamura, 1986). However, some species do not display any parental care at all and may eat young if they can gain access to them – in these cases, larvae may use plants or other shelters to avoid predation by adults. In a captive system, the aquarist will need to understand all of these natural tendencies of the species in order to provide fishes with the correct external cues and water conditions, sex ratios, amount of space, plants or substrates for nest building or laying eggs, and shelter for larvae.

Using the species listed in Table 18.1, examples of important behavioural ecology information that should inform captive care are provided in Table 18.2.

18.3 ENCLOSURE CONSIDERATIONS FOR FRESHWATER FISHES

Providing fishes with the correct water quality parameters for their species is fundamental to good welfare. As a result, one of the most important parts of enclosure design for fishes are the life support systems that help maintain water quality, such as filters, aeration equipment, and water pumps (Kasper, Adeyemo, Becker, Scarfe, & Tepper, 2021). These systems will need to be designed and routinely checked to provide water of the correct temperature, pH, hardness, salinity, flow rate, etc., and to ensure that waste (such as faeces or uneaten food) or toxic chemicals (such as nitrogenous compounds) are removed from the system. Some species are stimulated to breed in response to changes such as would be caused by the onset of rain, so life-support systems should be capable of mimicking this.

Another key part of enclosure design is tank size. Tanks that are too small can have detrimental impacts on behaviour and welfare, such as higher aggression, lower activity, less boldness, and poor stamina (Oldfield, 2011; Polverino, Ruberto, Staaks, & Mehner, 2016), but it is difficult to provide specific guidance on appropriate tank sizes, as many factors have to be considered. Larger tanks will generally be needed for large, active, or territorial species, or those that naturally live in very large groups, and consideration should be given to how much space is needed for planted or furnished tank areas vs open swimming space. Such decisions become even more complicated when a tank will hold a multi-species assemblage – do the different species interact well together? Do they naturally occupy the same or different parts of the water column? Might one species need refuges where it can retreat from the other species if necessary? Larger tanks may also be beneficial from a husbandry perspective, as they make it easier to maintain more stable water quality parameters.

The natural tendencies of fishes should be considered when deciding on tank shape as well as size. Species that inhabit deep lakes are likely to need deeper tanks than species that inhabit shallow rivers or streams, and different species make use of the water column in different ways. For example, Gimeno, Quera, Beltran, and Dolado (2016) compared swimming behaviour in zebrafish (*Danio rerio*) and black neon tetras (*Hyphessobrycon herbertaxelrodi*) and found that zebrafish seemed to make use of the whole water column, while tetras appeared to stay more in one stratum of the water column. Different tank shapes may also have important impacts on fish behaviour and welfare – male swordtails housed in tall, narrow tanks were less aggressive than those housed in wide, shallow tanks of the same volume (Magellan et al., 2012).

The interior of a tank may be furnished with substrates (such as rock, sand, or gravel), rockwork around the walls or in the centre of the tank, plants, dead wood, and other structures that provide refugia and structural complexity. Furnishings may be real or man-made – for example, some public aquaria include artificial trees or mangrove roots in some exhibits. Most tank furnishings are chosen with the intention of mimicking a particular ecosystem or habitat to at least some degree and this approach is likely to be the most beneficial for fish welfare, as well as helping to create exhibits that are more educational and interesting for visitors. Knowledge of how fishes use structures in the wild can help when designing aquariums for ornamental species. For example, young speckled piranhas (*Serrasalmus spilopleura*) use water hyacinths as shelter and a place to forage (Sazima & Zamprogno, 1985), while freshwater angelfish (*Pterophyllum scalare*) lay their eggs on large leaves (Brandão et al., 2021), so providing these plants is likely to benefit welfare and promote behavioural diversity. Some furnishings may have additional benefits, such as the use of dead leaves to increase tannin levels in the water as in blackwater streams of the Amazon – adding these increases the visual resemblance of the tank to the wild environment but also helps lower water pH and hardness (see Figure 18.1).

Table 18.2 Behavioural ecology of selected freshwater fishes common in the public aquarium.

Species	Example behavioural trait	Relevance to captive care
Red-bellied piranha 	Shoaling behaviour provides a protective function to individuals within a group, increasing levels of security against predators, e.g., giant otters (*Pteronura brasiliensis*).	Piranha in larger shoals have improved welfare (show reduced respiratory rates) and have an increased chance of performing breeding behaviours (spawning occurs as a group and individual fish darken in colour).
Discus 	Unique among the cichlids of the Americas, discus form static aggregations of many hundreds of individuals outside of the breeding season, with pairs moving away from their shoal to lay eggs and rear fry. Fry are fed on mucus, produced directly from the parent's skin. Discus have an elongated digestive tract, more similar to a herbivorous species, reflecting the variation in their diet caused by the seasonal rise and fall of the River Amazon.	Discus habitats are characterised by extreme seasonal fluctuations in food and shelter availability, dissolved oxygen, and risk of predation. Most of the anatomical, physiological, and behavioural adaptations of discus have been influenced by these fluctuating conditions. Captive discus are sensitive to incorrect feeding regimes, water parameters and social groups, and are susceptible to stress-related health issues.
Giant gourami 	Giant gouramis are predominantly herbivorous, feeding on algae when young and a wide range of plant material as mature adults. These large fish are active, with a high metabolism and have a high daily food intake.	As a mainly herbivorous species, giant gouramis in the aquarium can be kept occupied by "browse" (leaves and vegetation added to the aquarium for the fish to consume) in a similar manner to that used for browsing mammals.
Redtail catfish 	Redtail catfish have crepuscular and nocturnal foraging activities, being generally motionless during daylight hours. The preferred diet is live and dead fish, and the catfish's large mouth allows it to consume sizeable prey from an early age.	The large size of this species means only the largest aquariums can sufficiently provide space for a mature adult. Too regular feeding affects the health of this catfish, and tankmates need to be chosen carefully to avoid predation issues. The catfish's large mouth also puts it at risk of swallowing non-food objects within the aquarium.

Congo tetra 	A large and very active tetra that is always found in groups. Male Congo tetras have fin extensions and shimmering blue and gold scales that play an important role in courtship display.	Poor water quality negatively affects the development of fin extensions and colouration in males. A large aquarium is essential to maintain a sufficient number of individuals to allow appropriate social activities to be performed.
Koi carp 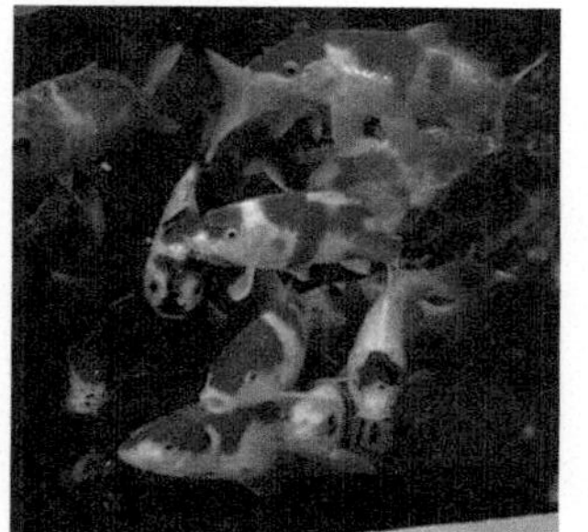	Sensitive barbels around the koi's mouth allow them to find food hidden in the substrate without needing to see where food is located. Koi have a downturned mouth and spend a large proportion of their time foraging in the substrate for food.	Foraging behaviour increases the turbidity of the water due to the increased suspension of particles from the substrate caused by digging and rooting behaviours. Strong filtration is required to ensure water quality remains high and oxygenation is needed in warmer weather to prevent respiratory distress.
Boeseman's rainbowfish 	Rainbowfish are always found in groups, and shoals of many hundreds of fish have been recorded. Such large shoals stimulate the performance of a wide range of social behaviours (affiliative, agonistic, and reproductive) and can enhance opportunities for social learning between individual fish.	Fish kept alone or in pairs or trios may display duller colours and be less active. The high activity levels of this species are promoted by providing an open area for swimming in the middle of the aquarium. Slow-moving tankmates are unsuitable as they can become stressed due to the rapid swimming speeds of the rainbowfish.
Guppy 	Like all members of Poeciliidae, guppies are livebearers. Internal fertilisation of the female by the male fish, using a gonopodium (modified anal fin) results in heavily gravid females giving birth to up to 200 fry. This reproductive strategy results in fry that are larger and studier than those from oviparous species, able to consume a wider range of foods and have an increased chance of survival.	The alternative name for the guppy ("millions fish") is not without meaning, as these fish can reproduce rapidly and a single pair can produce a large shoal in a short space of time. Population management and ethical control of breeding is required to prevent overstocking. Keeping multiple females to each male reduces female harassment.

(*Continued*)

Table 18.2 (Continued) Behavioural ecology of selected freshwater fishes common in the public aquarium.

Species	Example behavioural trait	Relevance to captive care
Cardinal tetra	A specialist of the acidic waters of the Rio Negro drainage system of Brazil, cardinal tetras are found in clearwater streams and blackwater flooded forests, moving in large shoals as water levels rise and fall seasonally. The distinctive colour pattern of the cardinal tetra reflects light in different angles, breaking up the fish's shape and confusing predators.	Provide open middle and top areas of the water column for this tetra to shoal in. Plants can be provided to fill the bottom of the aquarium, with some taller areas of refuge added. A mixture of shaded and brighter areas will show off the fish's colours and provide opportunities for display between individuals.
Clown loach	Complex social hierarchy with a shoal normally led by a dominant female who swims at the top of the group. Smaller and subordinate fish will press themselves close to larger fish and mimic their actions. Clown loach shoals will rest together, often in unusual positions such as lying on the substrates on their sides.	Provide hiding places in the aquarium for all fish to be able to shelter in together. Clown loach enjoy packing themselves into tight spaces, such as tubes. To replicate the social hierarchy documents in the wild, at least five loaches should be maintained in a shoal.
Silver arowana	A solitary predator that hunts at the water's surface of the flooded forest of the Amazon. Can wait in ambush for arboreal prey located above slow-moving water, then leap out of the water to grab prey.	Often develops downturned eyes ("droop eye") in aquaria. While the causal factor of this is unknown, encouraging the fish to look up for its food, reducing the amount of fatty foods provided, and encouraging activity at the water's surface may all help reduce this change in eye structure and position.

Congo tetra image credit Wikimedia Commons (creative commons licence).
Guppy image credit: Wikimedia Commons (creative commons licence).
All other images: P. Rose.

Figure 18.1 Examples of freshwater aquaria where ecology and behaviour influence tank design. Top left: marbled headstanders (*Abramites hypselonotus*) in an Amazonian blackwater set-up. Tannin-stained waters are caused by the use of bogwood, dead leaves, and blackwater extract to reduce pH and hardness. The characteristic "nose to the ground" swimming behaviour of headstanders is promoted by the open space provided in the aquarium for shoaling. Top right: heavy planting at the back of a clearwater Amazonian set-up for rummy-nose tetras (*Hemigrammus rhodostomus*), bleeding-heart tetras (*Hyphessobrycon erythrostigma*), and silver hatchetfish (*Gasteropelecus levis*) allows an open area for schooling but provides safe areas for refuge if needed. Planting in this manner calms fish and makes them more likely to be visible, as they will feel more safe and secure. Centre left: School of red-bellied piranha in breeding condition. Mature piranha darken in colour and chew down an area of aquatic plants ready for spawning. Mimicking a sudden flood of colder, clean water into the aquarium (as caused by the onset of the rainy season) can stimulate spawning. Centre right: A huge aquarium for arapaima and other such "tankbusters" (e.g., pacu, *Colossoma macropomum*) with minimal furnishing to allow maximum space for swimming, but robust refuge points available to the rear of the exhibit. The obligate atmospheric airbreathing nature of the arapaima is considered with the large, uncluttered area that allows the fish to access the water's surface. Bottom: A long aquarium for rainbowfish that provides an expanse of open swimming space, with patches of cover as well as an area of stronger water flow. All these features promote the natural high activity levels of the rainbowfish and enable them to move as a shoal, which reduces stress, enhances colouration, and increases behavioural diversity.

18.4 FRESHWATER FISH BEHAVIOURAL BIOLOGY AND WELFARE

Investigations into fish welfare have become more commonplace as we learn more about how various fish species perceive the environment around them, how they process information taken from the environment, and how they may perceive and experience distress and suffering. For the myriad of freshwater fish species kept in aquariums, specific indicators of welfare are still to be identified and developed. With this in mind, Figure 18.2 attempts to illustrate potential the physical, psychological, and behavioural indicators of welfare that could be used to infer the welfare states of freshwater fish in the aquarium.

The study of fish welfare has historically received less attention than the study of welfare in other vertebrates, and understanding of fish welfare has been complicated by debates over fish sentience and cognition (Brown, 2015). However, many authorities now state conclusively that fishes are capable of feeling pain, that they are sentient and experience different welfare states akin to other vertebrate species, and so consideration of their welfare, and how to improve it, is necessary (Sneddon et al., 2018). Despite this progress, assessing fish welfare can be a challenge, as there is still much to be learned about the subjective experiences of fishes, and how different stimuli affect behaviour and welfare.

Fish welfare may be assessed through physical indicators (such as eye and skin colour, the presence of injuries or disease symptoms), physiological indicators (such as cortisol release rates) or behavioural indicators. A combination of indicators will often provide the most complete picture of a fish's welfare state, but behavioural indicators may be particularly useful for day-to-day monitoring as they often happen rapidly in response to a stressor and may be relatively easy to observe. However, not all behaviour changes are relevant to

Figure 18.2 An example of potential identifiers of aquarium fish welfare. Physical indicators (green boxes) can relate to the fish's colouration and patterning, orientation (movement within the water column), overall health and condition of scales, fin integrity, and eyes. Behavioural indicators (blue boxes) include the movement of the fish, the normality of its behaviour patterns as befits the daily time budget of that species, and use of resources including the fish's interactions with environmental features. New measures of sensory perception and communication, for example sound production and acoustic signalling, will provide more information as to how fishes assess stimuli from their environment and present specific behavioural responses accordingly. Evaluation of behaviour patterns can indicate environmental appropriateness, which will influence the individual behavioural synchrony ("personality") of each fish kept – such parameters are indicated in the gold boxes. Social choice refers to the fish's activities within its shoal or social group (choosing to be with others or apart, and including synchronised collective behaviours, such as schooling). Individuality (e.g., bold or shy, neophobia or neophilia) and personality characteristics (e.g., excited, aggressive) refer to a behavioural expression common to that specific animal and can be a useful judge of internal state under prevailing environmental conditions.

all species or life stages, so specialist knowledge is needed to identify and interpret changes (Martins et al., 2012). Some examples of indicators an aquarist might look for are as follows.

18.4.1 Feeding behaviour

Feeding time often provides a convenient opportunity for doing health and welfare checks, and changes in feeding behaviour are an indicator of welfare issues in many species. Many species reduce foraging behaviour and/or overall food intake in response to stress (e.g., angelfish, Gómez-Laplaza & Morgan, 1993), which may cause reduced growth rates. Fishes may also increase or decrease how quickly they start feeding when presented with food (e.g., rainbow trout, *Oncorhynchus mykiss*, Øverli, Sørensen, Kiessling, Pottinger, & Gjøen, 2006; three-spined sticklebacks, *Gastereosteus aculeatus*, Magierecka, Lind, Aristeidou, Sloman, & Metcalfe, 2021). Other possible indicators of welfare related to feeding behaviour include poorer foraging efficiency and altered feeding patterns – European perch (*Perca fluviatilis*) were found to change their preferred feeding time from day to night after stress (Ferter & Meyer-Rochow, 2010).

18.4.2 Social behaviour

Many freshwater fishes are social, forming schools or shoals in the wild. For these species, shoaling behaviour can be a useful welfare indicator. Social fishes may shoal more tightly and cohesively in response to stressors, although some recent evidence from zebrafish has suggested "heightened shoaling" behaviours may be indicative of positive welfare (Franks, Graham, & von Keyserlingk, 2018). However, where welfare problems are associated with individuals rather than the whole group, a single fish may be observed being isolated or distant from the main shoal.

Being exposed to aggressive interactions is likely to have negative effects on fish welfare, but aggression can also be a consequence of other welfare issues. For example, in some species of Lake Tanganyika cichlid, aggression is more likely when there is high competition for resources, or in response to changing water conditions (Satoh, Ota, Awata, & Kohda, 2019). Aggression may also result from unstable social hierarchies, which may occur when there are too few individuals or too much disruption in a tank, and may be more common in more simple environments, perhaps due to there being fewer areas in which subordinate fishes can avoid dominant individuals (Oldfield, 2011). The aquarist should therefore use the presence of aggression in a tank as a possible indicator of compromised welfare, and may seek to minimise aggression by checking the water quality, ensuring the correct provision of food, space and shelter, and ensuring social groups are kept stable and the right size.

18.4.3 Activity

What counts as a "normal" activity pattern will depend strongly on the species in question – some fishes are generally active and spend much of the day foraging, while others spend little time moving. Whatever the "normal" state is, deviations from this may indicate welfare issues. Both increases and decreases in activity may be seen, varying according to both the species and the stressor in question, and fishes may display erratic movements or freezing behaviour (Blaser, Chadwick, & McGinnis, 2010). Decreases in activity may also be associated with greater use of shelters or a tendency to stay closer to the walls of the tank. Alongside changes in activity level, fishes may change their ventilatory activity, and monitoring this by counting opercular beats can be a useful measure of welfare.

18.4.4 Abnormal behaviours

A variety of abnormal behaviours may be seen in response to poor welfare, and some of these may be indicative of particular welfare issues. Many species will show "piping" behaviour (gasping at the surface) in response to low dissolved oxygen or "flashing" behaviour (sudden and rapid rubbing on surfaces) in response to parasite presence or poor water quality. Loss of equilibrium, buoyancy, or other difficulties swimming may indicate swim bladder issues or disease presence. Other abnormal behaviours, such as fin "clamping" (holding fins very close to the body) are more general indicators of stress. Stereotypic behaviours of freshwater fishes have not been characterised many times in the literature, but there is some evidence that at least some species display stereotypies in response to stress, e.g., fathead minnows, *Pimephales promelas* (Crane & Ferrari, 2016).

There is scope to add more behaviours to our repertoire of welfare indicators, but more research into these behaviours is needed. For example, numerous species of fish have been documented as producing sound. The red-bellied piranha produces sound as a form of interspecific communication (Millot et al., 2011), but also when handled, which might give insight into the fish's psychological state during handling and being removed from the water. Research into the acoustic communication of fish and how this corresponds to their behaviour patterns will provide aquarists with more information on which to base welfare judgements and husbandry needs.

18.5 SPECIES-SPECIFIC ENRICHMENT FOR FRESHWATER FISHES

The whole aquarium can be enriching if the right considerations of fish behaviour and ecology are used when furnishing, lighting, planting, and stocking. Water movement is also a source of enrichment, with different flow speeds and areas of still or flowing water helpful to increasing opportunities for a range of different movements or swimming patterns and speeds.

An example of how the whole aquarium can be enriching is provided in Figure 18.3. This mixed-species set-up provides stimulation at the intra- and interspecific level and fishes have been chosen to occupy different areas of the water column, thus avoiding competition and potential negative interactions. Territorial species, such as firemouth cichlids (*Thorichthys meeki*), have space at the bottom of the aquarium to form pairs and defend a specific patch, enabling them to perform important social behaviours without causing harassment. Surface-dwelling livebearers, such as splitfins (Goodiedae) will not encroach on the preferred space of the cichlids and instead use an area of the aquarium that enables them to avoid aggressive encounters. Planting and bogwood, and tight spaces behind rockwork, provide refuges

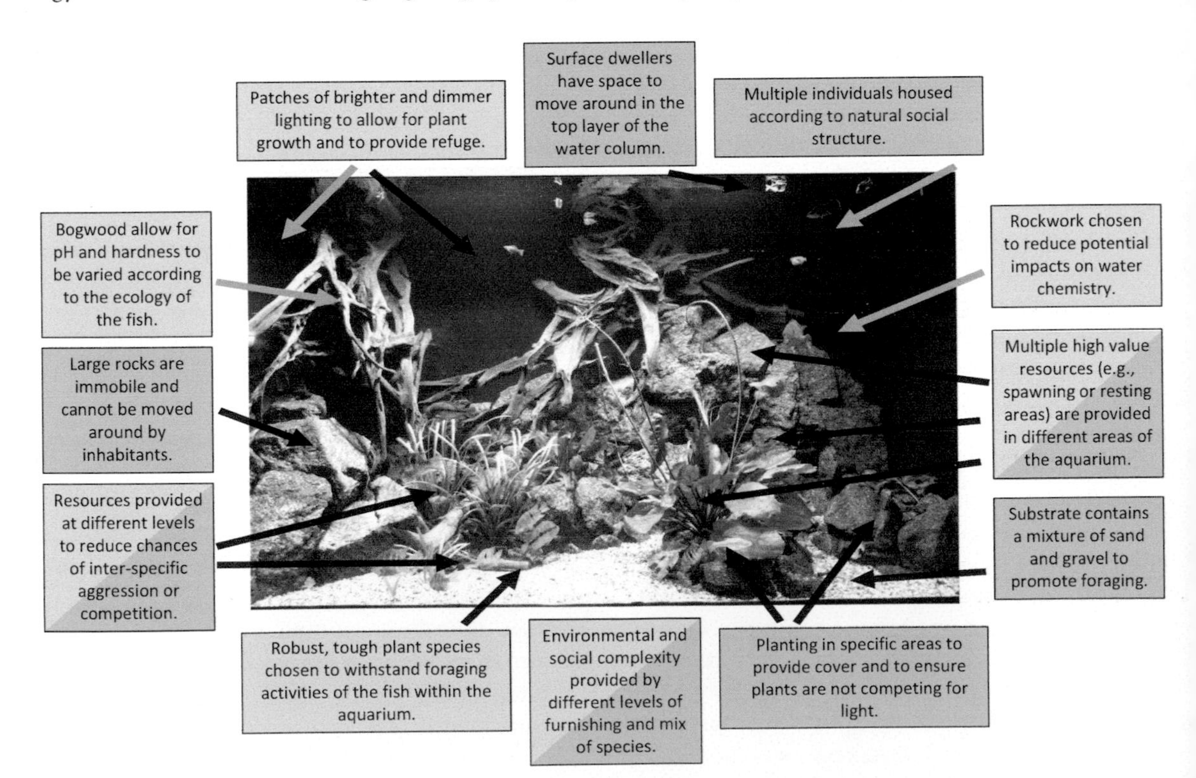

Figure 18.3 The entire aquarium can be enriching. Here is an example of a Central American community aquarium in a large public aquarium. Blue boxes show husbandry features based on fish behaviour, green boxes where ecological and environmental factors have been considered, and green and blue show application of behavioural ecology information.

for smaller, or more peaceful species or subordinate individuals.

Using Bloomsmith, Brent, and Schapiro (1991)'s categories of environmental enrichment, examples of fish behaviour and how to promote this activity in the aquarium using appropriate enrichment regimes are provided:

- Nutritional enrichment: providing increased opportunities for foraging over a longer time period and a wider area of the aquarium by changing the presentation of food or varying the type of food available to the fish. Providing opportunities to display natural hunting behaviours, e.g., by feeding live food, or presenting food in a way that mimics wild prey species, such as placing insects on leaves overhanging water to be shot at by archerfish.
- Occupational enrichment: providing stimuli that allow expression of exploratory, investigative, or other such behaviours that occupy the fish's time in beneficial activity. It may also be possible to provide occupational enrichment by training fishes to participate in daily husbandry, which may also make the provision of food and medication easier, as well as enable visual health checking to be conducted more easily (if fishes are target trained to a specific location in the aquarium).
- Physical: providing stimuli that increase environmental complexity and enable the performance of behaviours associated with the manipulation or direct engagement with aspects of the immediate environment. May also provide refuges from other fishes (e.g., to avoid negative social interactions) or from human disturbances.
- Sensory: providing biologically relevant stimuli that engage the fish's different senses (e.g., olfaction, vision, hearing) and increase the diversity of behaviours performed or increase the time spent on natural behaviour patterns.
- Social: providing the appropriate social group for the species being housed, including an ecologically relevant sex ratio and age structure, that allows for social choice as well as the performance of other social behaviours. The presence of fishes of other species may also be socially enriching, as long as they are compatible tankmates.

18.6 USING BEHAVIOURAL BIOLOGY TO ADVANCE FRESHWATER FISH CARE

As has been emphasised throughout this chapter, freshwater fishes are a phylogenetically, morphologically, and behaviourally diverse group, and it is difficult to provide a general summary of almost any aspect of their ecology and behaviour. Advancing their care is likely to rely on two major areas. The first is widening our understanding of the ecology and behaviour of the species of interest. In-depth knowledge of the wild ecology and behavioural biology of different species, both those commonly exhibited as well as those less familiar but in need of conservation action, provides a strong foundation for best practice care, and added educational and research value to captive fish populations. Aquariums that contain biologically relevant enclosures are also likely to be more inspiring and engaging to visitors.

The second area where behavioural biology can help to advance care is through a better understanding of welfare indicators in aquarium species. Fish welfare is a growing field of study, but our understanding of welfare in fishes still lags behind that of other taxa. There is much that is yet to be understood about the full behavioural repertoires of many species, the impacts different stimuli have on fish welfare, and behaviours that may be used as further welfare indicators. Some examples of relevant future research questions related to fish behaviour and welfare are provided in Box 18.1.

To guide further developments in our understanding of fish behavioural biology Box 18.2 provides information on useful sources of information that can help the reader learn more about freshwater fish biology, husbandry, and management.

18.7 CONCLUSION

This chapter has expanded on the importance of the evidence-based approach to freshwater fish husbandry and has illustrated why this is relevant using a range of examples of species, habitats, and fish behaviour patterns. When considering the husbandry and management needs of such a diverse, eclectic "group" of animals, it is clear that while some general trends in what fish need (e.g., an aquatic environment) will hold true, species-specific

BOX 18.1: Future research questions to further support evidence-based fish husbandry using behavioural information

How do social fish species use their space when housed in a single species and multi-species environment? Can compatibility of species mixes be determined by differences in collective behaviour and social assortment?

Effects of visitor presence (both direct and indirect) on fish behaviour and space use in public aquaria. How do changes in the sensory environment (e.g., differences in light and sound caused by aquarium visitors) affect what fishes do and where they are?

How much space is good space? Assessing quality rather than quantity space for fishes in public aquaria. What balance of planting, furnishing, and open space is required for activity and natural behaviour performance?

Do all "tankbusters" belong together? Is size the best consideration for mixing big fishes together, or should multi-species aquaria be themed more ecologically?

What forms of environmental enrichment work based in multi-species aquaria? How can enrichment be provided in a meaningful way to maximise its beneficial effects across all individuals?

How can positive challenges, e.g., through cognitive enrichment, be provided to provide opportunities for behavioural diversity and promotion of plasticity in long-term behaviour patterns?

BOX 18.2: Suitable references to help inform freshwater fish husbandry and care

An excellent example of how natural history information can be fully embedded into fish care to support animal welfare: Lee, C.J., Paull, G.C. & Tyler, C.R. (2022). Improving zebrafish laboratory welfare and scientific research through understanding their natural history. *Biological Reviews.* https://doi.org/10.1111/brv.12831

FishBase: www.fishbase.se

FishEthoBase: www.fishethobase.net

Roberts, H.E. (2011). *Fundamentals of Ornamental Fish Health.* Wiley-Blackwell.

Roberts, H., & Palmeiro, B.S. (2008) Toxicology of aquarium fish. *Veterinary Clinics of North America: Exotic Animal Practice, 11*(2), 359–374.

Saxby, A., Adams, L., Snellgrove, D., Wilson, R.W., & Sloman, K.A. (2010). The effect of group size on the behaviour and welfare of four fish species commonly kept in home aquaria. *Applied Animal Behaviour Science, 125*(3–4), 195–205.

Sloman, K.A., Baldwin, L., McMahon, S., & Snellgrove, D. (2011). The effects of mixed-species assemblage on the behaviour and welfare of fish held in home aquaria. *Applied Animal Behaviour Science, 135*(1–2), 160–168.

Walster, C. (2008). The welfare of ornamental fish. In E.J. Branson (Ed.), *Fish Welfare* (pp. 269–290). Blackwell Publishing Ltd, Oxford UK.

considerations are ultimately integral to long-term good animal welfare as well as to support ex situ population sustainability and conservation initiatives. Recreation of the whole or substantial elements of a wild habitat may be easier for some species of freshwater fish, enabling the whole environment of the aquarium to be enriching – promoting behavioural diversity and the performance of naturalistic time budgets. More research into freshwater fish behaviour is required, specifically to identify indicators of positive animal welfare states as well as to provide fundamental information on natural or normal behaviour patterns so that time-activity patterns of captive individuals can be judged as to their appropriateness. Given the huge populations of freshwater fish in ex-situ housing (from zoological collections to private aquaculture) small advances in our knowledge of their correct care benefit many thousands of individual animals.

REFERENCES

Blaser, R.E., Chadwick, L., & McGinnis, G.C. (2010). Behavioral measures of anxiety in zebrafish (*Danio rerio*). *Behavioural Brain Research, 208*(1), 56–62.

Bloomsmith, M.A., Brent, L.Y., & Schapiro, S.J. (1991). Guidelines for developing and managing an environmental enrichment program for nonhuman primates. *Laboratory Animal Science, 41*(4), 372–377.

Brandão, M.L., Dorigão-Guimarães, F., Bolognesi, M.C., Gauy, A.C.D.S., Pereira, A.V.S., Vian, L., Carvalho, T.B., & Gonçalves-de-Freitas, E. (2021). Understanding behaviour to improve the welfare of an ornamental fish. *Journal of Fish Biology, 99*(3), 726–739.

Brown, C. (2002). Do female rainbowfish (*Melanotaenia* spp.) prefer to shoal with familiar individuals under predation pressure? *Journal of Ethology, 20*(2), 89–94.

Brown, C. (2015). Fish intelligence, sentience and ethics. *Animal Cognition, 18*(1), 1–17.

Chen, M.J., Coss, R.G., & Goldthwaite, R.O. (1983). Timing of dispersal in juvenile jewel fish during development is unaffected by available space. *Developmental Psychobiology, 16*(4), 303–310.

Crane, A.L., & Ferrari, M.C.O. (2016). Uncertainty in risky environments: a high-risk phenotype interferes with social learning about risk and safety. *Animal Behaviour, 119*, 49–57.

Elliott, W.R. (2018). *The Astyanax caves of Mexico: Cavefishes of Tamaulipas, San Luis Potosí, and Guerrero*. Association for Mexican Cave Studies, Bulletin 26. Austin, TX, USA.

Farias, I.P., Leão, A., Almeida, Y.S., Verba, J.T., Crossa, M.M., Honczaryk, A., & Hrbek, T. (2015). Evidence of polygamy in the socially monogamous Amazonian fish *Arapaima gigas* (Schinz, 1822) (Osteoglossiformes, Arapaimidae). *Neotropical Ichthyology, 13*(1), 195–204.

Ferter, K., & Meyer-Rochow, V.B. (2010). Turning night into day: effects of stress on the self-feeding behaviour of the Eurasian perch *Perca fluviatilis*. *Zoological Studies, 49*(2), 176–181.

Franks, B., Graham, C., & von Keyserlingk, M.A.G. (2018). Is heightened-shoaling a good candidate for positive emotional behavior in zebrafish? *Animals, 8*(9), 152.

Gimeno, E., Quera, V., Beltran, F.S., & Dolado, R. (2016). Differences in shoaling behavior in two species of freshwater fish (*Danio rerio* and *Hyphessobrycon herbertaxelrodi*). *Journal of Comparative Psychology, 130*(4), 358–368.

Gómez-Laplaza, L.M., & Morgan, E. (1993). Transfer and isolation effects on the feeding behaviour of the angelfish, *Pterophyllum scalare*. *Experientia, 49*(9), 817–819.

Helfman, G., Collette, B.B., Facey, D.E., & Bowen, B.W. (2009). *The diversity of fishes: Biology, evolution, and ecology*. John Wiley & Sons, Chichester, UK.

Hesse, S., Bakker, T.C.M., Baldauf, S.A., & Thünken, T. (2012). Kin recognition by phenotype matching is family rather than self-referential in juvenile cichlid fish. *Animal Behaviour, 84*(2), 451–457.

IUCN. (2021). One third of freshwater fish face extinction, warns new report. https://www.iucn.org/news/water/202102/one-third-freshwater-fish-face-extinction-warns-new-report

Kasper, S., Adeyemo, O.K., Becker, T., Scarfe, D., & Tepper, J. (2021). Aquatic Environment and Life Support Systems. In L. Urdes, C. Walster & J. Tepper (Eds.), *Fundamentals of aquatic veterinary medicine* (pp. 1–27). John Wiley & Sons, Chichester, UK.

Kuwamura, T. (1986). Parental care and mating systems of cichlid fishes in Lake Tanganyika: a preliminary field survey. *Journal of Ethology, 4*(2), 129–146.

Magellan, K., Johnson, A., Williamson, L., Richardson, M., Watt, W., & Kaiser, H. (2012). Alteration of tank dimensions reduces male aggression in the swordtail. *Journal of Applied Ichthyology, 28*(1), 91–94.

Magierecka, A., Lind, Å.J., Aristeidou, A., Sloman, K.A., & Metcalfe, N.B. (2021). Chronic exposure to stressors has a persistent effect on feeding behaviour but not cortisol levels in sticklebacks. *Animal Behaviour, 181*, 71–81.

Martins, C.I.M., Galhardo, L., Noble, C., Damsgård, B., Spedicato, M.T., Zupa, W., Beauchaud, M., Kulczykowska, E., Massabuau, J.-C., Carter, T., Planellas, S.R., & Kristiansen, T. (2012). Behavioural indicators of welfare in farmed fish. *Fish Physiology and Biochemistry, 38*(1), 17–41.

Millot, S., Vandewalle, P., & Parmentier, E. (2011). Sound production in red-bellied piranhas (*Pygocentrus nattereri*, Kner): an acoustical, behavioural and morphofunctional study. *Journal of Experimental Biology, 214*(21), 3613–3618.

Oldfield, R.G. (2011). Aggression and welfare in a common aquarium fish, the Midas cichlid. *Journal of Applied Animal Welfare Science, 14*(4), 340–360.

Øverli, Ø., Sørensen, C., Kiessling, A., Pottinger, T.G., & Gjøen, H.M. (2006). Selection for improved stress tolerance in rainbow trout (*Oncorhynchus mykiss*) leads to reduced feed waste. *Aquaculture, 261*(2), 776–781.

Polverino, G., Ruberto, T., Staaks, G., & Mehner, T. (2016). Tank size alters mean behaviours and individual rank orders in personality traits of fish depending on their life stage. *Animal Behaviour, 115*, 127–135.

Satoh, S., Ota, K., Awata, S., & Kohda, M. (2019). Dynamics of sibling aggression of a cichlid fish in Lake Tanganyika. *Hydrobiologia, 832*(1), 201–213.

Sazima, I., & Zamprogno, C. (1985). Use of water hyacinths as shelter, foraging place, and transport by young piranhas, *Serrasalmus spilopleura*, *Environmental Biology of Fishes, 12*(3), 237–240.

Simon, K.D., & Mazlan, A.G. (2010). Trophic position of archerfish species (*Toxotes chatareus* and *Toxotes jaculatrix*) in the Malaysian estuaries. *Journal of Applied Ichthyology, 26*(1), 84–88.

Sneddon, L.U., Wolfenden, D.C.C., Leach, M.C., Valentim, A.M., Steenbergen, P.J., Bardine, N., Broom, D.M., & Brown, C. (2018). Ample evidence for fish sentience and pain. *Animal Sentience, 3*(21), 17.

Takeuchi, Y., Hori, M., & Oda, Y. (2012). Lateralized kinematics of predation behavior in a Lake Tanganyika scale-eating cichlid fish. *PloS One, 7*(1), e29272.

Valdesalici, S., & Cellerino, A. (2003). Extremely short lifespan in the annual fish *Nothobranchius furzeri*. *Proceedings of the Royal Society B: Biological Sciences, 270*(Suppl 2), S189–S191.

Wong, B., Rosenthal, G., & Buckingham, J. (2007). Shoaling decisions in female swordtails: How do fish gauge group size? *Behaviour, 144*(11), 1333–1346.

Wright, D., Ward, A.J.W., Croft, D.P., & Krause, J. (2006). Social organization, grouping, and domestication in fish. *Zebrafish, 3*(2), 141–155.

19

The behavioural biology of marine fishes and sharks

CHRISTOPHER D. STURDY
City College Norwich (Easton Campus), Norwich, UK

GEORGIA C. A. JONES
Bournemouth University, Poole, UK

JAKE SCALES
Meade Barn, Tadley, UK

19.1 INTRODUCTION TO MARINE FISH AND SHARK BEHAVIOURAL BIOLOGY

Marine fishes (Osteichthyes) and sharks (Chondrichthyes) are often the focus of public aquariums, with smaller exhibits present in zoos. Large aquariums exhibit sharks, rays, and shoals of tropical fish, creating immersive displays, attracting visitors, and providing an opportunity to educate about the oceans and their inhabitants. Zoos and aquariums have an important role to play in the education and conservation of marine species. Ex-situ programmes maintain captive stocks of endangered species with the possibility of reintroduction to the wild, while in situ projects receive practical advice and guidance (Hosey, Melfi, & Pankhurst, 2009). Examples include the Sustainable Aquariums Project (SNAP), a collaborative project between Bangor University and several UK zoos and aquariums, that focuses on ex-situ breeding of coral reef fish species that have not been previously bred in captivity (Marine Centre Wales, 2019).

Ex-situ captive breeding is important when it is considered that 90% of captive marine fish are wild-caught, especially if there are benefits to supporting

DOI: 10.1201/9781003208471-21

conservation programmes and reducing environmental damage (King, 2019; Tlusty, 2002). However, it is important to note that there can be advantages to sustainably caught wild fish, including environmental and socio-economic benefits (King, 2019). The Banggai Cardinalfish Action Plan, for example, includes the sustainable management of *Pterapogon kauderni*, the creation of marine protected areas and monitoring of the trade in *P. kauderni* (Ndobe & Moore, 2008). Zoos and aquaria have a responsibility to promote sustainability and conservation, either by collecting wild-caught fish from sustainable fisheries, or by developing ex-situ techniques to increase the number of captive-bred species kept for display. An understanding of diet, environment, and natural behaviour is essential for the success of keeping and breeding such species in ex-situ facilities.

The species covered in this chapter represent the orders Perciformes (perches, basses, sunfishes, bluefishes, remoras, jacks, pompanos, snappers, drums, angelfishes, cichlids mackerels, tunas, gobies, groupers, and swordfishes), Kurtiformes (nurseryfishes and cardinalfishes), Syngnathiformes (trumpetfish, pipefish, and seahorses), Anguilliformes (eels and morays), Carcharhiniformes (catsharks, hound sharks, weasel sharks, requiem sharks and hammerhead sharks), and Lamniformes (thresher sharks, basking sharks, mackerel sharks, megamouth sharks, goblin sharks, sand sharks, and the crocodile shark *Pseudocarcharias kamoharai*). Example species of these orders are detailed in Table 19.1 and have been selected based on their representation in zoological collections. Many of the species in the table are from the order Perciformes, which represents over 40% of all Osteichthyes. Within the Perciformes alone, there is a diversity in behaviours, including symbiosis, monogamy, and broadcast spawning.

The captive care of aquatic species has the added complexity of replicating the environmental conditions, considering how individuals interact with the environment, as well as other individuals.

19.2 ECOLOGY AND NATURAL HISTORY OF MARINE FISHES AND SHARKS RELEVANT TO THE ZOO

Marine displays are often mixed taxa, representing a particular habitat such as coral reefs, mangroves, or intertidal regions. An understanding of the natural history of all species in the exhibit is important to support captive management. A knowledge of environmental parameters is vital when designing an exhibit, from temperature and salinity to wave action, current strengths, and light intensity. These parameters not only influence the physiology of marine species, but the displayed behaviours. Size and morphology of species is indicative of the environment they inhabit and therefore can inform exhibit design, with hiding places provided for laterally compressed species or large open spaces provided for those that are dorsoventrally compressed. Foraging activity will inform species choice (predator and prey relationships; ability to provide diet in captivity; symbiotic and mutualistic relationships) and exhibit design (husbandry routine; filtration; flow and turnover rate). Literature on intra- and interspecific behaviours will inform species selection, the density of individuals, and sex ratios, while helping to avoid territorial and competitive behaviours, and increase reproductive success. Table 19.2 provides examples of behaviour traits of commonly kept marine fish and sharks, and how these behaviours can be applied to captive care.

Understanding interspecies behaviours can result in interesting mixed-species exhibits, while improving the welfare of individuals. A commonly exhibited species that demonstrates a symbiotic relationship is the common clownfish that inhabits anemones, utilising them for protection, while providing the anemones with nourishment from excreted ammonia and protection from anemone-eating fish. The anemone also benefits from the oxygenation of its tentacles by the movement of water created by the clownfish swimming. Although clownfish would not survive in the wild without an anemone, they are able to survive in captivity, utilising some coral species or hiding places such as rock. However, symbiosis with an anemone is a natural relationship that encourages the full repertoire of behaviours.

The role of cleaner wrasse from the genus *Labroides* has been observed in the wild and in captivity. Cleaner wrasse will perform mutualistic interactions with a range of species from morays and groupers to sharks, removing and feeding on ectoparasites from the oral cavity and gills. Species of *Labroides* exhibit different behaviours and social hierarchies, therefore exhibit design needs to be species-specific. *Labroides dimidiatus* are social wrasse, working in groups of 5–12 adults

Table 19.1 Representative species of marine fishes and sharks exhibited in zoological collections, including environmental conditions, to inform exhibit design.

Species	IUCN red list status (as of January 2022)	Distribution	Environment
Common clownfish (*Amphiprion ocellaris*)	Not Evaluated	Indo-West Pacific, including the eastern Indian Ocean.	Coral reefs, shallow, and calm lagoons. Associated with *Stichodactyla* anemones.
Regal tang (*Paracanthurus hepatus*)	Least Concern	Indo-Pacific, including East Africa, Japan, and the Great Barrier Reef.	On the seaward side of reefs, in clear, current-swept waters.
Banggai cardinalfish (*Pterapogon kauderni*)	Endangered	Western Pacific, restricted to the Island of Banggai.	Seagrass with silty sand bottoms. Associated with *Diadema* urchins.
Copperband butterflyfish (*Chelmon rostratus*)	Least Concern	Western Pacific, including the Andaman Sea.	Rocky shores, coral reefs, estuaries, and silty inner reefs.
Blue-green chromis (*Chromis viridis*)	Not Evaluated	Indo-Pacific, including East Africa, Japan, and the Great Barrier Reef.	Sheltered areas including subtidal reef flats and lagoons.
Cleaner wrasse (*Labroides dimidiatus*)	Least Concern	Indo-Pacific, including the Persian Gulf, Red Sea, and East Africa.	Coral reefs, inner lagoons, and subtidal reef flats.
Big-belly seahorse (*Hippocampus abdominalis*)	Least Concern	Southwest Pacific, mainly Australia and New Zealand.	Large rockpools at low tide, found amongst seaweed.
Green moray eel (*Gymnothorax funebris*)	Least Concern	Western Atlantic Ocean, typically Bahamas and Florida Keys.	Rocky tidal areas, coral reefs, mangroves, or sandy bottoms.
Black-tip reef shark (*Carcharhinus melanopterus*)	Vulnerable	Indo-Pacific, including Persian Gulf, Red Sea, East Africa, and Hawaiian Islands.	Shallow water, coral reefs, and intertidal reef flats, mangroves.
Sand tiger shark (*Carcharias taurus*)	Critically Endangered	Circumtropical, including the Indo-West Pacific, Red Sea, Western and Eastern Atlantic.	From inshore surf zone and shallow bays to outer continental shelves.

and juveniles; performing dancing movements on their cleaning stations to attract clients. Meanwhile, *L. bicolor* works alone, actively swimming to clients and then requesting permission to clean. Client species can seek and recognise cleaner wrasse due to the blue band on their lateral side, while cleaner wrasse are able to recognise clients and will prefer older, more experienced fish, especially if they have a higher parasite load. Studies into the personalities of cleaner wrasse (Wilson, Krause, Herbert-Read,& Ward, 2014) determine that they have a range of traits from aggression, risk-taking, exploratory, honesty, and deception. Deceiving wrasse gain higher nutrient intake by feeding on skin, mucus, and scales but potentially with fatal consequences, while honest wrasse, only feeding on ectoparasites, retain clients for longer. The range of personalities adds an extra layer of complexity, with the risk of some client fish being over-cleaned and injured as they cannot escape in captivity.

Table 19.2 Relevance of specific behavioural traits to the captive care of marine fishes and sharks.

Species	Example behavioural trait	Relevance to captive care
Common clownfish (*Amphiprion ocellaris*) 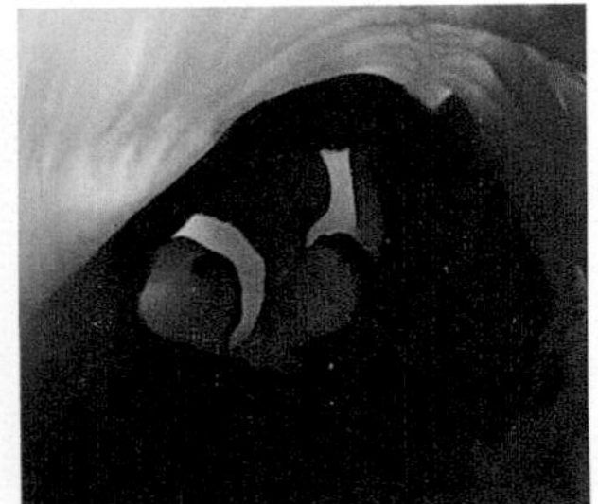	Classed as obligate symbionts, clownfish in the wild would not be able to survive without their host anemone. As protandrous hermaphrodites, the social hierarchy consists of a monogamous dominant female and male pair, surrounded by non-reproductive adolescents.	An anemone should be provided for the clownfish to express natural behaviour. In captivity, *Entacmaea quadricolor* is often used, however this relationship would not be seen in the wild. In the wild, *A. ocellaris* are associated with *Heteactis magnifica* or *Stichodactyla gigantea* (carpet anemones), both of which can be difficult to keep in captivity. When adding individuals to an established aquarium, it is vital that these are adolescents, otherwise competition for the dominant position will occur.
Regal tang (*Paracanthurus hepatus*) 	Often observed individually, Regal tangs can be found in pairs or small groups. They will form large breeding groups, broadcast spawning directly into the water. Associated with branching corals, regal tangs will hide within the coral, extending its caudal spine for stability. They graze on algae that grow on corals, therefore carrying out an important role in keeping the corals clear of growth.	Regal tangs are best kept in large displays where there is space to form social groups. They can become aggressive in smaller tanks and should be kept to one individual representing the species. Regal tangs can be kept in coral biotopes, where the corals provide refuge and algae to graze on. Artificial rocks and artefacts can be used as an alternative to live corals.
Banggai cardinalfish (*Pterapogon kauderni*) 	Banggai cardinalfish are paternal mouthbrooders, that are associated with *Diadem setosum* urchins. Groups of 2–60 fish will swim in seagrass habitats, in reach of the urchins. The male will incubate up to 40 eggs for approximately 20 days. Juvenile cardinalfish will return to the male's mouth for protection for approximately 10 days, after which the relationship with the male will end and the juveniles will remain close to the urchins. Juveniles will also use anemones.	Due to the social structure of this species, Banggai cardinalfish should be kept in small groups, ideally around 12. Kept individually they will spend a large amount of time hiding. The aquarium should be structurally complex, ensuring the cardinalfish are able to hide when they feel threatened. Paired individuals will breed if refuge spaces are provided, however, the presence of *Diadem sp.* urchins further encourages natural behaviour.

Copperband butterflyfish (*Chelmon rostratus*) 	A territorial species that will form monogamous pairs, however, juveniles will swim in shoals. Butterflyfish have a long pincer-like mouth that is adapted to feed on small benthic invertebrates, including anemones.	Consider aggression if attempting to form a breeding pair, as there is a high chance of aggression if a second is introduced. Introduction as juveniles is advised or within a large aquarium where there is the opportunity of escape. Plenty of live rock should be provided to encourage the copperband to swim among crevices and forage for food. Copperbands are often used in an aquarium to keep populations of *Aiptasia* sp. under control, an anemone with a high reproductive rate that often takes over displays.
Blue-green chromis (*Chromis viridis*) 	Chromis are a shoaling species and spawning typically occurs around the full and new moon. As demersal spawners, the males will establish nesting sites on suitable hard substrates and fragments of coral or rock. Males will guard the developing embryos until they hatch in a couple of days.	A shoal of 12 is ideal, with a minimum of six. Hiding will indicate not enough individuals in the group or an issue with the setup. Minor aggressive behaviour may be observed as the pecking order of the shoal is established. A photoperiod replicating the lunar cycle will encourage spawning.

(*Continued*)

Table 19.2 (Continued) Relevance of specific behavioural traits to the captive care of marine fishes and sharks.

Species	Example behavioural trait	Relevance to captive care
Cleaner wrasse (*Labroides dimidiatus*)	Renowned for establishing cleaning stations on reefs, cleaner wrasses often form monogamous pairs or harems consisting of a dominant male with females. These wrasses generally feed on ectoparasites of predatory species, a mutualistic relationship that seldom results in the wrasse being eaten.	When acquiring cleaner wrasse for the aquarium, it is advised to get one adult or a small number of adolescent individuals. Station feeding is more likely to occur with larger numbers of wrasse, though supplementary feeding for individuals can be done with a variety of invertebrate feeds and potentially more commercial feeds when food is scarce. Supplementary feeding should be given little and often to coincide with natural feeding patterns which would occur on feeding stations.
Big-belly seahorse (*Hippocampus abdominalis*)	Courtship is seasonal and initiation involves the male opening their brood pouch slightly, inflating the pouch and the pouch becoming lighter in colour. The body colour also becomes brighter. Once a mating pair has been formed, they will swim in tandem and entwine tails. Finally, they will swim upwards towards the water surface before transferring eggs.	Replication of seasonal photoperiod and temperatures are important; *H. abdominalis* will start courting when photoperiod is more than 11 hours light and less than 13 dark (Woods, 2000). Tank height is important, with a minimum depth of 90 cm to enable egg transfer.

Green moray eel (*Gymnothorax funebris*) 	An ambush predator which spends the majority of its time within hides between rocks or crevices in reefs. Moray eels predominantly utilise their sense of smell to detect passing prey as they are nocturnal predators.	Enclosure design must include a variety of cave-like areas allowing the preferred environment for these eels. When kept within a mixed species aquarium, other fish species that are considerably larger are more likely to be successful in housing with moray eels. However, if the eel is sufficiently fed regularly, the predation risk to those sharing its aquarium would be reduced. Eels may be kept together, but ensure significant space is offered and multiple hiding spots to reduce competition for food and territory.
Black-tip reef shark (*Carcharhinus melanopterus*) 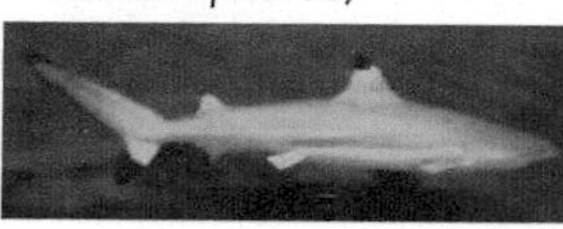	Pregnant females will leave their home range and swim to nursery areas to give birth. Females prefer shallower waters that are safe from predators and male harassment. They will become more rounded, with increased abdominal swelling. The lower abdomen will begin to bulge on both sides, about two weeks before birth (Porcher, 2005).	Enclosure design that has a shallower, connected pool has been observed as an effective method to prevent the risk of pup losses because of predation, as they can be separated from other predatory fish kept within the same enclosure (Hibbitt, Rees, & Brown, 2017). This can similarly be used to monitor breeding partners.
Sand tiger shark (*Carcharias taurus*) 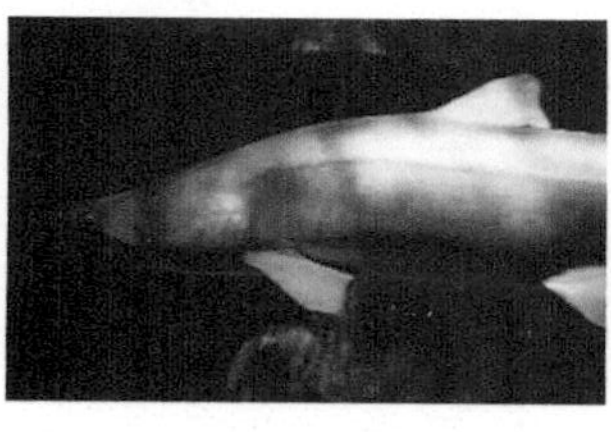	Sand tiger sharks are a slow swimming, migratory, colonial species, that exhibit social behaviours. As nocturnal hunters, they work together to hunt fish, small sharks, squid, and crustaceans. They are the only known shark species to gulp air from the surface, storing this in their stomachs, allowing them to remain neutrally buoyant. Sand tigers are typically active at night and tend to remain on the seabed during the day.	*C. taurus* demonstrate a range of behaviours, therefore exhibit design and enrichment programmes should be considered to reduce repetitive swimming behaviours, which influence the shape of the dorsal fin and can cause spinal deformities. Exhibit length should be at least 14 times the body length to enable to sharks to complete their full repertoire of swimming behaviours, which includes cruising, glide, recovery, and turning stages. The shape of the aquarium and the design of the exhibit should encourage complex swimming behaviours and reduce the amount of circular swimming (Tate, Anderson, Huber, & Berzins, 2013).

Photos: S. Adams & P. Rose.

Trophic activity and feeding behaviour should be considered when planning to exhibit a species. Establishing whether a species is carnivorous or herbivorous is important for species compatibility, while understanding feeding methods and diet will improve husbandry. Marine species' foraging activity can be dramatically different between juvenile and adult stages, and even within these different life stages, foraging patterns can alter according to time of day. For example, research on batfish (*Platax orbicularis*) shows wildly different foraging strategies between daylight and nocturnal time periods (Barros et al., 2008). Type of food selected and gastrointestinal tract anatomy and rate of passage also weigh heavily on foraging strategy, For example, grazing surgeonfish feed on microalgae film that grows on corals, which has low nutritional value, therefore they need to graze for long periods of time. In captivity, feeding should take up a large proportion of the activity budget of surgeonfish.

Seahorses exhibit a cryptic lifestyle, camouflaged to avoid predators due to their poor swimming ability, though *H. abdominalis* is much stronger than its relatives. They suction feed on small invertebrate species floating nearby through their toothless tubular mouth. Due to their increased metabolic rate, it is important that captive diets are fed regularly and ensure prey items are enriched as nutrient variability and uptake within the seahorse can vary.

Reproductive strategies of marine fish and sharks can be difficult to accommodate in captivity, especially those species that have a pelagic larval phase. Fish must overcome individual behaviours and join aggregates of reproductively active fish or find mating partners. Suitable spawning sites need to be found, either to build nests or to provide space for broadcast spawning. Territorial and courting behaviours are diverse and vary from species to species. Extra complexities need to be considered as some species will change sex, and can be parthenogenetic, monogamistic, promiscuous, or haremic.

Formation of mating groups requires an understanding of mating systems. The seahorse is relatively easy to breed in captivity and is often a key species in aquariums due to the unusual reproductive strategy of paternal care. Seahorses generally remain in monogamistic relationships, with egg transfer from the female to be fertilised in the male's incubator pouch. Females attract the male by performing swimming patterns above the eel grass and males select the bigger, most colourful female. In some instances, the female may seek out additional males to accommodate all her eggs. The Banggai cardinalfish is a paternal mouthbrooder, with females initiating breeding activity and forming a pair with a similar-sized male. Over a few days, the paired cardinalfish will defend a spawning area, where the female will deposit eggs, which are fertilised by the male, who then protects the eggs and larvae by incubating them in his mouth.

In addition to the previously mentioned monogamous systems, there are species that use polygamous systems, such as the clownfish which are protandrous, where small males will undergo a sex change when there is a vacant position of a dominant female. It is this larger female that will reproduce with selected males. The remaining individuals are under stress which causes a delay in their growth and causes sexual inhibition, in effect, causing temporary castration. The cleaner wrasse, however, is a haremic proterogyne, changing from female to male, but also returning to female in situations when they become subordinates to other males.

For some species, wild-type behaviours are not suitable for captivity or individual welfare, and an understanding can support management to reduce risk. The mating system for reef sharks, including the black-tip reef shark involves three to four males surrounding a female. One male will seize the female by the pectoral fin with his jaws. The male will hold the female's head down and push to the bottom, then inseminates by inserting a pterygopod (clasper) into the female's cloaca. The injuries caused by the males will leave scars on the females. There is also the risk that blood from the bites will cause more males to swim to the female and instigate predatory behaviour, putting the female at risk of being eaten. This is a behaviour that would not be desirable in captivity, often resulting in the sex ratio being switched to include more females, therefore defusing the aggressive male behaviour as males are only generally aggressive during the breeding period.

The sand tiger shark, however, has a complex reproductive strategy that is difficult to replicate in captivity. This species is aplacental viviparous and there is evidence that the mating system involves behavioural polyandry, but genetic monogamy, due to embryonic cannibalism, where the first hatchling will predate on subsequent hatchings (Chapman et al., 2013). A migratory species, pregnant females will move to warmer subtropical waters and then return to cooler waters to rest. In their resting year, they will remain in the cooler waters. Pre-copulation behaviour includes reduced feeding and increased swimming speed. Males will become more aggressive and will show interest in the females when the females slow down and present cupped pelvic fins. Competitive behaviours are exhibited between the males, including circling and tailing. Both male-male and male-female interactions will involve biting.

19.3 ENCLOSURE CONSIDERATIONS FOR MARINE FISHES AND SHARKS BASED ON BEHAVIOURAL EVIDENCE

Many marine species kept in aquariums can be found in inshore regions, from estuaries and bays to coral reefs, seagrass beds, and rocky tidal areas. The complexities of these habitats require the keeper to understand the environmental parameters and the structural complexities, from reef formations to mangrove roots. The morphology of a species can be used as an initial indicator of the species' niche in a selected habitat and the associated behaviours (Figure 19.1). The blue-green chromis is a small laterally compressed fish which has adapted to hide in reefs, particularly branching *Acropora* sp. of coral. They swim above the coral, feeding on phytoplankton, but will dart back into the coral to avoid predators or neighbouring aggressive fish. A setup for chromis should contain plenty of live rock and small polyp stony (SPS) corals to support their foraging and flight behaviours. As active swimmers, the middle and top areas of the display should be left clear, only filling the bottom section with rock and corals. As a small species, they have adapted to low current, sheltered waters; high turnover rates in aquariums will see chromis hiding more and displaying stress indicators. In contrast, the elongated anguilliform shape of eels and morays indicates their sedentary lifestyle as ambush predators. Hiding in crevices, reefs, or man-made structures such as wrecks, morays are camouflaged to their environment, waiting for their prey to swim past. Both species require places to hide within an aquarium, that replicates their natural habitat, for them to perform a full range of behaviours. Flow rate needs to be considered, in addition to light intensity, light cycle, and other abiotic parameters.

Seahorses are slow swimming with their dorsal fin providing the majority of the thrust. They inhabit seagrass meadows, coral reefs, and other sheltered habitats where the water movements are slow and tidal movements are repressed allowing them to drift between structures, using their prehensile tails for anchoring and control. It is essential that water flow is to a minimum within aquariums for these species (no faster than four to five times the tank volume turn around per hour), though allowing some change in tidal mimicking movements allows the expression of natural travelling behaviours as they swim amongst their surroundings. Depths and water conditions should be controlled for species-specific preferences as seahorse species vary globally.

Pelagic species are not influenced by structural barriers and territorial interactions, but abiotic factors such as temperature, salinity, and oxygen concentration alongside biotic factors such as feeding and spawning. Both abiotic and biotic changes can result in migration, both vertical and horizontal. It is possible to keep some pelagic species, like the sand tiger shark, in captivity, ensuring abiotic factors are adjusted to replicate seasonal migration. This could also be achieved when providing environmental complexity within an aquarium that suit the natural behavioural and biological needs of a species.

Stocking density is an important factor, often relating to water quality, ensuring the biological oxygen demand is not too high or negatively impacts on the nitrogen cycle. However, research into shoaling behaviours demonstrates that for many shoaling fish information flows faster between individuals, which can be seen by more social interactions. Demonstrated in the lab by the blue-green chromis where it was shown that these fish use sensory information from multiple individuals, including vision, olfactory, hearing, and mechanoreception, to solve behavioural problems, such as risk or feeding against predatory vigilance (Loannou, 2017). Other reef species such as damselfish, butterflyfish, and surgeonfish also utilise visual and olfactory systems to swim across the reef and return to their home habitat. Therefore, it's essential that stocking density considers social hierarchy as well as abiotic factors, when establishing a new setup.

Feeding strategy can influence management routine. For example, seahorses are ambush predators and continuously feed on live prey due to their lack

Figure 19.1 Examples of how morphology indicates niche. Top left: Laterally compressed shape of the yellow tang indicates it does not constantly swim, requiring bursts of speed as it swims amongst coral. Top middle: Depressiform shape of the turbot indicates a benthic lifestyle. Top right: The fusiform shape of the black-tip reef shark indicates a streamlined and fast swimming species. Bottom left: The unique S-curved shape of seahorses has adapted to a life of attachment to holdfasts and hunting. Bottom middle and right: The anguilliform shape of morays and garden eels enables them to hide amongst rocks or substrate (Photos: author contributed, P. Rose and S. Adams).

of a true stomach. Therefore, in captivity, keepers are required to continuously feed seahorses, which can impact water quality and enclosure cleanliness. It is common practice to siphon the substrate of seahorse exhibits twice a day to remove the high detritus load, therefore reducing negative impacts on welfare.

19.4 BEHAVIOURAL BIOLOGY AND WELFARE OF MARINE FISHES AND SHARKS

Fish present extra complexities when assessing welfare, due to the nature of their environment. When working with shoals of fish, individual health checks to assess physical indicators such as weight, are difficult. There is a reliance on visual health checks, observing both physical and behavioural indicators (Figure 19.2). Marine fish will present physical indicators of welfare stressors, mainly a change in colour, becoming pale when under stress. Behavioural indicators can further support assessment, observing ventilation rate (could be linked to poor water quality, including O_2 levels), aggression (resource competition), feeding (incorrect amounts, type of food or feeding method) and swimming (flashing due to parasite load) behaviours. Due to the diversity of behaviours expressed by marine species, it is important to understand species-specific behaviours, coping style, and whether they are normal or abnormal.

Feeding behaviours depend on the ecological niche of the species, for example, whether they are bottom or surface feeders, whether they are diurnal or nocturnal, or if they are scavengers, predators, or grazers. A change in feeding behaviour could indicate poor welfare. For example, in the UK, rays (*Raja sp.*) are commonly exhibited in open top displays and surface breaking behaviour is demonstrated, often associated with aquarists either target feeding from the surface or scatter feeding at set times for public feeds. Altering the feeding method to encourage benthic feeding and at random times, reduces the associated surface breaking behaviour.

Figure 19.2 Physical and behavioural indicators of welfare described in blue boxes, with corresponding enclosure design and husbandry considerations in green boxes.

A reduction in feeding amounts and normal foraging behaviour can be an indicator of stress. Environmental parameters should be investigated, such as water quality, sound, and vibration, as well as changes in social dynamics.

Aggressive and territorial behaviours vary on whether a species is generally solitary or social. Solitary species will exhibit territorial behaviours if kept with the wrong species or too many conspecifics, while social species can exhibit aggression if resources are limited within the enclosure. Competition for resources such as food and space can soon result in aggression between normally passive individuals. Martins et al. (2012) discusses the importance of individual social state and group social stability on the welfare of individuals. Factors such as an individual's boldness, size, or sex, as well as distribution of food, feeding frequency, amount of feed, and stocking density can influence the social hierarchy. Subordinate individuals tend to experience more aggression from dominant individuals, therefore being a subordinate may be a stressor.

Sound and vibration can be a deterring stimulus for elasmobranch species which may result in welfare implications within captivity where such disturbances are not recognised. Variations in unnatural sounds within the water can deter sharks and disrupt natural behaviours. If such noises can be heard during feeding this may lead to an individual or group ignoring that feed though they may be hungry enough to eat.

19.5 SPECIES-SPECIFIC ENRICHMENT FOR MARINE FISHES AND SHARKS

Marine species typically live in complex environments and individuals need to consider interactions with other individuals of the same and other species, impacts of currents, use of substrates and structures in the environment, as well as feeding strategies.

Enrichment for marine fish and sharks could include:

Figure 19.3 Examples of aquatic enrichment used for marine fish species (Photos: S. Adams & P. Rose).

- Enhancing the complexity of the aquarium to replicate the natural habitat and species-specific behaviours. Where an aquarium does not replicate the natural habitat, the exhibit design should be complex in physical enrichment consisting of stones, plants, kelp, sand, gravel, or artificial objects.
- Appropriate environmental parameters including temperature seasonality, currents, and varying photoperiod to simulate sunny or cloudy days, or seasonality.
- Novel items to change the environment, such as additional plants, rocks, or artificial items to allow fish to explore or defend items.
- Feeding live prey to encourage foraging behaviour. For example, using live artemia to feed seahorses.
- Dietary enrichment, including varying the time of day food is offered, alternating the area of the tank the fish are fed from, simulating movement of food by adding currents. This method has been used when feeding ambush predators; whole fish are dropped in front of internal pumps which propels the fish across the tank. This stimulates their ambush feeding behaviour.
- Operant conditioning with positive reinforcement has been successfully carried out with a variety of species, including sharks, rays, and morays. Behaviours include directing a fish to station, swimming into a stretcher or net, or moving from one exhibit to another.
- Auditory environmental enrichment. For example, Marchetto et al. (2021) highlight that musical auditory environmental enrichment reduced anxiogenic effects in zebrafish. This could lead to discussions as to whether public aquariums playing thematic music is a benefit to fish welfare.
- Shoal size appropriate for the species to exhibit a range of behaviours. Ensuring stocking density benefits behaviour, without impacting water quality or disease transmission.
- Enrichment devices to encourage feeding behaviours. For example, food placed within a cage to encourage foraging behaviours (Figure 19.3).

19.6 USING BEHAVIOURAL BIOLOGY TO ADVANCE CARE OF MARINE FISHES AND SHARKS

To provide further information on how behavioural biology should be used to support the development of species-specific care regimes for sharks and marine fishes, Box 19.1 discusses how wild shark

BOX 19.1: Case study on wild shark ecology with relevance to captive husbandry and a focus on sand tiger shark reproduction

Sharks have been swimming our oceans for over 450 million years – long before dinosaurs took over the land. They represent an incredibly diverse group (over 500 species), that come in all shapes and sizes, from the dwarf lantern shark (*Etmopterus perryi*) that reaches a maximum length of ~20 cm up to the whale shark (*Rhincodon typus*) which can achieve lengths of over 18 m. Due to commercial fishing, they are also one of the most threatened vertebrate groups, with one-third of sharks and their close ray relatives at risk of extinction (Dulvy et al., 2021). The generally negative public perception of sharks is a recognised barrier to their conservation and therein lies an excellent opportunity for aquaria to help change this perception through appropriate displays and interpretation.

Sharks are often portrayed and perceived as unintelligent, but this is far from the truth. In the wild, many species undertake accurate and long return-migrations, communicate with each other using body language, and can learn to exploit multiple food sources or indeed adapt to become specialists in a particular hunting strategy. It is even theorised that great white sharks (*Carcharodon carcharias*) may use star patterns to help them navigate across ocean basins (Bonfil et al., 2005). In captivity, multiple shark species have exhibited the ability to learn tasks and have been successfully trained using classical and operant conditioning (Smith et al., 2017). In addition to their intelligence, sharks are equipped with an impressive array of unique sensory organs and use hearing, smell, pressure changes, vision and electrosensory perception (through pores on their snouts called the Ampullae of Lorenzini) to hunt their prey. Given their intelligence and the likely benefits of allowing or encouraging sharks to display natural behaviours by using all their sensory organs (based on species-specific, wild behaviours), there is large scope for the development of innovative and appropriate enrichment and feeding activities.

In recent years, personality has been proved in sharks (Jacoby, Fear, Sims, & Croft, 2014) where individuals of multiple species have been found to exhibit consistent behavioural traits and even maintain long-term associations with individual conspecifics (Finger, Guttridge, Wilson, Gruber, & Krause, 2018). In black-tip reef sharks (*Carcharhinus melanopterus*), individuals divide to form multiple stable groups or communities at reef sites, partly influenced by sex and size but also by individual preferences (Mourier, Vercelloni, & Planes, 2012). Matching of individual personality types and maintenance of stable groups may help to reduce conflict and stress in captive settings and personality should be taken into account regarding diver and handler safety for some larger species.

As is typical for many predators, sharks exhibit sexual dimorphism – females attain larger sizes than males. The sexes also frequently segregate temporally and spatially, often aggregating only for mating or around especially rewarding food sources and even then, fine-scale segregation can occur. One key driver in the male/female segregation is thought to be female avoidance of male harassment. For example, female small-spotted catsharks have been found to sacrifice optimal environmental conditions in order to avoid year-round male sexual harassment (Wearmouth et al., 2012). Mating in sharks can be quite physically damaging for females, as males will typically grip them with their teeth in head, gill, and pectoral fin areas during copulation. This practice often results in "mating scars" which can be relatively severe and is considered a driver behind increased female size and thickness of the skin around the head. This is especially problematic when keeping male and female sand tiger (*Carcharias taurus*) sharks together, where repeated mating attempts in a short time period have proved fatal for females (personal communication reported in Smith et al. (2017)). Furthermore, male sand tiger sharks have been found to form dominance hierarchies, which can lead to aggressive interactions in both sexual and non-sexual contexts, sometimes requiring removal of either the most dominant or subordinate individual (Claus, Henningsen, Shivji, & Wetherbee, 2021). Despite the difficulties of keeping multiple sand tiger sharks together, especially in an attempt to breed them, aquaria are keen to try as the species is listed as critically endangered (Rigby, Carlson,

Dicken, Pacoureau, & Simpfendorfer, 2021) and maintaining a sustainable breeding stock rather than acquiring wild individuals is imperative.

Lessons from the wild are helping keepers to breed the species while avoiding conflict and fatalities. For example, female sand tiger sharks in the wild are known to reproduce on a biennial cycle, while males operate on an annual cycle (Gilmore, Dodrill, & Linley, 1983). Using hormone analyses, these patterns have been corroborated for captive individuals (Wyffels et al., 2020; Claus et al., 2021). A preliminary captive study suggests that sand tiger sharks preparing for a breeding year will increase their food intake by up to four times as much as in the non-breeding season and have suggested that monitoring individual food intake can give keepers valuable insight into the reproductive status of their sharks and therefore inform their management (Townsend & Gilchrist, 2017). However, breeding programmes for sand tiger sharks are still hampered by low success, with only four aquariums to date being successful, and as such, a detailed framework for monitoring behavioural and environmental cues of reproduction in successful aquaria was developed in 2017 (Henningsen et al., 2017). A recent study has revealed for the first time that drivers of this limited reproductive success are lower testosterone levels and poorer semen quality in captive vs. wild males (Wyffels et al., 2020). Mimicking the changes in photoperiod and water temperature that would occur during the long seasonal migration in wild is likely key to optimising male reproductive ability (Henningsen et al., 2017; Wyffels et al., 2020).

ecological data can improve reproductive success for ex-situ animals. This case study highlights the benefits that can be gained when intricate knowledge of shark physiology and interactions with their wild environments are replicated in the aquarium.

19.6.1 Further information to support the care of marine fishes and sharks

For information on how captive breeding of marine species can support the aquarium trade, as well as support the conservation of wild fish populations, including rearing protocols for the main marine species bred in captivity see:

Dominguez, A.M. & Botella, Á.S. (2014). An overview of marine ornamental fish breeding as a potential support to the aquarium trade and to the conservation of natural fish populations. *International Journal of Sustainable Development and Planning*, 9(4), 608–632.

19.7 CONCLUSION

This chapter has explored the diversity of behaviours amongst marine fish and sharks, and the importance of understanding these when designing an exhibit and selecting the required species. An increased amount of research has focussed on personality and social traits amongst groups of fish, including how vision, olfactory, hearing, and mechanoreception are utilised in communication and problem solving. Exhibit design needs to move away from just considering abiotic factors such as water quality, temperature, photoperiod, and light intensity, but consider biotic factors such as social hierarchy, foraging behaviours, reproductive strategy, territorial and competitive behaviours. Understanding species-specific behaviours will support husbandry routines, in addition to understanding the complexities associated with the maintenance of an array of complex environments.

Looking forward, there is an increased need to improve breeding success in captivity to maintain further sustainable populations of these marine species. Understanding wild ecology, personality, social behaviours, and reproductive strategies can further improve ex-situ captive breeding.

REFERENCES

Barros, B., Sakai, Y., Hashimoto, H., & Gushima, K. (2008). Feeding behavior of leaf-like juveniles of the round batfish *Platax orbicularis* (Ephippidae) on reefs of Kuchierabu-jima Island, southern Japan. *Journal of Ethology*, *26*(2), 287–293.

Bonfil, R., Meÿer, M., Scholl, M.C., Johnson, R., O'Brien, S., Oosthuizen, H., Swanson, S., Kotze, D., & Paterson, M. (2005). Transoceanic migration, spatial dynamics, and population linkages of white sharks. *Science*, *310*, 100–103.

Chapman, D.D., Wintner, S.P., Abercrombie, D.L., Ashe, J., Bernard, A.M., Shivji M.S., & Feldheim, K.A. (2013). The behavioural and genetic mating system of the sand tiger shark, *Carcharias taurus*, an intrauterine cannibal. *Biology Letters, 9*, 20130003.

Claus, E., Henningsen, A., Shivji, M., & Wetherbee, B. (2021). Sexual conflicts in sand tiger sharks *Carcharias taurus* (Rafinesque, 1810) in an artificial environment. *Journal of Zoo and Aquarium Research, 9*, 161–169.

Dulvy, N.K., Pacoureau, N., Rigby, C.L., Pollom, R.A., Jabado, R.W., Ebert, D.A., Finucci, B., Pollock, C.M., Cheok, J., Derrick, D.H., Herman, K.B., Sherman, C.S., Vander Wright, W.J., Lawson, J.M., Walls, R.H.L., Carlson, J.K., Charvet, P., Bineesh, K.K., Fernando, D., Ralph, G.M., Matsushiba, J.H., Hilton-Taylor, C., Fordham, S.V., & Simpfendorfer, C.A. (2021). Overfishing drives over one-third of all sharks and rays toward a global extinction crisis. *Current Biology*. https://doi.org/10.1016/j.cub.2021.08.062

Finger, J.S., Guttridge, T.L., Wilson, A.D.M., Gruber, S.H., & Krause, J. (2018). Are some sharks more social than others? Short- and long-term consistencies in the social behavior of juvenile lemon sharks. *Behavioral Ecology and Sociobiology, 72*, 17. https://doi.org/10.1007/s00265-017-2431-0

Gilmore, R.G., Dodrill, J.W., & Linley, P.A. (1983). Reproduction and embryonic development of the sand tiger shark Odontapsis taurus (Rafinesque). *Fishery Bulletin, 81*(2), 201–225.

Henningsen, A., Street, E.P., Claus, E., Littlehale, D., Choromanski, J., Parkway, K., Gordon, I., & Willson, K. (2017). Reproduction of the Sand Tiger Sharks, Carcharias Taurus, in Aquaria: A Framework for a Managed Breeding Program. In Smith, M., Warmolts, D., Thoney, D., Hueter, R.M.M., & Ezcurra, J. (Eds.), *The elasmobranch husbandry manual II: Recent advances in the care of sharks, rays and their relatives*. Special Publication of the Ohio Biological Survey Inc., Columbus, Ohio, pp. 375–390.

Hibbitt, J.D., Rees, E., & Brown, C. (2017). Blacktip Reef Shark Reproduction and Neonate Survivorship in Public Aquaria. In Smith, M., Warmolts, D., Thoney, D., Hueter, R.M.M., & Ezcurra, J. (Eds.), *The elasmobranch husbandry manual II: Recent advances in the care of sharks, rays and their relatives*. Special Publication of the Ohio Biological Survey Inc., Columbus, Ohio, USA, p. 443.

Hosey, G., Melfi, V., & Pankhurst, S. (2009). *Zoo animals: Behaviour, management, and welfare* (2nd edition). Oxford University Press, Oxford, UK.

Jacoby, D.M.P., Fear, L.N., Sims, D.W., & Croft, D.P. (2014). Shark personalities? Repeatability of social network traits in a widely distributed predatory fish. *Behavioral Ecology and Sociobiology, 68*, 1995–2003.

King, T.A. (2019). Wild caught ornamental fish: A perspective from the UK ornamental aquatic industry on the sustainability of aquatic organisms and livelihoods. *Journal of Fish Biology, 94*, 925–936.

Loannou, C.C. (2017). Swarm intelligence in fish? The difficulty in demonstrating distributed and self-organised collective intelligence in (some) animal groups. *Behavioural Processes, 141*, 141–151.

Marchetto, L., Barcellos, L.J.G., Koakoski, G., Soares, S.M., Pompermaier, A., Maffi V.C., Costa, R., da Silva, C.G., Zorzi, N.R., Demin, K.A., Kalueff, A.V., & de Alcantara Barcellos, H.H. (2021). Auditory environmental enrichment prevents anxiety-like behaviour, but not cortisol responses, evoked by 24-h social isolation in zebrafish. *Behavioural Brain Research, 404*, 113169.

Marine Centre Wales. (2019). *SustaiNable aquariums project (SNAP)*. Bangor University, Bangor, UK http://snap.bangor.ac.uk/index.php.en

Martins, C.I.M., Galhardo, L., Noble, C., Damsgård, B., Spedicato, M.T., Zupa, W., Beauchaud, M., Kulczykowska, E., Massabuau, J.C., Carter, T., Planellas, S.R., & Kristiansen, T. (2012). Behavioural indicators of welfare in farmed fish. *Fish Physiology and Biochemistry, 38*, 17–41.

Mourier, J., Vercelloni, J., & Planes, S. (2012). Evidence of social communities in a spatially structured network of a free-ranging shark species. *Animal Behaviour, 83*, 389–401.

Ndobe, S., & Moore, A. (2008). Banggai cardinalfish: Towards a sustainable ornamental fishery. In *Proceedings of the 11th International Coral Reef Symposium*, Ft. Lauderdale, FL, 7–11 July.

Porcher, I.F. (2005). On the gestation period of the blackfin reef shark, *Cacharhinus melanopterus*, in waters off Moorea, French Polynesia. *Marine Biology, 146*, 1207–1211.

Rigby, C.L., Carlson, J., Dicken, D., Pacoureau, N., & Simpfendorfer, C. (2021). IUCN red list of threatened species: *Carcharias taurus. IUCN Red List of Threatened Species*. https://doi.org/10.2305/IUCN.UK.2021-2.RLTS.T3854A2876505.en

Smith, M., Warmolts, D., Thoney, D., Hueter, R., Murray, M., & Ezcurra, J. (Eds.) (2017). *The elasmobranch husbandry manual II: Recent advances in the care of sharks, rays and their relatives.* Special Publication of the Ohio Biological Survey Inc., Columbus, USA.

Tate, E.E., Anderson, P.A., Huber, D.R., & Berzins, I.K. (2013). Correlations of swimming patterns with spinal deformities in the sand tiger shark, *Carcharias taurus*. *International Journal of Comparative Psychology*, 26: 75–82.

Tlusty, M. (2002). The benefits and risks of aquacultural production for the aquarium trade. *Aquaculture, 205*, 203–219.

Townsend, R., & Gilchrist, S. (2017). Preliminary Evidence for a Biennial Feeding Strategy Related to Reproduction in Female Sand Tiger Sharks, *Carcharias taurus* (Rafinesque, 1810). In *The elasmobranch husbandry manual II: recemt advances in the care of sharks, rays and their relatives.* Ohio Biological Survey, Columbus, USA, pp. 153–157.

Wearmouth, V.J., Southall, E.J., Morritt, D., Thompson, R.C., Cuthill, I.C., Partridge, J.C., & Sims, D.W. (2012). Year-round sexual harassment as a behavioral mediator of vertebrate population dynamics. *Ecological Monographs, 82*, 351–366.

Wilson, A.D.M., Krause, J., Herbert-Read, J.E., & Ward, A.J.W. (2014). The personality behind cheating: Behavioural types and the feeding ecology f cleaner wrasse. *Ethology, 120*, 1–9.

Woods, C.M.C. (2000). Preliminary observations on breeding and rearing the seahorse *Hippocampus abdominalis* (Teleostei: Syngnathidae) in captivity, New Zealand. *Journal of Marine and Freshwater Research, 34*(3), 475–485.

Wyffels, J.T., George, R., Adams, L., Adams, C., Clauss, T., Newton, A., Hyatt, M.W., Yach, C., & Penfold, L.M. (2020). Testosterone and semen seasonality for the sand tiger shark Carcharias taurus. *Biology of Reproduction, 102*, 876–887.

20

The behavioural biology of invertebrates

JAMES E. BRERETON
University Centre Sparsholt, Winchester, UK

20.1 INTRODUCTION

Invertebrates are undeniably the most abundant animal taxonomic group in terms of numbers of species, numbers of individuals, and biomass (Chapman, 2009). Invertebrates appeared on Earth millions of years before (and ultimately gave rise to) vertebrate animals. It is believed that 95.5% of all animal species are invertebrates. However, the true proportion of invertebrate species is likely far higher, as many more invertebrate species are likely to be discovered. Small body sizes, difficult-to-access habitats, and challenges with identification present barriers to invertebrate taxonomists.

Invertebrates are not a monophyletic (i.e., descended from one common ancestor) group. Some invertebrate taxa share more of their evolutionary heritage with vertebrates than other invertebrates. For example, insects have very little in common with cnidarians (e.g., jellyfish and sea anemones). Evolutionarily, invertebrates are paraphyletic, which means they are grouped together because of their shared features (i.e., lack of a backbone) rather than due to any shared evolutionary relationships. This paraphyletic nature of invertebrates means that there is enormous variation in terms of their biology and behaviour. Even aspects of biology that we take for granted, like the respiratory system, vary widely. In arachnids, breathing is achieved through book lungs, a series of thin, vascularised plates that are held together in a pouch and found just below the abdomen. In the insects, breathing is achieved instead through a series of small holes called spiracles, which are found in lines along the thorax and abdomen. As for jellyfish, oxygen enters the body through the slow process of diffusion. This bewildering variation in invertebrate biology poses challenges, especially when identifying best practice husbandry and management techniques across multiple species.

Invertebrates are also rarely the first taxonomic group that one associates with a trip to the zoo. Admittedly, invertebrates rarely feature in the "top ten" list of animals to see. However, invertebrates permeate all areas of zoo and aquarium animal husbandry. For example, invertebrates are a common sight in aquarium tanks, in the form of starfish (Phylum Echinodermata), shrimp (Class Malacocostrans), and corals (Phylum Cnidaria), for example. They may also appear in the traditional "tropical house" or butterfly garden as key attractions. In terms of collection plans, zoos keep

DOI: 10.1201/9781003208471-22

more species of invertebrate than they do reptiles or amphibians (Brereton & Brereton, 2020).

Invertebrates play further roles in zoos and aquariums. Many species are regularly cultured or purchased as live foods, such as mealworms (*Tenebrio molitor*) and fruit flies (*Drosophila melanogaster*). Invertebrates can also be unwanted pests, posing challenges for food safety and disease risk. Some invertebrates play a key role in ecosystem services (such as pollination of key crops or protection of browse forests) and they are often used in visitor education programmes.

Despite their common presence in the zoo, there remains much to be learned about invertebrate biology and behaviour. This poses challenges for the zoo biologist, who must build from existing research to understand the needs of captive animals (Harvey-Clark, 2011). With few standardised biological features across invertebrate taxonomic groups and many gaps remaining in our understanding of behaviour, welfare assessments are a challenge to complete.

Invertebrates are not always afforded the same level of protection as other taxonomic groups. For example, in the UK, the Animal Welfare Act (2006) provides protection to vertebrates and only cephalopod invertebrates (Class Cephalopoda). More recently, lobsters and crabs (Family Malacostracans) have been considered candidates for greater protection under the proposed Animal (Welfare) Sentience Bill in the UK. It is often suggested that this differentiation in law between vertebrates and (most) invertebrates focuses on sentience and the ability to feel pain. There is a growing body of evidence to suggest that some invertebrates are sufficiently complex to be afforded some protection (Keller, 2017).

By necessity, this chapter will focus on a few key examples, to show how the concepts of behavioural biology can be applied across invertebrate species. Where relevant, signposting to other species-specific sources of information is supplied for further reading.

20.2 NATURAL HISTORY OF INVERTEBRATES

The natural history of some invertebrates has been the subject of intense study. However, other species have rarely been documented in the literature. For example, the social structure and communication of eusocial insects (e.g., termites (Order Isoptera) and ants (Order Hymenoptera)) have been exceptionally well studied. In contrast, research on tenebrionid beetles, a taxon containing over 20,000 species, is relatively scarce. In some respects, the sheer diversity of species of arachnids, beetles, and molluscs may present an overwhelming choice of species to focus research on. As such, researchers must prioritise those species relevant to answering the most useful or impactful questions, particularly those related to conservation, health and disease, or food production.

To add a further layer of natural history complexity, the habitat and dietary preferences of some invertebrates transform as they age. Many insect species undergo a process of complete metamorphosis as they develop. Complete metamorphosis involves transformations in body size and shape, often accompanied by changes in behaviour and biology. The Atlas moth (*Attacus atlas*) is a perfect example. After feeding voraciously for weeks following hatching, the larvae of the Atlas moth spin a papery cocoon, in which they live dormant for around four weeks. During this dormancy, the biology of the insect reshuffles, so that the resulting adult moth is able to fly using wings that the caterpillar lacked. The moth also swaps its original, chewing mouthparts for a short but non-functional proboscis, meaning that the adult moth is unable to eat. Insects that undergo complete metamorphosis may fill several different roles in an ecosystem. For the zoo, this means that the collection needs to consider the needs of the larva (sufficient access to food plants), cocoon (a safe, humidity-controlled environment, free from disturbance), and the adult moth (sufficient space to fly and access to mates in a very short time period).

Another example of complete metamorphosis can be found in the sun beetle (*Pachnoda marginata*). This sun beetle is well known to animal keepers in its guise as a live food item for birds and reptiles. The larval form of the beetle is white and subterranean; it spends its time deep in soil, rotting logs, or leaf litter. The grub uses large jaws and a hind gut fermentation system to break down high-fibre food items such as leaves and rotten wood. Just before metamorphosis, the behaviour of the grub changes rapidly. It starts to collect chunks of soil from its environment, and it seals these items together using its own secretions to form a cocoon (Figure 20.1). Once the cocoon is complete, the

Figure 20.1 The three life stages of the sun beetle (*Pachnoda marginata*). Sun beetle larva undergo several molts until they reach a size in which they can metamorphose. They build a cocoon to protect them during this period. Several weeks later they emerge from their cocoon as a beetle.

grub sheds its skin and pupates. In its new form, it has only limited scope for movement and no legs, so is vulnerable to attack. However, the cocoon it formed as a larva gives it a measure of protection, preventing attack from others. During its time within the cocoon, the pupa prepared for its final metamorphosis.

When the beetle finally emerges, it is totally different to the original grub. The beetle possesses a brightly coloured exoskeleton, along with wings. The beetle's gut has also transformed; rather than relying on hindgut fermentation, a much simpler gut has taken its place and the adult beetle now actively searches for fruit items as its primary food source.

Clearly, the three forms of sun beetle have entirely different biological needs. In the zoo, the requirements of all life stages need to be considered. For the larval form, this will include the provision of a deep (and edible substrate). Food substrates can be identified from the literature. Suitable food items include rotting leaves and wood, although the larvae are sufficiently versatile to survive in substrates as challenging as bat guano! A deep substrate and rotten wood should also meet the needs of the pupa, as it will provide them with materials to construct a cocoon. The beetle form, by contrast, will need access to fruit items for food, and a range of elevated branches for climbing and flying.

There are many occasions where little is known about the natural history of captive invertebrates. In some cases, the zoo can be used as a research facility to better understand the natural history and habitat requirements of the species. An example is the Desertas wolf spider (*Hogna ingens*), a Critically Endangered species from Deserta Grande island near Madeira (Rowlands, Capel, Rowden, & Dow, 2021). In its native habitat, the Desertas wolf spider is an ambush predator, making use of a burrow to hide from unsuspecting prey. However, little is known about burrow construction or spider behaviour in the wild. In a series of preference tests, zoo scientists were able to identify preferred substrates for burrow building, preferred substrate depth, and requirements for anchor points. The results were used to further inform the captive care of the species and are being used to inform plans for future reintroductions (Figure 20.2).

20.3 FEEDING ECOLOGY

There are a huge variety of feeding strategies among the invertebrates. From blood and haemolymph-drinking parasites to filter feeders, decomposers

Figure 20.2 The Desertas wolf spider (*Hogna ingens*). Studies on captive specimens were used to identify preferred substrates, substrate depth and aspects of burrow construction.

to carnivores, invertebrates have developed almost every conceivable strategy for obtaining and digesting food. There are wasps that feed only on honey, but whose larva feed exclusively on the internal organs of unfortunate cockroach hosts. There are some herbivorous species which partner with gut bacteria to help digest fibrous material. Other species bypass the symbiotic relationship entirely by producing the enzyme cellulase, allowing them to break down the plant fibre cellulose into simple sugar.

It can often be a challenge, not only to find foods that are an adequate substitute for wild food items, but also that allow the selected invertebrate species to express their natural feeding behaviour. For some invertebrates, there may be only limited information available about the animal's natural diet. However, aspects of the animal's physical anatomy may hint at suitable food types.

Tarantulas (Family Theraphosidae) are a good example. On very close inspection, it becomes clear that tarantulas are not equipped with chewing mouthparts. While the animals do possess fangs (which are technically called chelicerae), they are not able to chew their food. This suggests that tarantulas will struggle with shelled prey, or food with a thick exoskeleton. As a result, prey items should have a relatively soft exoskeleton (such as crickets). After capturing prey, the tarantula injects it with venom. This venom breaks down the internal organs, transforming the prey into a liquid. This liquid can then be sucked up by the tarantula for digestion.

For a handful of commercially reared invertebrates such as shrimps, there are now pellets available that address key nutrient requirements. However, attention must still be paid to the wild ecology of the species and their prey acquisition strategies. For example, investigations of pellet size for the Southern brown shrimp (*Penaeus subtilis*) revealed that the shrimps often discarded large pellets (Nunes & Parsons, 1998). This is a reflection of the prey size of the shrimp in the wild. Large food items could easily overpower the shrimp or result in excessive competition against other shrimps for food. Careful attention to prey size, resulting in a smaller pellet or crumb, could improve food acceptance.

For some species, it is simply not possible to source the wild diet in captivity. The *Portia* sp. Spiders, for example, feed almost exclusively on other spiders. Many stick insects (Order Phasmatodea) specialise in eating the leaves of a single species of plant. In these cases, a substitute food item is necessary. This food should be similar nutritionally to the original diet, and ideally, should allow the animal to engage in its natural feeding behaviour. An example of successful substitution is seen in the Peruvian stick insect (*Peruphasma schultei*), a Critically Endangered species that was first discovered by scientists in 2002 and is found only in a five-hectare section of the Cordillero del Condero region of Peru (Conle, 2005). In the wild, this stick insect is a specialist and feeds only on the leaves of *Schinus* spp. Trees. However, a similar and common food item, privet (*Ligustrum vulgare*) has been identified as a suitable substitute. As this species feeds on only a single plant species in the wild, there is little need to vary food types throughout the year.

20.4 MATING SYSTEMS AND REPRODUCTION

The full spectrum of mating systems can be seen in invertebrates. Some species are monogamous, whereas others are polygamous. Other species have reproductive strategies that are rarely seen in vertebrates. For example, the Indian stick insect (*Carausius morosus*) can reproduce sexually, but also can reproduce asexually through a process known as parthenogenesis. Others, such as the immortal jellyfish (*Turritopsis dohrnii*), can revert to a polyp form during times when food become scarce. Egg laying is common among many invertebrates. However, there are also many species, for example, the Malagasy hissing cockroach (*Gromphadorhina portentosa*) that give birth to live young through a process known as ovoviviparity.

Knowledge of an invertebrate's mating system is essential to successful and viable captive breeding. Some species cope well in colonies, though others are highly territorial. Others pose a risk to conspecifics during mating. The praying mantids (Order Mantodea) have achieved infamy in this capacity, with the females of many species incapacitating the male during the act of copulation. While this is a concerning sight for any animal manager, it should also be considered that this cannibalism does occur in nature, and it is believed that female reproductive performance is enhanced as a result. No such luck for the male, however!

For some invertebrates, the process of rearing young is especially complex. The jewel wasp (*Ampulex compressa*) is an example in which egg-laying is a sinister activity. This wasp is largely nectivorous but still possesses a sting. When preparing to lay an egg, the wasp searches for a host, such as an American cockroach (*Periplaneta americana*). Upon finding a cockroach, the wasp delivers stings to paralyse the front legs of the cockroach, followed by a sting to the brain to prevent escape reflexes. Once the cockroach is incapacitated, the wasp drags its victim to its burrow, where it lays eggs between the host's legs. These eggs hatch after a few days and the resulting larva use the host cockroach as a food source throughout their development.

While a rather grisly process, this complex incubation process can be easily replicated in the zoo. After the jewel wasp has mated, a cockroach can be released into the exhibit. This should be combined with the addition of several fake wasp burrows, consisting of large test tubes. Once the cockroach has been successfully stunned, the entire test tube can be removed and incubated at a suitable temperature for optimal larval development. The results of this behavioural biology knowledge embedded into wasp management are successful cultures of the jewel wasp in zoos.

Where possible, invertebrates should be kept in ways that allow them to engage in natural breeding behaviour. However, this is not always the case. For example, the African field cricket (*Gryllus bimaculatus*) is commonly kept in captivity. In nature, the male field cricket defends a territory against other males. Females are attracted through stridulation – the male rubs his wings together to produce the characteristic "chirping" sound. In the wild, the cricket that can stridulate the loudest and longest is normally rewarded with increased breeding opportunities. In captivity, African field crickets are often used as a live food and as a result, are rarely kept as exhibits. The insects may be kept in mixed-sex groups, often with many individuals sharing small spaces. While this may be justified to maximise space use, it may also reduce the health of individual crickets and result in competition over resources and mates.

20.5 MEASURING INVERTEBRATE WELFARE

Invertebrates pose many challenges when assessing welfare. Many of the pre-prepared assessment tools that are available for vertebrates cannot be applied to invertebrates. For example, body condition scoring cannot yet be applied to insects or arachnids on account of their inflexible exoskeletons. Behavioural indicators of poor or excellent welfare are still largely unknown for most species.

While it is essential that welfare assessments are developed and welfare issues are identified, invertebrate researchers must conduct foundational work on "What is invertebrate welfare?" before it is fully understood and assessment methods can be applicable.

With physiological indicators of stress being in an early stage of validation for vertebrates, there is limited scope to apply these indicators to invertebrates. However, some initial studies are available. For example, new research detected cortisol in spider and scorpion haemolymph (Somerville, Baker, Baines, Trim, & Trim, 2021). Cortisol levels were elevated when king baboon spiders (*Pelinobius muticus*) and Indian giant scorpions, (*Heterometrus swammerdami*) were maintained under full-spectrum lighting, versus low lighting. This suggests not only that these species would benefit from lower light levels in captive husbandry, but also that cortisol assessment might be a viable tool for arachnid welfare assessments.

One of the major impediments to invertebrate welfare is the limited enforcement of good welfare in law. For example, in the United Kingdom, invertebrates (except for cephalopods) are excluded from two key legislative Acts that were developed to safeguard welfare. These are the Animal Welfare Act (2006) and Animals (Scientific Procedures) Act 1986. These two Acts place a duty of care on animal keepers and protect animals against suffering (such as invasive procedures). While cephalopods (and more recently several crustacean taxa) are afforded some protection through the legislation, most invertebrates are not considered. This means that even if welfare issues are identified in invertebrates, the law has limited ability to prosecute or fine offenders.

The justification for the exclusion of invertebrates from welfare legislation centres around their ability to feel pain. It has been suggested that some invertebrates are unable to feel pain (as reviewed in Keller, 2017; Drinkwater, Robinson, & Hart, 2019), and therefore these taxa have limited capacity for suffering. However, research suggests that not only cephalopods, but also decapod crustaceans such as lobsters are able to feel pain (Elwood, 2012). Furthermore, there are suggestions that the capacity for suffering may not be based purely on the perception of pain, but also on cognitive abilities. Many invertebrate taxa have been shown to solve complex problems (Drinkwater et al., 2019), the likes of which were thought to be only possible for vertebrates. Given the growing body of evidence, ethical treatment of invertebrates is advised, even if it is not yet a requirement by law in many countries.

Live food cultures pose a particular challenge for animal welfare assessments. Like traditional farm livestock, live foods are reared under conditions that maximise growth and reproduction. Rearing under these conditions may sometimes result in live feed animals being housed at high stocking densities. However, unlike farm livestock, there is limited research available on the potential welfare ramifications of high stocking density or husbandry practices.

The mealworm is a typical example. Mealworms are used as a live food item for many other zoo-housed species. Orders of mealworms can be purchased online, with some vendors providing sacks of over 1,000 individual larvae. Although the mealworms are normally provided with substrate (e.g., oat flakes or bran), the animals are closely packed during transit. If a shipment is delayed, particularly during warm weather, there is the possibility that some individuals will become dehydrated or overheat. While there remains a debate as to whether invertebrates can suffer, there is the potential for many animals to be impacted by relatively minor occurrences (e.g., live food orders delayed in transit).

Clearly, there is a need for animal practitioners to be able to assess invertebrate welfare. One potential avenue that has shown some promise for assessing vertebrate welfare is the concept of Qualitative Behavioural Assessment (QBA) (Wemelsfelder, 2008). When using QBA to investigate welfare, an assessor reviews the physical and behavioral traits of the animal to gain a holistic overview of the animal's current psychological welfare state. The assessment may include descriptions of the animal's features (e.g., brightly coloured exoskeleton, evidence of scratches), behaviour (e.g., withdrawn into shell), and environment (is it appropriate for the species). A good understanding of the natural environment, and physical indicators of health for the species are therefore required if QBA is to be meaningful (Riley & Rose, 2020). However, if assessors have sufficient experience, QBA is promising as it allows welfare to be assessed rapidly and at minimal cost.

Using the Malagasy hissing cockroach (*Gromphadorhina portentosa*) as an example species, QBA

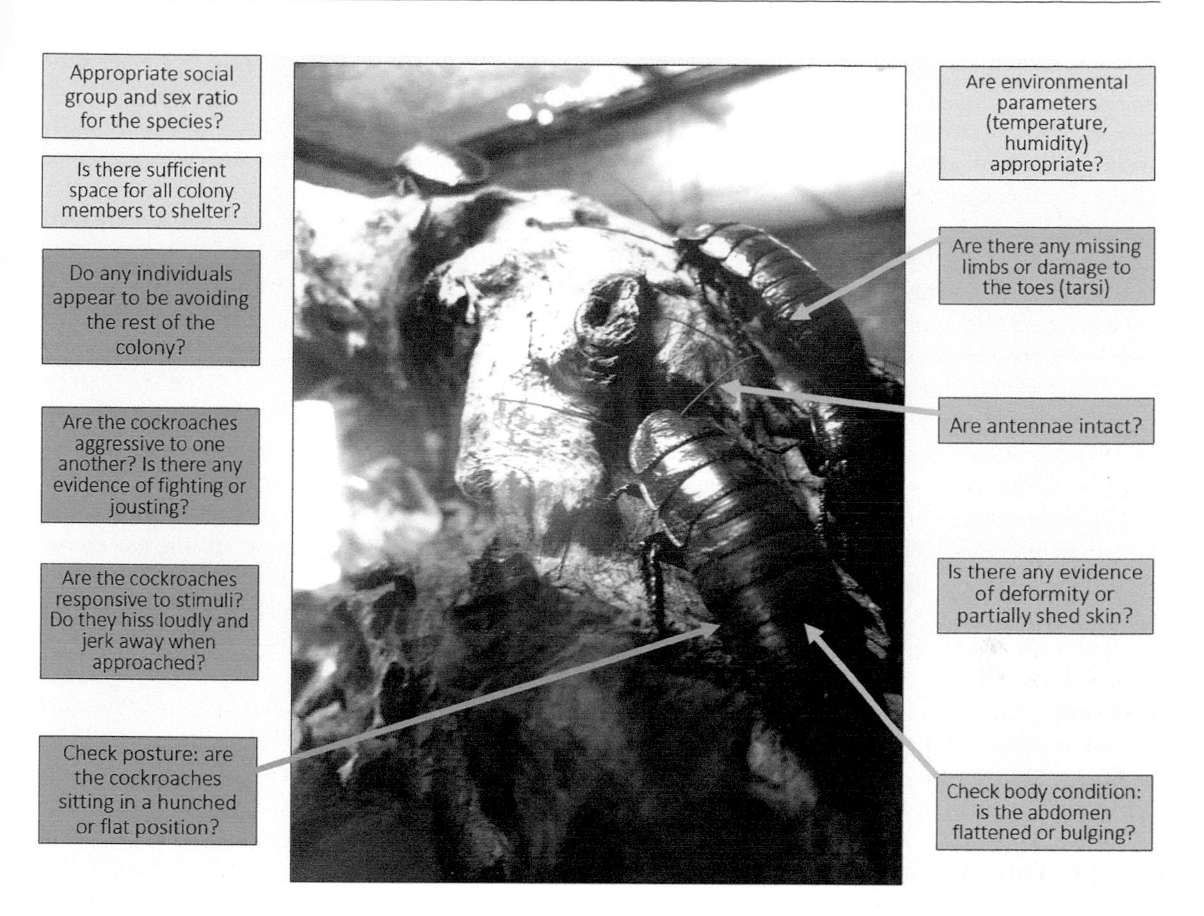

Figure 20.3 An example of a QBA assessment for a colony of Malagasy hissing cockroaches. Here, QBA involves an assessment of the physical health (green boxes) of the insect, along with its behavioural states and examples of body language (behavioural expression) (blue boxes) and social grouping (yellow boxes). Attention is also paid to the environment of the animal and its suitability as a major influence over behavioural expression.

assessment could be used on a colony of animals (Figure 20.3). Before beginning the QBA, the assessor would evaluate the temperature and humidity provision for the species, to ensure it is in line with the species' needs. Here, reference might be made to husbandry or "Best Practice Guidelines" such as those developed by the European Association of Zoos and Aquaria (EAZA) if available. If species-specific guidelines are not available, the assessor may instead use published information on the species' wild habitat to determine optimal temperature and humidity ranges to inform zoo care.

Next, social grouping should be considered. Malagasy hissing cockroaches form colonies, so group housing is generally suitable. However, high stocking densities or male-skewed sex ratios often result in heightened aggression. In this case, signs of unsuitable social grouping might manifest as broken antennae or missing tarsi (toes) for some colony members. Behaviour may also be affected, with subordinate individuals spending long periods of time hidden in sheltered locations. The assessor would take note of both the behaviour of the cockroaches and the animal's behavioural expression (i.e., its body language). For example, a cockroach that is sitting in a hunched position, and jerks away rapidly in response to stimuli, may be uncomfortable or experiencing distress due to unsuitable social grouping.

Care must be taken when interpreting behavioural expression. To understand the welfare inferences from behavioural performance, the assessor should be well versed in the normal behaviour of the species. For the hissing cockroach, the species is generally more active by night and takes shelter during the day. Inactivity, therefore, does not indicate poor welfare. Subtle signs such as the location of the cockroach (e.g., far away from the rest of the group) and its quality of behaviour (running rapidly when disturbed versus moving slowly while exploring using antennae) would be used to determine the overall welfare of the individual.

One species-specific behavioural trait that the assessor may listen for is hissing. The Malagasy hissing cockroach, as its name implies, can make a loud, hissing noise when threatened. The noise is produced by the insect forcing air out throughout its spiracles. While hissing is a natural behaviour, excessive evidence of hissing tends to suggest that individual cockroaches are threatened.

While there is a level of subjectivity to QBA, the assessments can be standardised if assessors are sufficiently trained. QBA may yet require further methodological refinement if it is to be applied to invertebrates more generally. However, with further study and validation of descriptors of behavioural expression, QBA could be an excellent method for evaluating the welfare of individual invertebrates, even when housed socially.

20.6 BEHAVIOUR

Historically, it was believed that invertebrate behaviour was "hard-wired." Invertebrates hatched with a set of instincts to help them cope with their respective environments and had little to no capacity for learning. New research suggests this is not the case, and some invertebrates have some capacity to respond to, and learn from, their environment. This capacity for learning is especially well developed in some invertebrate taxa. Bees (Order Hymenoptera), for example, are especially capable learners, and can develop comprehensive maps of their local surroundings, along with information on local food resources and their hive. The bee's ability to memorise the local landscape is so advanced that they are also able to communicate this information to conspecifics. Using their "waggle dance," the bee can communicate both the location and distance of potential food resources (Grüter & Farina, 2009). This suggests that some invertebrates have a considerable capacity to learn and therefore adjust their behaviour in line with environmental change. For the zoo, creative enclosure design cold help in highlighting the learning and communication abilities of this taxon.

Many invertebrate species are reliant on complex patterns of behaviour for their continued survival. Leafcutter ants (*Acromyrmex* spp.), for example, rely on a mutualistic relationship with fungi to help them break down their food. This relationship requires worker ants to locate and cut fresh leaves, and then to transport them back to the fungus. The ants must also take the role of farmers, ensuring that the fungus remains healthy and pathogen-free. Consistent care of the fungus is needed if the ant colony is to thrive. This complex behaviour can be manipulated in captivity. Technically, freshly cut leaves could be deposited daily near to the ant colony. This would reduce the amount of work required by the ants in cutting up leaves and transporting them. However, the challenge of cutting and transporting leaf matter may be biologically valuable for ants. *Acromyrmex* have evolved over millions of years to improve their ability to cut and transport leaves; the ants are now able to lift and carry more than their own body weight. Rather than reduce the workload for the ants, physical challenge could be provided, requiring the ants to transport the leaves several metres back to the colony.

Some zoos now actively showcase the abilities of the ant and develop enclosures in which the ant leaf transportation process is showcased to the public. This often comes in the form of rope bridges for ants to travel across, or in even more creative designs, through glass handrails, as seen in the Haus des Meeres (2021) in Vienna. Not only does the ant benefit from positive challenges it would receive in the wild, but visitors are able to see the communication and perseverance of this small insect.

Wild invertebrates may engage in even more complex behaviours, some of which can be challenging to incorporate into captive management. Termites, for example, are an example of a eusocial insect that work cooperatively to build a protected home. These termite mounds are structurally complex, and have thermoregulatory and gas exchange capabilities (Korb, 2010). The mounds also vary according to locale to ensure that the internal mound environment remains stable. Selection pressures have been applied over millions of years to termites for such complex mound-building behaviours to develop. Providing suitable substrates and opportunities for termites to engage in mound-building is therefore important for welfare. Relatively few zoological collections keep termites, but Saint Louis Zoo's "Insectarium" is an example of a collection that has successfully set up a colony (Stevens, 2006).

Invertebrates are no longer considered to be automatons, and there is clear evidence that invertebrates can engage in complex behaviours such as tool use, social learning, and environmental modification. Many of these traits were historically believed to be found only in higher mammals. However, if provided with barren and unengaging environments, invertebrates will be unable to showcase their abilities. Zoos should therefore consider the behavioural biology of each species, to ensure that opportunities are provided for animals to engage in natural behaviour. One way in which this can be easily achieved is through the use of environmental enrichment.

20.7 ENRICHMENT

Environmental enrichment is commonly used to enhance the welfare of zoo-housed animals (Mather & Anderson, 2007). Enrichment may aid in reducing stereotypy and self-directed behaviour, while potentially improving cognitive function. According to Bloomsmith, Brent, and Schapiro (1991) there are five types of environmental enrichment, which consist of physical, nutritional, sensory, social, and cognitive. Some authors also suggest there is a sixth category, training (e.g., Melfi, 2013). It is well acknowledged that enrichment is essential for good welfare for many zoo-housed species. However, invertebrates have typically been the subject of less enrichment research. This does not necessarily mean that enrichment is not being practised. Instead, enrichment is simply not being reported as commonly in scientific literature for invertebrate taxa. It has been suggested that enrichment practices are used much more widely, and for a greater range of species, than is reported in the literature (Rose & Riley, 2019). However, the depth of evidence is a challenge, as there are fewer available studies for practitioners to use to inspire their enrichment practice. To overcome this limitation, several potential enrichment options are detailed in Table 20.1.

The need for enrichment by some invertebrate taxonomic groups has been well acknowledged. Cephalopods, well known for being adaptable and capable problem solvers, are commonly provided with enrichment items that require them to solve problems or manipulate their environment (Mather & Anderson, 2007). For many cephalopods such as octopi (Order Octopoda), enrichment, and environmental complexity is not just a luxury, but essential. Research on the tropical octopus (*Callistoctopus aspilosomatis*) suggested that octopi kept in enriched conditions were more likely to respond well to an approach stimulus (Yasumuro & Ikeda, 2011).

Enrichment for octopi is varied, with practices encompassing all six enrichment categories. Food enrichment is a common example, in which tough invertebrates such as crabs and clams are provided. To feed, the octopus must break down the exoskeleton or shell of its prey (Anderson & Wood, 2001). This form of enrichment is both nutritional and potentially cognitive, especially where the octopus must use objects in its tank to break apart its prey's shell. Other potential enrichment strategies include regular training. Octopi respond well to training regimes and can rapidly learn new behaviours.

The behavioural biology of the species in question should be used to develop enrichment practices. Reference to the underlying diet, habitat, and lifestyle of the species can be used to develop meaningful enrichment. While enrichment practices do not necessarily need to be naturalistic, they should be developed with the natural history of the species in mind. Animals may be more motivated to engage in behaviours that they have evolved to do. Similarly, motivation may be higher when they are provided with enrichment items that mimic valuable resources from the animal's wild habitat. For example, enrichment for a sun beetle could make use of the adult beetle's sense of smell. The beetle is highly sensitive to olfactory cues in the wild, and it uses these abilities to find food (Stensmyr, Larsson, Bice, & Hansson, 2001) (Figure 20.4.). In addition to the normal provision of fruit, enrichment items could include a range of flowers. To increase the challenge associated with the enrichment, the flowers could be provided in elevated branches, requiring the beetles to climb or fly to access the resource. This allows the enrichment to fit into two enrichment categories – physical and nutritional.

While there remain gaps in the knowledge of effective invertebrate enrichment, provision of enrichment in the form of preference tests could be used to inform captive husbandry. Preference tests can be used to identify which of several stimuli is

Table 20.1 Potential enrichment ideas for a range of invertebrate taxa.

Species	Enrichment type	Description
Tarantulas (Family Theraphosidae)	Scent trail	Large tarantulas may occasionally be fed on rodents (e.g., "pinky" mice). Mouse-scented water could be sprayed around the enclosure prior to feeding, to simulate the presence of rodents and prepare the tarantula for feeding.
Leafcutter ants	Insect shed skin	In the wild, leafcutter ants regularly encounter other invertebrate species. Inserting an insect shed skin onto the normal travel paths of the ant colony may result in the ants responding to an "invading" invertebrate.
Malagasy hissing cockroach	Social enrichment	Many cockroach species natural form social groups. Introduction of new individuals to an established group can result in increased social interaction in the form of greetings (with antennae) and breeding, though also fighting (among males).
Praying mantis	Live prey	Live prey are commonly provided for captive praying mantids. However, fast-moving live prey could be provided that require the praying mantis to actively hunt down its meal. Example of fast, flying live foods include the fruit fly and house fly (*Musca domestica*).
Hermit crab (Family Malacostracans)	Shells	Hermit crabs use the discarded shells of other invertebrates as a form of protection. As they grow, hermit crabs regularly need to upgrade to a larger shell. Providing novel shell shapes and sizes can encourage hermit crabs to regularly try or swap their inhabited shell.
Octopus	Puzzle feeder	Octopi have already shown themselves to be competent problem solvers. Enrichment devices can be developed that require the octopus to manipulate a puzzle to receive a food reward.

most preferred by an animal. From this information, judgements can be made as to which stimulus is considered the most valuable. For example, sun beetles, when provided with a choice of scents, consistently chose fruit odours over green leaf volatiles and alcohols (Larsson, Stensmyr, Bice, & Hansson, 2003). This reinforces the point that fruit is a valuable commodity for this species over other potential food sources.

20.8 EXHIBIT DESIGN

With careful attention to exhibit design, public perception of invertebrates can be changed. Use of interpretation and signage to construct a message that might cause the zoo visitor to reconsider their views of invertebrates. Traditionally, invertebrates are kept in terrariums or aquariums. While an attempt is often made to recreate the animal's natural environment (e.g., use of planting or substrate), the result is not always inspiring. Showcasing invertebrates in novel exhibit styles may give visitors food for thought.

An example of novel practice is the Zoological Society of London's (ZSL) London Zoo's "*In with the Spiders*," exhibit. This exhibit is in the "Tiny Giants" building, which was originally developed to show visitors a glimpse of the variety of animal species encompassed within biodiversity. The "*In with the Spiders*" exhibit displays a range of arachnids which are showcased in a unique, walk-through experience (ZSL, 2021). At first, the visitor encounters spiders in a familiar scenario – in the bathtub. Supporting signage helps to discount some common misconceptions about house spiders. Next visitors encounter endangered

Figure 20.4 The sun beetle (*Pachnoda marginata peregrina*) uses its antennae to detect olfactory cues, and shows preference for the scent of food types including banana.

spiders, such as the (native to the UK) fen raft spider (*Dolomedes plantarius*), which is housed in a naturalistic enclosure. The aim here is to show visitors that spiders also are threatened with extinction, and therefore need support. Finally, visitors are introduced to a walk-through, in which Madagascar orb weavers (*Nephila inaurata madagascariensis*) and golden orb weavers (*N. edulis*) *have made their webs among live plants. Visitors can walk directly below these webs and re-evaluate their beliefs about spiders. This should allow visitors to consider the wider context of spiders within their respective ecosystems.*

Invertebrates can also be used to tell key conservation stories. For example, *Partula* sp. are tree snails that originate from Polynesia. With their small size and relatively dull markings, Partula are not the traditional, crowd-drawing celebrities of the zoo. However, Partula are in urgent need of conservation breeding because of an introduction of giant African land snails (*Achatina fulica*) to their habitat, followed by a poorly managed biological control attempt involving rosy wolf snails (*Euglandina rosea*) (Haponski et al., 2017). With the last few species reliant on captive breeding and reintroduction, the zoo is critical in the conservation for these snails.

The problem remains that Partula itself remains an unremarkable exhibit despite its key conservation story. Some zoos make up for this shortfall by using creative exhibit design. In some collections (e.g., ZSL London Zoo), the Partula exhibits are set up in a "scientific study" room, which is set up in the style of a laboratory. Exhibits are on show to the public, along with keepers as they practice daily husbandry. This gives the public a "behind the scenes" view of the husbandry for the species. Complimentary signage around the exhibit, making use of a comic book style, is used to dramatise the plight of the snail. While this exhibition technique does not attempt to recreate the wild environment of the Partula snail, it may have some value in conservation education.

There is also scope in the realm of habitat recreation. With attention to detail, immersive exhibits can be developed that allow the invertebrate to express almost all its natural behaviour. If attention is paid to environmental parameters, planting, and inclusion of other species, the invertebrate may not be aware it is in captivity at all. For terrestrial invertebrates, bioactive substrates can be used. A bioactive substrate may contain a range of bacteria and fungi that would normally. These are normally accompanied by additional invertebrate populations, such as springtails and woodlice (Rizzo, 2014). Together, the populations of microbes and invertebrates can help to maintain a more hygienic exhibit, reducing the need for regular enclosure maintenance. The populations of small "cleaner" invertebrates might also provide opportunities for hunting in carnivorous species.

In an exhibit containing plants, microbes, and invertebrates, it is technically possible to reach a point of equilibrium, in which minimal maintenance is required. In this scenario, plant growth is kept in check by small invertebrates, whose populations are in turn controlled by larger invertebrates, the "stars" of the exhibit. In this scenario, maintenance might only consist of glass cleaning, regular spraying, and topping up of water, along with occasional pruning of plants. This type of exhibit may have enormous educational value to visitors, while also allowing the inhabitants to live in a naturalistic way.

20.9 USING BEHAVIOURAL BIOLOGY TO ADVANCE CARE

The sheer diversity of invertebrate species, plus a relative scarcity of wild research across a wide range of species, provide challenges when advancing captive care. However, the scarcity of information also provides opportunities. For example, where limited research is available, zoo studies can help better understand the behavioural biology of specific species. Examples include the preference tests on substrate consistency and depth, as developed by Rowlands et al. (2021). Similar studies could be used to "ask" the invertebrate which enrichment types or enclosure styles best meet their needs. These can then be used to inform conservation strategies or reintroduction attempts.

Invertebrates underpin ecological services and healthy biodiversity, yet their roles are sometimes underappreciated by the public. A better understanding of the behavioural biology of invertebrates may aid scientists in identifying the importance of particular taxa. Carefully designed exhibits that show the natural behaviour of invertebrates, when combined with carefully designed zoo interpretation, can help to raise public awareness that invertebrates have a key role to play in biodiversity and planetary conservation.

BOX 20.1: The behavioural biology of parasites

So far in this chapter, we have not considered the behavioural biology of parasites, or their role in zoological collections. Many invertebrate taxonomic groups specialise in parasitising other animals: these may vary from blood-drinking ticks to the helminths of the digestive system. Often, parasites are considered as species of concern to health. However, if invertebrates are in the standing for some level of protection, questions could also be asked as to whether parasitic invertebrates also have welfare needs. Parasites play a key role in natural ecosystems, and where host species are threatened with extinction, the parasite may also be at risk. This is particularly true where parasites have co-evolved alongside one specific host species, as co-extinction can take place. Several avian lice species, including the Guadalupe caracara louse (*Acutifrons caracarensi*), are already extinct, owing to the extinction of hosts (in this case the Guadalupe caracara, *Caracara lutosa*). These lost parasites represent missing biodiversity and may also have value in terms of improving host immune function.

The behavioural biology of parasites often centres around one specific host species, or a group of biologically similar hosts. If one hypothetically planned to protect endangered parasites such as lice, they would need to consider the blood biochemistry of the host, alongside louse breeding and migration. This may mean that conservationists would be required to maintain populations of host species, for the sole purpose of culturing their parasites. Considering the current threats faced by wild vertebrate populations, attention needs to be given to this "parasite question" before further endangered parasite species are left to go extinct.

20.10 CONCLUSION

In summary, invertebrates provide an opportunity to push the boundaries of zoo science. Not only are invertebrates ubiquitous in zoo and aquarium collections, but they also present a wide range of different husbandry, feeding, and breeding styles. While the behavioural biology of some invertebrate species is well understood, many species would benefit from further study. Progress in understanding the behavioural biology of invertebrates could allow the zoo to exhibit their animals in more biologically informed enclosures, helping to showcase key messages regarding their diversity and conservation value to their visitors.

REFERENCES

Anderson, R.C. & Wood, J.B. (2001). Enrichment for giant Pacific octopuses: Happy as a clam?. *Journal of Applied Animal Welfare Science*, *4*(2), 157–168.

Bloomsmith, M.A., Brent, L.Y., & Schapiro, S.J. (1991). Guidelines for developing and managing an environmental enrichment program for nonhuman primates. *Laboratory Animal Science*, *41*(4), 372–377.

Brereton, S.R., & Brereton, J.E. (2020). Sixty years of collection planning: What species do zoos and aquariums keep? *International Zoo Yearbook*, *54*(1), 131–145.

Chapman, A.D. (2009). *Numbers of living species in Australia and the world*. Australian Government, Department of the Environment, Water, Heritage and the Arts.

Conle, O.V. (2005). Studies on neotropical Phasmatodea I: A remarkable new species of Peruphasma Conffile & Hennemann, 2002 from northern Peru (Phasmatodea: Pseudophasmatidae: Pseudophasmatinae). *Zootaxa*, *1068*, 59–68.

Drinkwater, E., Robinson, E.J., & Hart, A.G. (2019). Keeping invertebrate research ethical in a landscape of shifting public opinion. *Methods in Ecology and Evolution*, 10(8), 1265–1273.

Elwood, R.W. (2012). Evidence for pain in decapod crustaceans. *Animal Welfare*, *21*(1), 23–27.

Grüter, C., & Farina, W.M. (2009). The honeybee waggle dance: Can we follow the steps?. *Trends in Ecology & Evolution*, *24*(5), 242–247.

Haponski, A.E., Lee, T., & Foighil, D.Ó. (2017). Moorean and Tahitian Partula tree snail survival after a mass extinction: New genomic insights using museum specimens. *Molecular Phylogenetics and Evolution*, *106*, 151–157.

Harvey-Clark, C. (2011). IACUC challenges in invertebrate research. *Ilar Journal*, *52*(2), 213–220.

Haus des Meeres (2021). Our animals. https://www.haus-des-meeres.at/en/Our-Animals/Zoo-animals/Aqua-Terra-Zoo.htm

Keller, M. (2017). Feeding live invertebrate prey in zoos and aquaria: Are there welfare concerns?. *Zoo Biology*, *36*(5), 316–322.

Korb, J. (2010). Termite mound architecture, from function to construction. *Biology of Termites: A Modern Synthesis*, 349–373.

Larsson, M.C., Stensmyr, M.C., Bice, S.B., & Hansson, B.S. (2003). Attractiveness of fruit and flower odorants detected by olfactory receptor neurons in the fruit chafer *Pachnoda marginata*. *Journal of Chemical Ecology*, *29*(5), 1253–1268.

Mather, J.A., & Anderson, R.C. (2007). Ethics and invertebrates: A cephalopod perspective. *Diseases of Aquatic Organisms*, *75*(2), 119–129.

Melfi, V. (2013). Is training zoo animals enriching?. *Applied Animal Behaviour Science*, *147*(3–4), 299–305.

Nunes, A.J., & Parsons, G.J. (1998). Food handling efficiency and particle size selectivity by the southern brown shrimp *Penaeus subtilis* fed a dry pelleted feed. *Marine & Freshwater Behaviour & Physiology*, 31(4), 193–213.

Riley, L.M., & Rose, P.E. (2020). Concepts, applications, uses and evaluation of environmental enrichment: Perceptions of zoo professionals. *Journal of Zoo and Aquarium Research*, *8*(1), 18–28.

Rizzo, J.M. (2014). Captive care and husbandry of ball pythons (*Python regius*). *Journal of Herpetological Medicine and Surgery*, *24*(1–2), 48–52.

Rose, P.E., & Riley, L.M. (2019). The use of Qualitative Behavioural Assessment to zoo welfare measurement and animal husbandry change. *Journal of Zoo and Aquarium Research*, 7(4), 150–161.

Rowlands, A.N., Capel, T., Rowden, L., & Dow, S. (2021). Burrowing in captive juvenile Desertas wolf *spiders Hogna ingens*. *Journal of Zoo and Aquarium Research*, 9(2), 66–72.

Somerville, S., Baker, S., Baines, F., Trim, S.A., & Trim, C. (2021). Full Spectrum Lighting Induces Behavioral Changes and Increases Cortisol Immunoreactivity in Captive Arachnids. *Journal of Applied Animal Welfare Science*, 24(2), 132–148.

Stensmyr, M.C., Larsson, M.C., Bice, S., & Hansson, B.S. (2001). Detection of fruit-and flower-emitted volatiles by olfactory receptor neurons

in the polyphagous fruit chafer *Pachnoda marginata* (Coleoptera: Cetoniinae). *Journal of Comparative Physiology A, 187*(7), 509–519.
Stevens, J. (2006). Building the Monsanto Insectarium at the St. Louis Zoo. In *2006 Entomological Society of America Annual Meeting*, New York, USA, 10–13th December 2006.
Wemelsfelder, F. (2008). Qualitative Behaviour Assessment (QBA): A novel method for assessing animal experience. *Proceedings of the British Society of Animal Science, 2008*, 279–279. 10.1017/S1752756200028246
Yasumuro, H., & Ikeda, Y. (2011). Effects of environmental enrichment on the behavior of the tropical octopus *Callistoctopus aspilosomatis. Marine and Freshwater Behaviour and Physiology*, *44*(3), 143–157.
ZSL (2021). In with the spiders. Zoological Society of London. https://www.zsl.org/zsl-london-zoo/exhibits/in-with-the-spiders

PART III

For the future

DOI: 10.1201/9781003208471-23

21

Behavioural biology and zoo animal welfare: For the future

LISA M. RILEY
University of Winchester, Winchester, UK

MARÍA DÍEZ-LEÓN
Royal Veterinary College, Hatfield, UK

PAUL ROSE
University of Exeter, Exeter, UK
Wildfowl & Wetlands Trust, Slimbridge, UK

21.1 INTRODUCTION

Animal welfare is a complex (e.g., what to assess), contentious (e.g., how to assess, interpret indicators and gauge importance of husbandry improvements), and challenging (e.g., animals judge their own welfare state) issue to grabble with when determining the best way to enhance the husbandry, management, and lives of captive wild animals. While animal welfare primarily refers to the subjective states of individuals (Duncan, 1993), original definitions of welfare, "the state of the individual as it attempts to cope with its environment" (Broom, 1986) are still helpful, especially when combined with expanded approaches to guide assessment on how welfare is impacted by both environmental and animal-based variables (Mellor, 2012). Consideration of the impact of human interactions on an animal's ability to attain positive affective states is also important (Mellor et al., 2020). It is imperative to remember that welfare assessment is a continual cycle of observation, data collection, review, and evaluation, change to practice (if needed), further observation, evaluation, and so

DOI: 10.1201/9781003208471-24

on. Welfare assessment in the zoo must not be a simple "tick-box" exercise conducted only once or twice throughout the course of an individual's life. Without validated and accessible tools, welfare measurement will be difficult to do well and in a relevant manner for the species being observed. Fortunately, zoos and zoo organisations are working together to promote positive animal welfare practices and share research outcomes and methods that can support this continual cycle of welfare review, for example, the BIAZA Animal Welfare Toolkit, published by BIAZA's Animal Welfare Working Group (Harley & Clark, 2019).

Distinctions around definitions and terminology used are also important. "Quality of Life" (QoL) is often referred to when considering zoo animal welfare, and it is important to remember what this means as it is grounded in human psychology. QoL is defined by the World Health Organization (2022) as "*an individual's perception of their position in life in the context of the culture and value systems in which they live and in relation to their goals, expectations, standards and concerns.*" From an animal welfare perspective, QoL can be likened to a life worth living and a good life (where the balance sways towards positive, good, enjoyable experiences) rather than a bad life and a life not worth living (where the opposite balance is seen) (Green & Mellor, 2011). Zoos have an obligation to ensure all their animals experience a life worth living and a good life, that animals experience "great welfare" and are thriving, not just surviving (Melfi, 2009).

Covering all possible aspects of zoo animal welfare – from the development and validation of welfare indicators to species-specific assessments – is well beyond the realm of a single book chapter. Consequently, this chapter aims to look to the future and examine areas of animal welfare science and practice that can help reduce the complexities, clarify the contentious, and overcome the obstacles. We provide an overview of the importance of behaviour and cognition, the essence of behavioural biology when considering animal welfare, as well as review some of the limitations and challenges posed by measuring welfare across zoo taxa. We briefly summarise established and novel behavioural welfare indicators and focus on the use of Qualitative Behavioural Assessment (QBA) as a case study of advantages and challenges when applying this metric to assess zoo animal welfare. We end the chapter by providing a list of future areas where the study of behaviour would be critical to advance zoo animal welfare research and practice.

21.2 CONVERGING SCIENTIFIC RESEARCH AND PRACTICE IN ANIMAL WELFARE

The evidence-based approach empowers practitioners to use data and substantiated protocols when designing and implementing all areas of zoo and aquarium animal care. Welfare assessment and evaluation methods are not exempt from the evidence-based approach, and in fact should be rooted in it to be truly effective. In Chapter 6, the autonomy approach is advocated. This novel approach on the husbandry of managed animals puts the individual and sense of self at the centre of our philosophy of animal care. Given the cognitive expansion, behaviour, and close relatedness of primates to humans it is perhaps inevitable that we focus on primates for this approach. Since the early 1970s experiments have shown us that chimpanzees (*Pan troglodytes*) have a sense of self. The Gallup Jr (1970) mirror self-recognition task has since been tested with a wide range of animals and several primate species, elephants, and cetaceans show direct evidence of self. Of course, species that have failed this test may still have self-recognition and this shows the importance of developing tests with ecological validity. Chimpanzees also show a second-order theory of mind (understanding what another sees, knows, and believes), during psychological tests (Hare, Call, & Tomasello, 2001) and thus have a well-developed sense of self and "others" too.

Given the theory of mind, tool-use, primate cultural traditions, and the social complexities of primate living (e.g., Machiavellian Intelligence) it is unsurprising that we consider primates to possess human-like cognitive skills, but this can mean many things and does not occur spontaneously without cause. In primates, as in many other species, brain expansion evolved through ecological constraints – directly due to foraging niche and food acquisition strategies and indirectly due to sociality and complex social interaction (sociality is determined by food abundance/distribution and predation risk – see Chapter 6 for further explanation). In general, science and societies across the globe are increasingly recognising animals as sentient and consider animals to have moral significance; in 2022 the legal system in the United Kingdom reviewed the need to include animal sentience in their animal welfare legislation, which will now include Cephalopod Molluscs and Decapod Crustaceans. Further, an increasing number of countries are recognising individual animals of some species as legal persons (chimpanzees, orangutans, *Pongo* sp., and, depending on the outcome of current litigation, elephants, *Elephantidae*) (Nonhuman Rights Project, 2022).

Given continuing advancements in our understanding of animal consciousness, and sentience particularly, it makes logical sense that one aspect of the future of zoo keeping is to centralise the themes of cognition and autonomy into the care and husbandry regimes of captive animals. Future cognition-based research must urgently determine the extent and boundaries of sentience in animals, and therefore diversify the scope of such research to include a wider variety of metrics (e.g., beyond tool-use or mirror-test) in: (i) a wider range of primate species; (ii) other phylum to include invertebrates beyond Cephalopoda; (iii) other classes within the phylum Chordata, to include birds (beyond well-studied species within the Psittaciformes and Passeriformes), reptiles, amphibians, and fish. This centralisation of the individual into husbandry planning and practice means we are maximising the opportunity to improve and maintain good welfare. The justification for keeping wild animals in captivity is reduced greatly if we cannot evidence such good practice.

As this book advocates, one way to achieve this is using behavioural biology to inform husbandry practice. Throughout the taxonomic chapters, case after case has been made in support of this approach. Aligning husbandry with the physiological and behavioural adaptations an animal possesses means to provide for and challenge them in biologically appropriate ways. Historically, animal welfare has focused on the avoidance of negative experiences – fear, distress, pain, and discomfort – which must all be avoided under a Five Freedoms framework and in the UK, this forms the basis of the Animal Welfare Act 2006. This has created a culture of caution regarding the provision of any challenge assumed to lead to negative welfare. Conversely, researchers and practitioners alike, motivated by the best of intentions (not to cause animals harm or distress) often overuse enrichment strategies they know to be effective (e.g., promote positive welfare states, short- or long-term), typically feeding enrichment (Brereton & Rose, 2022) and enrichment is often applied as a generally effective treatment for abnormal behaviour. Yet negative states are only half the welfare story. In their seminal review of positive affective states in animals, Boissy et al. (2007) conclude that good welfare arises not simply from a lack of negative affective states but due to the presence of positive affective states, like pleasure, and advocate the application of physical, social, and cognitive enrichment to provide pleasurable experiences. Yet in real-world applications, enrichment is seldom used to promote pleasure; reducing abnormal repetitive behaviours for example is important for welfare but does not mean the animal has experienced a positive affective state.

21.2.1 Banishing the boring

By considering a scientific concept of animal welfare which includes naturalness, a captive animal should be given opportunity to behave in ways they are adapted to do, in ways that align with their anatomical, physiological, and behavioural adaptations. This would lead to pleasure and positive welfare, as frustration and the negative states associated with it is likely to arise when evolved, highly motivated behaviours are thwarted (Fraser, Weary, Pajor, & Milligan, 1997). It follows that behavioural biology-informed enrichment and wider husbandry should encourage animals to experience a life worth living (Mellor, 2016). At a basic level, this approach is evidenced in zoo practice; a fish is adapted to life in the water and therefore is kept in an aquarium, a wading bird is given shallow water in which to wade, but adaptations are about problem-solving, overcoming the challenge of everchanging wild ecosystems, extreme weather, and predation risk, food scarcity, and the threat from competitors, all rarely considered in husbandry practice currently. The chapters of this book explore the diverse opportunities we can capitalise upon to appropriately challenge zoo animals and stimulate their cognition and anatomy to overcome ecologically relevant issues and improve their welfare. Such challenge (when ecologically relevant and appropriately monitored) provides opportunities for animals to make choices over how they behave and control over use of their environment. Actively making decisions about daily actions, feeding times, rest times, reproduction, and play, for example, help an animal meet their motivational needs which are strongly driven by their wild ecology and evolutionary history.

Static environments inhibit decision-making and action – predictable is boring. Yet, it is essential to provide balance – too little predictability is difficult for animals to adapt to, it is frightening and disorientating. In addition, it is inevitable that staff and resource limitations need to be considered in real-world applications. Thus, husbandry informed by behavioural biology should provide some choice and control but also provide a level of predictability without being overly boring (Figure 21.1). It is then that behavioural biology-informed husbandry can improve or lead to good welfare. With a focus on good welfare, the remainder of this chapter investigates links between welfare assessment and behavioural biology, and the challenges, and solutions, to extending welfare research and assessment to "non-traditional" taxa.

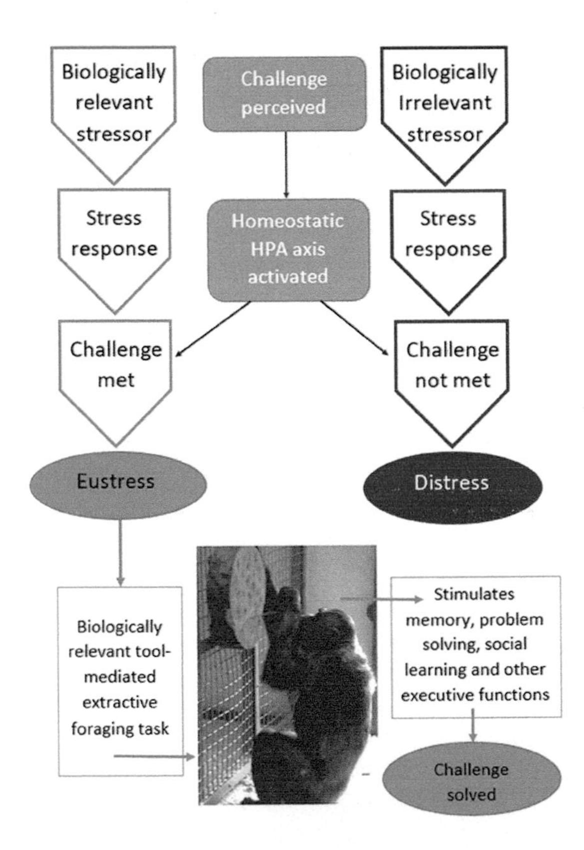

Figure 21.1 A simplistic representation of how biologically relevant challenges can activate the hypothalamic-pituitary-axis (in the example given via social competition for food rewards) and lead to positive stress. Both the challenge and solution align with the chimpanzee's behavioural ecology. It is important to note that eustress can become *distress* in chronic situations, and therefore the provision of challenge needs to be regularly evaluated.

21.3 BEHAVIOURAL BIOLOGY AND WELFARE ASSESSMENT

Given Fraser et al. (1997)'s concept of welfare, that welfare is comprised of naturalness, biological functioning, and feelings, it is reasonable to assume that aligning husbandry with behavioural biology would promote natural behaviour expression, biological fitness, and adaptive benefits, and a range of positive affective states. It follows that assessment of welfare would then need to focus equally on biological function, behaviour, and feelings, and prioritise positive welfare states. QBA can purportedly achieve this, at least for some species. QBA assesses emotional expression of an animal via evaluation of the animal's demeanour and expressive quality of their behaviour, going beyond behavioural assessment into a holistic evaluation of individual welfare (Wemelsfelder, 2007). This process sees a list of emotional descriptors (chosen from the literature or via free-choice profiling) applied to the animal's body language and expression, e.g., "happy," "joyful," "bold," "depressed," "fearful," "distressed." The extent to which the assessed individual displays each demeanour is then judged using a visual analogue scale by humans who typically have years of experience working with the species and individuals being assessed. The potential application of QBA is vast, and Chapter 20 expands on how QBA could be applied to inferences of captive invertebrate welfare.

What is essential for QBA to work well, is to ensure that those collecting welfare-focused data know who-is-who. Knowing individual animals, use of markings or tags, recognition of behaviour patterns or ways of moving, observation by caregivers, and accurate records, will allow for quality data on behavioural and physical characteristics that will help infer welfare state (Figure 21.2). When QBA is applied to easy-to-follow individuals, it is likely to result in better quality data collection than if individuals all look the same and are harder to follow. It is turning the "population into the individual" that is a specific challenge to welfare measurement for many zoo-housed species. Ultimately, it is important to acknowledge that QBA relies on the human assessment of the animal's welfare state and should thus be validated and/or used in conjunction with animal-based indicators.

Figure 21.2 "Charismatic" individuals can exist in non-mammalian zoo populations. A single blue-striped angelfish (*Chaetodontoplus septentrionalis*), top left, can have its behaviour easily followed and recorded by caregivers and become a "known personality," in the same way as the personality traits and characteristics of the zoo's easy-to-identify bull elephant, Elephas maximus (top right) enables easier welfare assessment. Similarly, mammals maintained in large herds where all individuals look similar at a passing glance, e.g., a herd of red lechwe (Kobus leche), bottom right, pose the same challenge for welfare assessment as a flock of budgerigars (Melopsittacus undulatus) in a mixed-species aviary. How do you follow the individual, consistently? How do you measure welfare without influencing individual behaviour patterns?

21.4 ASSESSING THE AWKWARD: HOW TO MOVE FORWARD WITH WELFARE MEASUREMENT FOR A DIVERSE ARRAY OF SPECIES?

While the basis for welfare assessment for zoo mammals can be founded on the results available from work on domestic and lab mammals, many of the other taxonomic groups have patchy to limited reliable or valid supporting information with which to structure zoo welfare assessment methods. Despite research showing that Mammalia is not the most speciose of Orders housed by zoos (Rose, Brereton, Rowden, Lemos de Figueiredo, & Riley, 2019), our familiarity with mammals and our inherent understanding of many of their needs and behaviours, and the perception that they could be more "deserving" of improved welfare, has pushed the development of welfare assessment tools to the forefront for these species. Many mammals also show facial expressions, which provide an observer with clues about mood and emotion, as well as allowing prediction of future behaviours and can provide evidence for

QoL estimation. Many understudied taxonomic groups are considered difficult to investigate empirically because we may know little about their natural history, evolutionary ecology, or behaviour patterns and their physiology sometimes lacks familiarity to an observer. Birds, reptiles, amphibians, fish, and invertebrates do not show easy-to-decipher facial expressions like mammals do (see limitations of e.g., QBA, earlier). Consequently, this barrier between the observer (struggling to decipher the meaning behind the behaviour performed) and the animal (its actions and behaviour patterns) could cause a collective reluctance to develop holistic welfare assessments for non-mammalian taxa.

If we first consider the natural behaviour of these "difficult to assess" taxa, we can provide a useful starting point for what we need to provide in the zoo to meet their ecological and behavioural requirements. Such evidence-based husbandry, population management and enclosure design will be the strongest foundation for the eventual attainment of positive welfare states in such taxa. Providing positive challenges in the zoo environment that enable animals to experience eustress (see Figure 21.1), further promotes beneficial welfare states by providing opportunities for behavioural plasticity and flexibility, enhances problem-solving capacities and can ultimately be cognitively enriching (Rose & Riley, 2019; Villalba & Manteca, 2019). A captive environment conducive to this eustress approach may be relatively easy to construct for many species of reptile, amphibian, fish, and invertebrate as the enclosures for these species are more likely to be microcosms of their overall natural habitat or ecological niche. The challenge is still to identify behaviours or activity patterns that can be observed (and therefore measured) that demonstrate positive welfare experiences by the individual, as well as to provide challenges in a way that animals can control (to e.g., avoid acute eustress becoming unavoidable chronic stress).

For many birds, reptiles, amphibians, fish, and invertebrates, "traditional" behavioural indicators of welfare may not work effectively when: (i) Animals live in large groups and behaviour may be hard to follow at the individual level; (ii) Individuals are sedentary or stationary for long periods of time and assessing behaviour change is complicated due to the timeframes for the observation that may be needed; (iii) Little information on behavioural responses to an impoverished environment are available to guide the evaluation of captive behaviour (with the exception of some bird species). Data collection across facilities to compare responses of populations (e.g., leafcutter ants, *Acromyrmex* and *Atta* sp., or neon tetras, *Paracheirodon innesi* where large group sizes and lack of individual identification may prevent individual welfare assessment) or individuals (e.g., Nile crocodile, *Crocodylus niloticus* or Galapagos giant tortoise, *Chelonoidis niger* complex, where individuals can be marked and followed at the specific animal level) to the prevailing environment provided by each zoo can then be evaluated to see any correlations or relationships between similar environmental features noted across institutions and the behavioural responses of the animals.

For example, space use by crocodiles and tortoises can be a biologically relevant metric for estimating welfare state. As ectotherms, these species need access to basking areas for correct thermoregulation and maintenance of homeostasis. Inappropriate enclosure design that impacts on an individual's ability to thermoregulate and/or a social grouping that causes competition or excessive agonistic behaviour around key resources (such as basking spots) can be measured based on the individual's behavioural responses compared to data from zoos where animals have choice and space to bask and thermoregulate when they desire without undue aggressive encounters.

Promotion of behaviours that maintain colony function (e.g., behaviour associated with defined roles of castes in an ant colony or the schooling behaviour of tetras) could be scored against the size, shape, and complexity of the environment provided. Estimation of welfare at group level is probably more biologically relevant for such species as the adaptive benefits of colonial living are essential to the health and fitness of the individuals themselves. A basis for welfare scoring using group size, e.g., an inappropriate social group of one or two neon tetras instead of a large school of fish, may be a suitable trajectory for data collection. Space use, position in the water column and activity patterns compared for fish in different school sizes would identify what is optimal for behavioural performance. Colour change, social dynamic, changes to body condition alongside behavioural responses such as lethargy or inactivity, orientation in the water column (e.g., at the surface or resting on the bottom), ease of movement, and ability to maintain equilibrium are all potential measurements for assessing the welfare of fish (and potentially aquatic amphibians, e.g., newts, too). Chapter 18 provides an illustration of physical, behavioural, and psychological measures of freshwater fish welfare that could be useful for the practitioner to use when judging the welfare state of the individual within its shoal.

Visibility of cryptic species, e.g., camouflaged species of frogs and toads, could also indicate the welfare state at a given time. An environment that makes cryptic animals feel comfortable to be on view, knowing that hiding opportunities exist, may be a useful baseline approach to commence a more detailed estimate of the welfare state. Amphibians are shown to respond differently, at the species-specific level to the presence and absence of visitors (Boultwood, O'Brien, & Rose, 2021), with aposematic (bright, warning colours to signal toxins) species responding differently to those species whose antipredatory response is to remain hidden. The more we understand about underlying ecology and biology, the more we can craft meaningful ways of determining the welfare states of these species in the captive environment.

21.4.1 Mammals still matter!

While mammals are often the subject of welfare-focused zoo research (Binding, Farmer, Krusin, & Cronin, 2020; Brereton & Rose, 2022; Rose et al., 2019), not all mammals are treated equally, and it is important to look across this taxonomic class when deciding on research subjects. Mammalian species can display subtle or hard-to-read behavioural, outward signs of welfare compromise and we need to look more deeply into the behaviour patterns of all the mammalian species that we house in zoos to ensure that we understand each individual's prevailing welfare state. Rodents in large colonies, or ungulates and small carnivores that can be perceived as less behaviourally diverse than "higher" primates are worthy candidates for bespoke welfare assessment measures to be constructed. As is noted in Chapter 7, the specific evolutionary adaptations and ecological requirements of ungulates can cause tremendous husbandry challenges and welfare assessment is needed to enable measurement and evaluation of improvements to practice, to ensure changes are actually improvements for the animals. Overall, for such welfare research and assessment to move forward, we need to take an ecological view of what the animal needs and why, remembering how important environmental parameters are to physiological functioning, so that we base our welfare assessments on what the animals can and cannot do in the environment created for them.

21.5 FURTHER WAYS BEHAVIOURAL BIOLOGY CAN ENHANCE WELFARE

Animal welfare research in zoo animals enjoys steady attention (Walker, Diez-León, & Mason, 2014) and we predict it will increasingly do so. To illustrate the application of behavioural biology to welfare enhancements, we provide some areas of future research and recommendations for further study:

- Include behaviour and welfare in population planning decisions (i.e., not just consider individual genetic characteristics but individuals that develop full potential in terms of behavioural skill).
- Increase use of multi-zoo and/or global zoo record databases to assess welfare-relevant parameters, that can then inform individual case studies.
- Develop and validate indicators generally, but particularly for "less-well-researched" species, particularly those where borrowing from other fields in the first instance is not possible.
- Refine existing behavioural indicators of welfare and lead research on novel ones (e.g., play) or often discussed but rarely examined states (e.g., boredom) that would add to our knowledge of welfare.
- Develop practitioner-led welfare assessment that considers how we can determine welfare metrics that are accessible for zoo personnel to implement quickly.
- Consider the positive impacts on conservation and evidence of the International Union for Conservation of Nature (IUCN) One Plan Approach – better welfare is associated with behavioural plasticity needed if zoo animals are to be incorporated into reintroduction programmes.
- Enhance "pure science" research at the zoo – animals acting and developing in ways they are adapted to do, allowing fundamental knowledge of natural phenomenona to grow.

21.6 CONCLUSION

The welfare of all taxa at the zoo is important. Promotion of positive welfare and affective states rather than avoidance of negative states should be a priority. This chapter helps the reader to identify the challenges of welfare assessment, from identifying emotions such as happiness in charismatic mammals to identifying the individual in social insects or schools of fish. Yet this chapter offers some solutions and argues that sentience is important regardless of taxonomic division, advocating that to enhance the welfare of zoo animals, consideration of behavioural biology and behavioural expression are important across all zoo animals. As zoo husbandry evolves, so the evolution and ecology of the animals kept should become a central focus of both husbandry practice and welfare assessment.

REFERENCES

Binding, S., Farmer, H., Krusin, L., & Cronin, K. (2020). Status of animal welfare research in zoos and aquariums: Where are we, where to next? *Journal of Zoo and Aquarium Research*, *8*(3), 166–174.

Boissy, A., Manteuffel, G., Jensen, M.B., Moe, R.O., Spruijt, B., Keeling, L.J., Winckler, C., Forkman, B., Dimitrov, I., & Langbein, J. (2007). Assessment of positive emotions in animals to improve their welfare. *Physiology & Behavior*, *92*(3), 375–397.

Boultwood, J., O'Brien, M., & Rose, P.E. (2021). Bold frogs or shy toads? How did the COVID-19 closure of zoological organisations affect amphibian activity? *Animals*, *11*(7), 1982.

Brereton, J., & Rose, P.E. (2022). An evaluation of the role of 'biological evidence' in zoo and aquarium enrichment practices. *Animal Welfare*, *31*(1), 13–26.

Broom, D.M. (1986). Indicators of poor welfare. *British Veterinary Journal*, *142*(6), 524–526.

Duncan, I.J.H. (1993). Welfare is to do with what animals feel. *Journal of Agricultural & Environmental Ethics*, *6*(2), 8–14.

Fraser, D., Weary, D.M., Pajor, E.A., & Milligan, B.N. (1997). A scientific conception of animal welfare that reflects ethical concerns. *Animal Welfare*, *6*(3), 187–205.

Gallup Jr, G.G. (1970). Chimpanzees: Self-recognition. *Science*, *167*(3914), 86–87.

Green, T.C., & Mellor, D.J. (2011). Extending ideas about animal welfare assessment to include 'quality of life' and related concepts. *New Zealand Veterinary Journal*, *59*(6), 263–271.

Hare, B., Call, J., & Tomasello, M. (2001). Do chimpanzees know what conspecifics know? *Animal Behaviour*, *61*(1), 139–151.

Harley, J., & Clark, F.E. (2019). Animal welfare toolkit. Retrieved from London, UK.

Melfi, V.A. (2009). There are big gaps in our knowledge, and thus approach, to zoo animal welfare: A case for evidence-based zoo animal management. *Zoo Biology*, *28*(6), 574–588.

Mellor, D.J. (2012). Animal emotions, behaviour and the promotion of positive welfare states. *New Zealand Veterinary Journal*, *60*(1), 1–8.

Mellor, D.J. (2016). Updating animal welfare thinking: Moving beyond the "Five Freedoms" towards "a Life Worth Living". *Animals*, *6*(3), 21.

Mellor, D.J., Beausoleil, N.J., Littlewood, K.E., McLean, A.N., McGreevy, P.D., Jones, B., & Wilkins, C. (2020). The 2020 five domains model: Including human–animal interactions in assessments of animal welfare. *Animals*, *10*(10), 1870.

Nonhuman Rights Project. (2022). Nonhuman rights project. https://www.nonhumanrights.org/. (Access Date 12/02/2022).

Rose, P.E., Brereton, J.E., Rowden, L.J., Lemos de Figueiredo, R., & Riley, L.M. (2019). What's new from the zoo? An analysis of ten years of zoo-themed research output. *Palgrave Communications*, *5*(1), 1–10.

Rose, P.E., & Riley, L.M. (2019). The use of qualitative behavioural assessment to zoo welfare measurement and animal husbandry change. *Journal of Zoo and Aquarium Research*, *7*(4), 150–161.

Villalba, J.J., & Manteca, X. (2019). A case for eustress in grazing animals. *Frontiers in Veterinary Science*, *6*, 303.

Walker, M., Diez-León, M., & Mason, G.J. (2014). Animal welfare science: Recent publication trends and future research priorities. *International Journal of Consumer Studies*, *27*(1), 80–100.

Wemelsfelder, F. (2007). How animals communicate quality of life: The qualitative assessment of behaviour. *Animal Welfare*, *16*(2), 25–31.

World Health Organization. (2022). WHOQOL: Measuring quality of life. https://www.who.int/toolkits/whoqol. (Access Date 12/02/2022).

22

Behavioural biology and animal health and wellbeing

MICHELLE O'BRIEN
Wildfowl & Wetlands Trust, Slimbridge, UK

22.1 INTRODUCTION

The health and wellbeing of all species held in zoological collections is intrinsically linked to their behavioural biology and the opportunity for the individual to perform natural behaviours. Legislation in the UK instructs keepers to monitor "condition, health and behaviour of all animals" (Secretary of States Standards of Modern Zoo Practice, 2012) and mentions provision of "an environment well adapted to meet the physical, psychological and social needs of the species to which it belongs" (Zoo Licensing Act, 1981). In recent years there has been a transition from resource- to animal-based indicators in welfare assessments, with animal-based indicators measuring a combination of physiological, behavioural, and health variables. Assessment at the individual animal level (instead of the group overall) provides a greater opportunity for health monitoring for a particular animal, and therefore the chance to intervene if welfare appears compromised.

The more positive the welfare state of an animal, the more likely it is that the immune system will be more robust, providing the animal with greater reserves to combat disease or survive certain levels of trauma when compared to an animal with a compromised welfare state and a corresponding weakened immune system. Differences in individual personality (e.g., a shy octopus will constantly hide in the presence of visitors) causes differential vulnerability to stress and thereby also a different susceptibility to disease (Mather & Carere, 2019). Behavioural assessment can also help to predict the outcome of some disease outbreaks. The intraspecific or interspecific interaction of individuals can affect the likelihood of transmission within or between enclosures. Behavioural research into enclosure use can then be used to inform some requirements for disease screening or prophylactic medication.

DOI: 10.1201/9781003208471-25

Behavioural biology is used as an integral part of animal training procedures to increase voluntary participation in husbandry or veterinary procedures in zoo animals. This training allows procedures to be carried out and samples to be taken to monitor the health status or reproductive state (for example) of an animal without the requirement for sedation or other methods that may increase the stress.

22.2 ANIMAL HEALTH AND WELFARE ASSESSMENTS IN ZOOS

To monitor the health of an individual factors such as weight, the performance of behavioural abnormalities, the extensiveness of "normal" behavioural repertoire, and engagement with the environment should be monitored (Ward, Sherwen, & Clark, 2018). Reduction in enclosure use or a diminished behavioural repertoire can be a sign of pain, disease, or injury. So too can changes in gait, posture, facial expression, or vocalisation, which may be outside of the human auditory range (Wolfensohn et al., 2018). However, these changes can also be due to reproductive behaviour or seasonal variation, so knowledge of the behavioural biology of each species is required to judge the motivation behind behavioural performance (Table 22.1). Research has historically tended to focus on mammalian taxa, so it is likely that the diversity and complexity of social behaviour in other taxa has probably been underestimated. For example, social systems are currently known in less than 1% of lizard species (Benn, McLelland, & Whittaker, 2019). The class Aves covers an extremely diverse group of species and assessment of the health of individuals needs to be tailored specifically to the species or group of species within a zoo enclosure. Birds are particularly good at masking signs of injury and illness until they are extremely unwell or debilitated. Carrying out health checks, such as regular weighing, can enable monitoring of individual bird health non-invasively on a regular basis.

Using positive reinforcement training and strengthwening the animal's trust in care staff can help to reduce the need for fear-driven management of zoo species (e.g., herding animals away from staff to lock them inside a house) (Wolfensohn et al., 2018). This can help reduce the traumatic injuries that can occur as part of any fear response. Due to evolutionary and ecological differences between taxa, general welfare assessment models, which can be used to monitor health, need to be adapted for use at a specific taxonomic level (Rose & O'Brien, 2020). The ability to assess the behaviour of a species accurately provides the chance to intervene if negative health indicators are reported.

22.3 EXTERNAL FACTORS AFFECTING THE HEALTH OF ZOO ANIMALS

Several activities related to the work of zoos can potentially affect the health and wellbeing of captive animals. Veterinary interventions in zoological collections may involve capturing large groups of animals for prophylactic treatments such as vaccination. Measurement, via welfare assessment, of the effects of those interventions could help to determine methodologies that minimise the potential stress-related health effects of these procedures.

Travelling internationally to participate in breeding programmes involves crating and restraint, change in social hierarchy, change in personnel delivering care, a new environment and numerous stressors which could negatively affect the health and welfare of an animal (Wolfensohn et al., 2018). Human presence impacts on animal health in other ways too. The Covid-19 pandemic enabled zoos to carry out research related to the effect visitor absence and presence has on zoo animal health and wellbeing. Interactions with visitors may be auditory, visual or vibration related, and the way in which that interaction is perceived by the animal determines whether it is a positive or negative experience (Sherwin & Hemsworth, 2019). A lack of visitors has been shown to have a negative effect on some species and a neutral or positive effect in others (Boultwood, O'Brien, & Rose, 2021). Those species showing any negative responses (directly on behaviour but likely on health too) should be housed differently or provided with space away from visitors.

22.4 TAXONOMIC-SPECIFIC HEALTH ISSUES

22.4.1 Invertebrates

Health assessment in invertebrates can be challenging. Invertebrates that undergo an ultimate moult show lesions associated with general "wear and tear"

Table 22.1 Examples of health issues with behavioural causation.

Health indicator	Behavioural cause	Taxonomic example
Traumatic injury	Fear response	Deer (Cervidae)
	Intraspecific aggression	Hyrax (Hyrocoidea)
	Stereotypy	Giraffe and okapi (Giraffidae)
	Digging behaviour	Armadillos (Cingulata, Dasypodidae)
	Unusual resting behaviour	Bats (Chiroptera)
	Collisions due to naturally fast movement	Hummingbirds (Trochiliformes)
	Exploratory behaviour	Parrots (Psitttacines)
	Running territory boundary (corners)	Canidae
	Climbing smooth or slippery surfaces and wedging into very small crevices	Caudata
	Cavity nesting behaviour in inappropriate environments	Passerines
Capture myopathy	Excitable disposition	Macropoda
Over aggression towards females	Stress and fear responses to keepers	Gorilla (Redrobe, 2008)
	Flight restraint preventing escape behaviour	Anseriformes
Consumption of inappropriate objects – heavy metal toxicity	Pica and consumption of rocks	Crocodilia
	Investigative behaviour	Penguins (Sphenisciformes) – young and nesting females
Consumption of inappropriate objects – foreign body impaction	Specific feeding strategies	Amphibians
	Abnormal compensatory picking	Ratites – juveniles
	Investigative behaviour	Walrus (Odobenidae)
Voluntary emesis	Displacement behaviour	Pinnipeds
Ocular disease	Inability to hide from excessive sunlight	Pinnipeds
Stress responses due to undesirable acoustics/ sound	Acute hearing	Insectivorous mammals
	Increased sound exposure	Zebrafish, *Danio rerio* (Jones et al., 2021)
Skin disease	Inability to hide from excessive sunlight	Tapiridae
	Intraspecific aggression	Agoutis (Dasyproctidae)
Thermal burns	Wrapping around structures	Caecilians
Dental disease	Chewing on or gnawing at enclosure structures	Pinnipeds
Reproductive disorder – egg binding	Chronic stress due to husbandry or environmental factors	Lizards (Benn et al., 2019)
Reproductive disorders, e.g., infant mortality	Chronic stress due to husbandry or environmental factors	Elephants (Mason, 2010)
Obesity	Chronic stress and seasonal energy requirements	Chelonia (Benn et al., 2019)
Digestive disease	Inability to maintain preferred optimum temperature	Crocodilia
Pododermatitis	Intense seasonal migratory restlessness	Migratory birds (Mason, 2010)

(*Continued*)

Table 22.1 (Continued) Examples of health issues with behavioural causation.

Health indicator	Behavioural cause	Taxonomic example
Feather plucking	Stress secondary to the inability to perform natural behaviours	Psittacines (Rose & Riley, 2019)
Opportunistic infection secondary to immunosuppression	Chronic stress secondary to the inability to perform natural behaviours	Cheetah (*Acinonyx jubatus*) -*Helicobacter sp.* Forest duikers (*Cephalophus*) – jaw abscesses (Mason, 2010)

For more information on each example, without a citation, see the corresponding chapter in Miller, M.E. & Fowler, R.E. (Eds.) (2015). *Zoo and Wild Animal Medicine* (8th edition). Elsevier, Missouri, USA.

whereas signs in those that undergo moulting throughout their lives are more subtle (e.g., lessening in body condition). Age has been shown to affect the ability of the invertebrate to regenerate (Pellet, O'Brien, & Kennedy, 2020), which has a bearing on health and welfare. Each animal must be considered as an individual, although invertebrate studies relating to personality and stress are limited, their stress response can be similar to vertebrates, so similar personality-related differences in coping could be evident in invertebrates (Mather & Carere, 2019).

Invertebrates showing indicators of ill health may be at risk of being cannibalised within a population, so may be more appropriately managed alone or within a geriatric group. Although there is still debate regarding whether invertebrates feel pain, recent behavioural and physiological work has suggested that there may be some evidence for consciousness in invertebrates (Drinkwater, Robinson, & Hart, 2019). Euthanasia must be taken into consideration for invertebrates with age-related disease and unmanageable conditions. Adequate analgesia and/or anaesthesia must be provided before euthanasia.

22.4.2 Fish

Research on fish behaviour has often been concentrated on those species either bred for food or those often found in the ornamental pet trade. In European sea bass (*Dicentrarchus labrax*), muscle activity was significantly higher at high densities (50 kg/m^3) resulting in higher use of reserves, which reduces the ability of the fish to manage additional stressors (Carbonara et al., 2015), including disease. Behaviours used to measure health and welfare in ornamental fish include feeding/foraging behaviour, aggression, neophobia, gasping at the water surface and locomotor activity. Changes in ventilation rate and observations of physical damage also represent non-invasive measures that could be combined with behavioural observations to monitor the health of individuals (Jones et al., 2021). Behavioural and social traits differ between species, e.g., territoriality and dominance/size-based hierarchies (Jones et al., 2021), therefore it is essential that the person observing aggressive behaviour is familiar with the species and able to distinguish between normal and abnormal levels of aggression.

Fish can be monitored individually or in groups as most signs of infections and diseases (e.g., damage to fins, flashing – rubbing the body on objects around the fish), are usually evident through visible indications and/or behavioural anomalies and therefore, would be easily noticeable and treatment or management changes put in place (Jones et al., 2021).

22.4.3 Amphibians

If amphibians are unable to use the full range of habitat zones (and access their species-specific optimum temperature range) within their captive environment, behavioural performance may be restricted, and welfare and health could be negatively affected (Boultwood et al., 2021). Amphibians are sensitive to sound and vibration and abnormal behaviour such as "balling" in newts can be seen in instances where building work (high vibration level) is carried out. Some species of amphibians can show reduced enclosure use in response to visitor presence (Boultwood et al., 2021), which can potentially lead to an increased likelihood

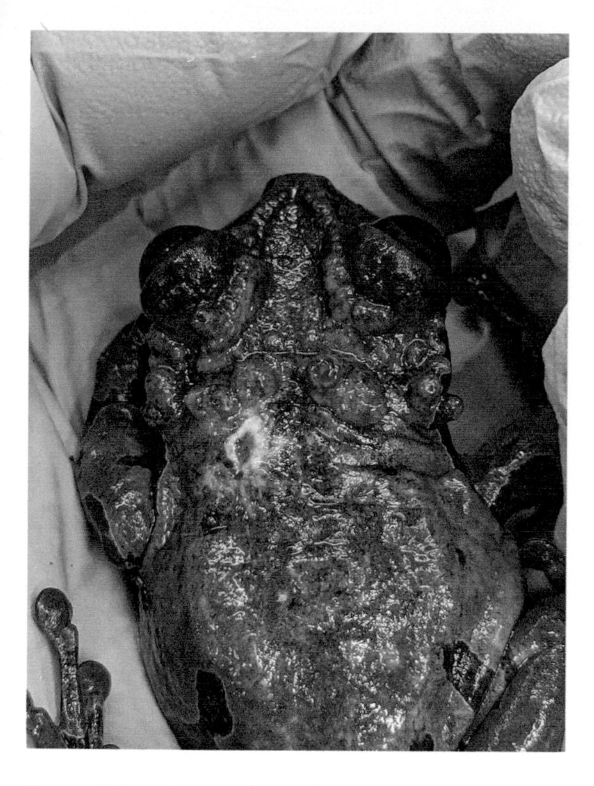

Figure 22.1 Crowned tree frog (*Triprion spinosus*) showing injuries related to intraspecific aggression between males (photo: Freya Boor).

of health-related conditions secondary to stress. Intraspecific injuries can also be caused by territorial behaviour in some species (Figure 22.1).

22.4.4 Reptiles

Chronic stress in reptiles has been shown to lead to aggression, anorexia, redirected activities, stereotypies, and displacement behaviours (Benn et al., 2019) all of which affect the health of the animal and lead to secondary disease processes. Compared with mammals, the humoral immune response of reptiles is weaker and slower and can be affected by season, reproductive state, and overall environmental conditions. It can be optimised within the species-specific optimum temperature range (Benn et al., 2019).

Stress has also been found to lead to egg binding, which can be a relatively common issue in gravid lizards (Yeates & Main, 2008) and may require medical or surgical intervention to remedy. This can then affect further reproductive ability, which is particularly important in individuals who are of conservation concern or part of breeding programmes. Reptiles can show signs of chronic stress in a variety of ways – from muscle degeneration and growth suppression (leading to emaciation) to obesity (Benn et al., 2019). Understanding the behavioural biology of a reptile can help to manage its seasonal energy requirements and any causes of stress which can exacerbate these issues.

22.4.5 Birds

The keeping of birds in zoos can require flight restraint unless birds are kept free flying in an aviary. Although flight restraint is generally considered to have a negative effect on health and wellbeing due to restriction of a natural behaviour, this should be assessed in relation to the behavioural biology of an individual species and the nature of the flight restraint method involved (Rose & O'Brien, 2020). Training to allow regular feather trimming can reduce stress and the need for permanent flight restraint measures in certain species. Telemetry is occasionally used in captive birds, and health effects should be regularly monitored as there can be a noticeable difference between individuals in the way they react to its application.

Physiological changes linked to breeding will also affect both bird behaviour and their interaction with other individuals or species housed within the same environment. Some species can be managed in a mixed exhibit during most of the year but during the breeding season, their territorial behaviour may lead to stress and injury in other individuals (Figure 22.2). Behavioural aggression in breeding males of some species can lead to injury in females of both intraspecific and interspecific groups.

Birds also display social networks within groups and relationships should be considered when removing individuals in order to minimise stress to both the individual and the group. Individuals of flock species should not be housed singly to reduce stress, which could lead to immunocompromise (Morishita, 2015).

One of the most prevalent health issues in captive parrots is feather plucking – a complex and multifactorial condition. Individuals with increased behavioural flexibility, using a wider range of coping strategies, may manage better in a captive environment (Rose & Riley, 2019) and be less likely to display this condition. Understanding the health reasons behind bird behaviours is vital, as some

Figure 22.2 Orinoco geese (*Neochen jubata*) can display territorial behaviour during the breeding season leading to stress and potential trauma to other species housed in the exhibit with the geese (photo: Jonathan Beilby).

behaviours can represent non-specific health problems, e.g., diurnal feeding in kiwi, *Apteryx* sp. (Boardman, 2008) whereas some natural behaviours in curassows (Cracidae) can be mistaken for neurological disorders (Tocidlowski, 2008).

Environmental changes (e.g., housing to mitigate for diseases such as Avian Influenza) are becoming increasingly relevant to zoological collections and understanding the potential stress and health issues (including intraspecific aggression) they may cause is essential to managing the wellbeing of the birds involved. Many avian diseases (e.g., candidiasis, aspergillosis) are caused by opportunistic pathogens associated with immunosuppression resulting from chronic stress. Stress may also result in hepatic dysfunction (Smith, 2015).

22.4.6 Mammals

Behavioural biology of any species is affected by their natural intraspecific social interactions. Where mammals are housed individually, or a single aged animal is left from what was previously a group, it is important to regularly monitor health and welfare as behavioural abnormalities may become more likely and must be managed using a variety of mechanisms to maintain wellbeing in such circumstances (Figure 22.3).

Health issues related to abnormal repetitive behaviours/stereotypies in mammals are many and varied and have been noted in bears (Ursidae), elephants, giraffe, prosimians, pinnipeds, and cetaceans (Mason, 2010). Being naturally wide-ranging predicts poorer welfare in carnivores (Mason, 2010). Overcrowding or separating compatible individual animals from each other may result in stress-related behaviour, e.g., the movement of individual female giraffe between herds has a greater chance of being disruptive than when moving males, and this should be considered when determining the placement of animals for breeding purposes (Gage, 2019).

Some species may require a period of seasonal torpor or hibernation, for example raccoon dogs (*Nyctereutes procyonoides*) (Padilla & Hilton, 2015)

Figure 22.3 Conducting a health assessment in a geriatric North American river otter, *Lontra canadensis* (photo: Jack Boultwood).

and bat species (Buckles, 2015) to maintain their health, which can be difficult to achieve in captivity.

In multispecies exhibits caution should be exercised related to behavioural interactions, e.g., tapirs have been known to eat birds, and zebras chasing new-born antelopes or giraffe can lead to myopathy and potentially death (Kaandorp, 2012).

22.5 VETERINARY CONSIDERATIONS FOR OLDER ANIMALS

As many individuals housed in zoological collections, when provided with good quality nutrition, veterinary care, and comfortable housing, can reach lifespans greater than free-living counterparts, the health and wellbeing of geriatric individuals needs to be considered. Assessing behavioural differences between young, mature, and older animals can be a useful way of identifying any potential health and wellbeing issues in geriatric zoo residents. Box 22.1 and Table 22.2 describe and explain some of these behavioural considerations for the aged zoo animal.

BOX 22.1: Health and wellbeing of older animals

As improvements in animal husbandry and veterinary care have led to longer life spans of numerous taxa in captivity, the health of animals suffering from the diseases of old age can become an issue for many collections. Providing easily accessible food sources as well as regularly monitoring appetite, weight, activity level, ability to navigate enclosure and skin condition can all provide information on the behaviour of geriatric animals that can help to determine health status. Behaviour patterns change as an animal ages and medical, nutritional, exhibit layout, and enrichment protocols may need to be modified for these geriatric challenges to maintain health, e.g., management of cardiac disease or osteoarthritis in great apes (Murphy, 2015).

In ageing invertebrates, the ability to navigate around an enclosure can become reduced, especially if limbs have been lost. Reducing vertical space can reduce any damage should the invertebrate fall, e.g., for elderly snails and New World theraphosid spiders. Ensuring that food is provided within easy reach for those with reduced flying or climbing ability is also an important consideration (Pellet et al., 2020).

Reptiles generally have long lifespans and the onset of age-related ailments may be insidious and only seen in subtle reductions in performance and vitality. Many species of squamates show gradual senescence with age whereas testudines and crocodilians show almost negligible senescence (Benn et al., 2019), see Table 22.2. Behavioural monitoring can objectively assess the health of older animals and in the case of slow deterioration, can help to ensure that if welfare declines, options can be considered such as increasing medication and eventual euthanasia. Chronic lifelong stress can predispose older animals to opportunistic infections or increase the visibility of behavioural issues related to these diseases, e.g., lameness and weight loss in birds suffering from chronic *Mycobacterium avium* infection.

Table 22.2 Examples of age-related behavioural changes affecting health of older animals.

Species	Behaviour issue in older animals	Health issue
Orb spiders	Reduced web building ability	Reduced ability to catch prey
Lepidoptera	Reduced flight due to wing damage	Decreased ability to find food
Phasmids	Reduced movement due to limb loss	Decreased ability to find food
Fish	Reduced mobility Inappetence Failure to control swim bladder volume	Spinal issues Lipid keratopathy
Predatory fish	Inability to close mouth	Age-related joint disease Difficulty in eating
Elasmobranchs	Rubbing against tank wall	Chronic ulceration
Multiple reptile species	Loss of range of movement Reluctant to move	Osteoarthritis Gout Renal disease Muscle atrophy
Reptiles and amphibians	Inability to find prey	Ocular disease
Snakes	Reluctance to move Stiffness on movement	Spondylopathy
Raptors	Reluctance to move Altered stance Weight loss Reluctance to fly Increased collisions	Osteoarthritis Ocular disease
Rodents	Reduced movement Reduced hygiene	Dental disease Osteoarthritis Neoplasia
Felids	Reduced appetite Weight loss	Dental disease Osteoarthritis Chronic renal failure (Longley, 2012)

For more information on each example, without a citation, see *Veterinary Clinics of North America Exotic Animal Practice, 23*(3). 2020.

22.6 CONCLUSION

This chapter has summarised some of the many behavioural indicators of health, wellbeing, and disease in captive wild animals. It illustrates more unknowns than knowns and highlights how research into behaviour and health needs to continue to fully advance zoo animal husbandry.

Ultimately, the question needs to be considered – if we cannot provide environments suited to the behavioural biology of certain species in captivity, likely leading to an increase in stress and reduction in health, should we be keeping these species at all? Especially if we are not doing so for conservation or where there is little chance of returning individuals to the wild in the future…

There is a need for greater development of taxon-specific health and welfare assessment tools with consideration of species-specific biology (Benn et al., 2019). Understanding and therefore being able to increase the behavioural repertoire of zoo animals can help solve numerous health issues. For example, increasing activity, stimulating the brain to reduce the likelihood of stereotypies, and providing more opportunities for natural behaviour – some of which may stimulate breeding or increased social interactions that can have positive health associations. Further research is also needed into those aspects of life in the zoo that modify behaviour and potentially have a detrimental effect on health (e.g., flight restraint in birds, the effect of visitors, and health issues related to sound levels and vibration). It is these unknowns that need our attention to ensure that health and welfare assessments are as comprehensive and relevant as possible.

REFERENCES

Benn, A.L., McLelland, D.J., & Whittaker, A.L. (2019). A review of welfare assessment methods in reptiles, and preliminary application of the welfare quality® protocol to the pygmy blue-tongue skink, *Tiliqua adelaidensis*, using animal-based measures. *Animals, 9*, 27.

Boardman, W. (2008). Veterinary Care of Kiwi. In M.E. Fowler, & R.E. Miller (Eds.), *Zoo and wild animal medicine* (6th edition). Elsevier, Missouri, USA, pp. 214–221.

Boultwood, J., O'Brien, M., & Rose, P. (2021). Bold frogs or shy toads? How did the COVID-19 closure of zoological organizations affect amphibian activity? *Animals, 11*, 1982.

Buckles, E.L. (2015). Chiroptera (Bats). In R.E. Miller, & M.E. Fowler (Eds.), *Zoo and wild animal medicine* (8th edition). Elsevier, Missouri, USA, pp. 281–290.

Carbonara, P., Scolamacchia, M., Spedicato, M.T., Zupa, W., McKinley, R.S., & Lembo, G. (2015). Muscle activity as a key indicator of welfare in farmed European sea bass (Dicentrarchus labrax L. 1758). *Aquaculture Research, 46*(9), 2133–2146.

Drinkwater, E.D., Robinson, E.J.H., & Hart, A.G. (2019). Keeping Invertebrate research ethical in a landscape of shifting public opinion. *Methods in Ecology and Evolution, 10*(8), 1265–1273.

Gage, L.J. (2019). Giraffe Husbandry and Welfare. In R.E. Miller, N. Lamberski, & P. Calle (Eds.), *Zoo and wild animal medicine* (9th edition). Elsevier, Missouri, USA, pp. 619–622.

Jones, M., Alexander, M.E., Snellgrove, D., Smith, P., Bramhall S., Carey, P., Henriquez, F.L., McLellan, I., & Sloman, K.A. (2021). How should we monitor welfare in the ornamental fish trade? *Reviews in Aquaculture*. https://doi.org/10.1111/raq.12624

Kaandorp, J. (2012). Veterinary Challenges in Mixed Species Exhibits. In R.E. Miller, & M.E. Fowler (Eds.), *Zoo and wild animal medicine* (7th edition). Elsevier, Missouri, USA, pp. 24–31.

Longley, L. (2012). Aging in Large Felids. In R.E. Miller, & M.E. Fowler (Eds.), *Zoo and wild animal medicine* (7th edition). Elsevier, Missouri, USA, pp, 465–469.

Mason, G.J. (2010). Species differences in responses to captivity: Stress, welfare and the comparative method. *Trends in Ecology & Evolution, 25*(12), 713–721.

Mather, J.A., & Carere, C. (2019). Consider the Individual: Personality and Welfare in Invertebrates. In J.A. Mathers, & C. Carere (Eds.), *The welfare of invertebrate animals*. Springer, Cham, Switzerland, pp. 229–245.

Morishita, T.Y. (2015). Galliformes. In R.E. Miller, & M.E. Fowler (Eds.), *Zoo and wild animal medicine* (8th edition). Elsevier, Missouri, USA, pp. 143–155.

Murphy, H.W. (2015). Great Apes. In R.E. Miller, & M.E. Fowler (Eds.), *Zoo and wild animal medicine* (8th edition). Elsevier, Missouri, USA, pp. 336–354.

Padilla, L.R., & Hilton, C.D. (2015). Canidae. In R.E. Miller, & M.E. Fowler (Eds.), *Zoo and wild animal medicine* (8th edition). Elsevier, Missouri, USA, pp. 457–566.

Pellet, S., O'Brien M., & Kennedy B. (2020). Geriatric invertebrates. *Veterinary Clinics of North America Exotic Animal Practice, 23*(3), 595–613.

Redrobe, S.P. (2008). Neuroleptics in Great Apes. In M.E. Fowler, & R.E. Miller (Eds.), *Zoo and wild animal medicine* (6th edition). Elsevier, pp. 243–250.

Rose, P.E., & O'Brien, M.F. (2020). Welfare assessment for captive anseriformes: A guide for practitioners and animal keepers. *Animals, 10,* 1132.

Rose, P.E., & Riley, L.M. (2019). The use of qualitative behavioural assessment in zoo welfare measurement and animal husbandry change. *Journal of Zoo and Aquarium Research, 7*(4), 150–161.

Secretary of States Standards of Modern Zoo Practice (2012). Available at: https://www.gov.uk/government/publications/secretary-of-state-s-standards-of-modern-zoo-practice

Sherwin, S.L., & Hemsworth, P.H. (2019). The visitor effect on zoo animals: Implications and opportunities for zoo animal welfare. *Animals, 9,* 366.

Smith, J.A. (2015). Passeriformes (Songbirds, Perching Birds). In R.E. Miller, & M.E. Fowler (Eds.), *Zoo and wild animal medicine* (8th edition). Elsevier, Missouri, USA, pp. 236–246.

Tocidlowski, M.E. (2008). Medical Management of Curassows. In M.E. Fowler, & R.E. Miller (Eds.), *Zoo and wild animal medicine* (6th edition). Elsevier, Missouri, USA, pp. 186–190.

Ward, S.J., Sherwen, S., & Clark, F.E. (2018). Advances in applied zoo animal welfare science. *Journal of Applied Animal Welfare Science, 21*(sup1), 23–33.

Wolfensohn, S., Shotton, J., Bowley, H., Davies, S., Thompson, S., & Justice, W.S.M. (2018). Assessment of welfare in zoo animals: Towards optimum quality of life. *Animals, 8,* 110.

Yeates, J.W., & Main, D.C. (2008). Assessment of positive welfare: A review. *The Veterinary Journal, 175,* 293–300.

Zoo Licensing Act (1981). Available at: https://www.gov.uk/government/publications/zoo-licensing-act-1981-guide-to-the-act-s-provisions

23

Behavioural biology and enhancing visitor education and experiences

BEAU-JENSEN MCCUBBIN
Natural History Museum, London

23.1 INTRODUCTION TO BEHAVIOURAL BIOLOGY AND ENHANCING VISITOR EDUCATION AND EXPERIENCES

Modern zoological organisations hold valuable knowledge on behavioural biology, fuelling advances in research, husbandry, conservation, and conservation education. The dissemination of this information boosts the experience of visitors and elevates well-being, empathy for, and engagement in the natural world. This, in turn, can support the behavioural changes that conservation education programmes desire. The future of aquariums and zoos (hereafter zoos) relies on their contribution to conservation education (Thomas, 2016) and currently, WAZA members only reach 10% of the population (WAZA, 2021). Increasing educational reach by catering to the needs of the remaining 90% can ensure zoos function, stay relevant, and contribute to and in the future. Equitable programmes are essential to ensure conservation messages are available, accessible, wanted, effective, and expand their reach. Conservation education's rich holistic tapestry should utilise the empathetic attributes of behavioural biology (how biology affects behaviour and vice-versa) to achieve this.

Empathy is a key element in communication and a desirable trait, or one to be developed, in those that are yet to be reached by conservation education. Behavioural biology widens the pool of endearing educational foci (such as eating habits, ecological niches, or social structures) a zoo has at their disposal for possible avenues towards empathy. Especially when sharing important "big picture" information, while still being relatable. By bridging the gap, these endearing behaviours can assist to form bonds by underlining shared traits, experiences, and problems. By using behavioural biology in conservation education, zoos can capitalise on their unique offering to cater to a visitor's (learner's) education, enjoyment, well-being, and role as a custodian of the planet. For example, flamingos (*Phoenicopteridae*), a familiar zoo animal, can help deliver the important message of how the

DOI: 10.1201/9781003208471-26

climate crisis can influence where a species (inclusive of humans) lives, while also relating to a learner's welfare with information on how diet directly affects health (Rose, 2018).

To develop and maintain relationships with learners, modern zoos must lead by example to align themselves with (and help shape) the societal zeitgeist. Zoos continuously modernise and now must evolve again as the next biggest change in their history from living museums into environmental resource centres transpires (Rabb & Saunders, 2005). Past exclusionary practices should be identified and rectified to guarantee increased uptake of educational offerings (Dawson, 2019). The variables in organisations and learners are vast and no perfect formula exists, but animals, and their endearing behaviours, are the enduring bridge between nature and peoples of all ages, abilities, and cultures (Fraser & Switzer, 2021). This chapter will explore how education programmes can be more meaningful and accessible using examples of integrating animal behaviour that unify the human experience with the natural world and deliver the curriculum equitably. When used proficiently, animal behaviour can enjoyably, ethically, and equitably engage and educate learners. It is from these ideals that a new era of conservation education programmes can be nurtured.

23.2 EQUITABLE PROGRAMME DESIGN

Though modern principles and practices should permeate throughout an entire organisation (Thomas, 2020), it is in the pedagogy (methods/practices of teaching) of practitioners (facilitators) of conservation education that should accelerate towards the next big shift in zoos as so the modern zoo can exemplify what it means to be humane (Kagan, Allard, & Carter, 2018). These modern programmes, new and existing, should be assessed to ensure equitability. Equitable programmes consider each learner's individual needs, aim to reach everyone, and ensure all stratifications (distinct social and cultural groups) and resource requirements are catered for. The objective of equitable education is an equal outcome for all.

Designing/redesigning programmes, using behavioural biology, that contribute to an ecoholistic approach is no small challenge and equity and ethics must oversee these designs in responsibility to animals, learners, and society. Key aspects of this challenge are:

1. Developing outcomes, equitably, in the production and reproduction of pedagogy.
2. Ensuring conservation education is ready for and participates in the future.
3. Positioning conservation education as ethical and productive.
4. Ensuring the animal behaviours used in conservation education are relatable to an individual, curriculum, and cause.
5. Researching and reaching as much of the community as possible.

It is essential to evaluate current practices and ensure programmes are designed with and by those they are intended for. This will alleviate appropriation, ensure appropriate delivery, boost engagement, and evolve a learner along with the organisation. The "equity compass" (Figure 23.1) can assist outputs in being socially just, including talks, signage, recruitment, gift-shop design, language, social media posts, etc. Facilitators can identify ways to support critical STEM (science/technology/engineering/mathematics) agency for learners to take action on issues that are meaningful to them (YESTEM Project Team, 2021). Evidence suggests pedagogies in zoos which challenge elite STEM practices and representations can have a direct effect on the equitable outcomes of learners (Archer et al., 2020).

Behavioural biology has many equitable pedagogical avenues to explore. Being out of the classroom lends itself to a multitude of learning styles, appealing to a wide range of learners. Kinaesthetic (active) learning is inclusive and allows individuals independence and ownership (agency) over their education. Something as simple as an activity based around observing a species travelling through its natural habitat can illicit inclusive and agent critical thought.

- What species would you like to observe?
- What equipment do we need?
- How does it use its limbs?
- What adaptations have this species evolved to suit its environment?
- Why is it travelling?
- Would it be able to travel in the same way in a different habitat?
- Is it similar or different to how we use our limbs?

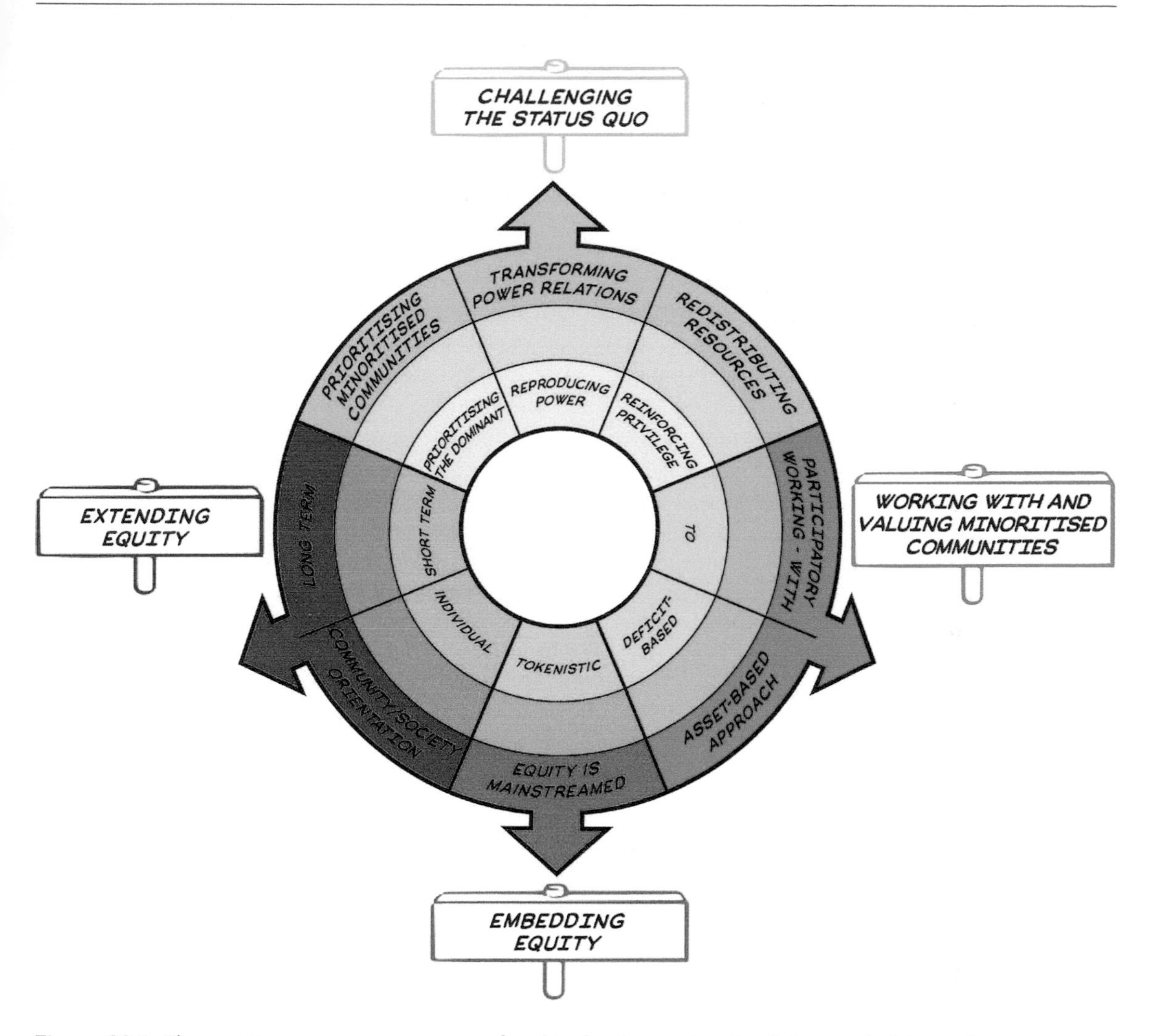

Figure 23.1 The equity compass, a resource for developing and maintaining equitable pedagogy.

- Can you move in the same way?
- How can human activity affect the way this species travels?
- How can we help this species achieve its goals?

Animal behaviour can be used as a learning resource for any topic in conservation education (deforestation, nutrition, lifecycles, plastic pollution etc.) and it lends itself to variety, equity, and empathy. Using these qualities, a learner has more opportunity to engagingly reach an objective in their preferred way. Proximity to live animals when observing behaviour provides a connectivity that the classroom cannot and increases the value of the animal being observed – therefore enhancing the role of the zoo.

How conservation education is taught is crucial to its viability. Weight should lean towards iterative development, as opposed to "this is how we have always done things." The ideal is producing exciting, evaluated programmes that enjoyably, ethically, and equitably engages and educates to elicit satiable behavioural change. The aims using behavioural biology is to engage learners in the natural world by instilling agency to evoke empathetic, meaningful, transferable, and lifelong behaviours.

In evaluating suitability, a facilitator might ask of their programme: Is it …

- **… Equitable**? Current programmes are seemingly designed for the masses, yet only serve a consistent few (Bourdieu, 1979). Though the premise may state the contrary, spaces of informal science learning are not for everyone, and incite personal feelings that certain spaces are "not designed for us" (Dawson, 2014). Equitable efforts are needed for the 90% that are not currently engaged. Are all learners being reached and is their individual journey tailored to their needs? Providing an arboreal species with climbing apparatus while furnishing an aquatic species with swimming space to exercise their behaviours, is an equitable measure. Is this emulated in the programme?

Are taxa represented equitably? If using given names, then all animals should be named, not just the "charismatic" few. These names should inclusively honour an animal's geographical/cultural/evolutionary background, and unhelpfully gendered, stereotypical, or anthropocentric names avoided. When disseminating biodiversity, facilitators ensure an individual species' importance to an ecosystem is recognised. This ethos needs to be supported universally. How can facilitators encourage empathy for an insect, if other areas of a zoo constantly focus their attention on "fluffier megafauna"? As with biodiversity, is diversity celebrated? Do learners of all stratifications have opportunities to see themselves represented in the material and the zoo?

- **... Enjoyable?** In what ways do the intended learners enjoy themselves? Fun is essential to the acquisition of knowledge and benefits different learning styles. Trialling sessions, market research and creating content with (not "for") learners is essential if predicting enjoyment, and therefore agency and uptake on understanding.
- **... Ethical?** Have all ethical considerations been assessed (for all animals including the learners)? Practising what is preached, all contradictions must be eliminated or justified. It is a touchy subject, but to move forward with the aims of the modern zoo, how do traditions of "touch" fit in? Some evidence shows that touching animals increases positive attitudes, but this is accompanied and compromised by increasing safety, hygiene, and ethical issues. Ethical positive attitudes towards wildlife can be achieved through approximated scenarios (bioartefacts, observing animal behaviour etc.) (Kidd, Kidd, & Zasloff, 1995). "Charismatic" animals (e.g., elephants, sharks, tigers) achieve deep, lifelong affinities with learners without close contact. The negative connotations and contradictions with "touch" are tangible, and the shift in the culture of zoos must include seeing species as subjects, not objects (Kimmerer, 2015).

  In any interactive scenario, an animal must have a choice (WAZA, 2020) to avoid unwanted stress (Salas & Manteca, 2017). Having no choice in interactions takes away the element of respect zoos are consistently trying to impart. When being allowed to hold, touch, or take a selfie with an animal and concurrently being told not to do this in the wild, or at another facility where the animals may be mistreated or drugged, is confusing at best, and hypocritical at worst. The contradictory practice is no longer conducive as zoos must adapt, influence, and honour the highest integrity and ethical standards (AZA, 2021) at all times.

  Through observing natural behaviours, complemented with bioartefacts, bioresources, and interacting with the environment and enrichment, is consistent with how species should be treated in the wild, when trained educators are not there to facilitate safe practices. Much can be learned from museums and other spaces of informal science learning, where bioartefacts and multimedia approaches are routinely employed.
- **... Engaging?** Will all learners be engaged? What can be provided to ensure a range of engagement methods can be accessed? Observing and expediting behaviours, incorporating Art into STEM practices (creating STEAM), and involving young people in the dissemination of the aims are all engaging methods (Figure 23.2).
- **... Education**al? Does it cover the local curriculum and the aims of the zoo? Does it elicit behavioural change? Is the educational outcome useful to the learner?
- **... Empathetic?** Are there relatable subjects making empathetic connections? Has the facilitator worked with learners to research what is relatable, resulting in sources of empathy?

Relying on a "build it and they will come" attitude is moot. It has been built, and they're not all here yet. Future programmes need to be built with learners. Traditionally, conservation education was something presented to an audience. More and more, it is now done for an audience. But now is the time to do it with them. All of them. It makes rudimental sense (whether in terms of business, altruism, social

Figure 23.2 Top: Increasing empathy with close proximity to wildlife needs to be ethical, respectful, and the animal must have complete choice in participation. Removing the possessive "touch" element does not mean less engagement. Middle: Programmes that include agency can also impart longer relationships with nature. Building bug hotels promotes engagement, celebrates local native wildlife, and encourages repeat visits to see how their contribution to conservation is serving. Bottom: Programmes need to be run WITH learners. Here, a member of a young governance committee (Direction Board, Hanwell Zoo, UK) is demonstrating an ethical learning experience.

responsibility, or conservation) to ensure everyone is reached. Currently, zoos have a huge reach, but it is only 10% of the population. If zoos wish to achieve the goal of reaching everyone, and impart on them a conservation capital (set of useful tools), they must first achieve the objective of changing a visit to the aquarium or zoo from a privileged pastime of the few, to a powerful rite of passage for all.

23.3 INTERPRETING EXISTING BEHAVIOURS

Interpreting biological behaviours need not be an expensive task. By highlighting the demonstrable behaviours of zoo animals, interpretation and dissemination can be simply achieved. This also enhances a visitor's experience by supplying them with the science capital to decipher science that is happening now. Explaining behaviours rarely seen (hunting, courting, fighting, breeding etc.) has its value, but behaviours the visitor will probably see (sleeping, eating, preening, social engagement, interacting with an environment etc.) increase engagement and promote citizen science. How and why these are portrayed is essential, with heavy gravitas given to empathy, equity, and ethics. There are already many modern zoos that celebrate and capitalise on certain behaviours. For example, in feeding demonstrations of the Eurasian otter *Lutra lutra* in Ueno Zoo, an underwater viewing device is used to showcase natural feeding behaviours (Figure 23.3). Even between feeding demonstrations, the apparatus lends itself to many informal learning opportunities.

Each species, and often more appropriately each specimen, has observational behaviours that learners can be taught to observe. This act can help retrain visitors (who are evolving with the organisation) on how to behave around animals. Specific nods towards tapping the glass, being noisy, feeding, or touching zoo animals, can be addressed and learners taught that these actions will erode natural behaviours and lessen the experience for all concerned. Existing examples include:

- Nurse sharks (*Ginglymostoma cirratum*), a nocturnal species, are relatively slow and sluggish. Their behaviour often contrasts with diurnal species within their space. This creates an opportunity to orchestrate learners to look for moving gills as the shark pumps water through

Figure 23.3 Top: Incorporating behavioural biology into animal space design to maximise conservation education at the Eurasian otter space in Ueno Zoo, Japan. Bottom: Interpretation of biological behaviours for Asiatic black bears (Ursus thibetanus) with an interactive enrichment device on display at Sapporo City Maruyama Zoo, Japan.

them. Unlike some other sharks, nurse sharks can breathe while lying still. Behaviours of more sedentary animals can be complemented with interactive items, such as sharing enrichment devices with learners, mirroring those within the animal's space (Figure 23.3).

- Garden eels (*Heteroconger hassi*) oscillate with the water's current and retreat into a burrow if disturbed. Garden eels feed on plankton and face the current while anchored in the substrate, waiting for food to drift within reach. A learner will not be able to observe these behaviours if the aquarium glass is tapped.
- Bittern (*Botaurus stellaris*), with patience and quiet, can be seen hiding among the reeds. Teaching learners about native species' behaviours (and how to observe them) bridges gaps between conservation education in zoos and its fruitful applications into the wild. Celebrating and conserving local fauna (and flora and fungi to assuaging the ecoholistic approach), living wild in a zoo, by creating spaces and programmes based on their behaviours, increases the number of identifiable species within a learning space and provides opportunities for more relatable conservation practices (NZP Delhi, 2022).
- Space for Life, offers an interactive experience that relies on behaviours to enhance a visitor's experience. The app assists learners through spaces while teaching natural history, behaviours, and conservation. It provokes engagement through observing specie specific behaviours to engage exploration. When in the Canada lynx (*Lynx canadensis*) space, learners are encouraged to critically think about specific adaptations the lynx has for its habitat and how to observe them. Equitably, iPads are available for those who may not have a compatible device (Space for Life, 2022).

The numbers of behaviours are obviously far too numerous to list, and the individuals working with specific animals are far more qualified to highlight them, which reiterates the importance of cross-organisation parity. Every specimen can have their behaviours interpreted in this more ethical and engaging approach.

23.4 SPECIFIC BEHAVIOURS

Reminding visitors that they too are animals living on a fragile planet can achieve empathy that leads to behaviour change. Highlighting differences to enthuse, and commonalities to engage will re-establish links (or create links that have not yet formed) to the natural world.

Though this paragraph highlights the links between zoo animal and human behaviour (which of course is still animal behaviour) it is important not to anthropomorphise species. Clear links should be given, but subjectivities (such as objective human traits that currently cannot be scientifically attributed to other animals) should be avoided. Subjectivities serve as a keystone to facilitate humans as the default, and defaults are not equitable and should be avoided.

Biological behaviours that will serve as a commonality with the human experience include: feeding, gender, movement, nurturing, sexuality, social structure, and symbiotics. Relating to specific social skills are beneficial on many levels. Exploring traits such as altruism, teamwork, and symbiotics will affirm empathy and cater to the personal needs of learners. Such positive attributes can be complimented with kinaesthetic activities for learners, echoing the social responsibilities and key skills that humans desire and require. Examples can range from the rotational leadership of Canada geese (*Branta canadensis*) flying in "V" formation, to more cognitive device management (Figure 23.3).

Emphasising the pressures that the climate emergency puts on species, demonstrated through their behaviour, can engage empathy and connectivity. Such as the evolution of successful biological traits that cope with rising temperatures as seen in the "Allen's affect" in birds (Ryding, Klaassen, Tattersall, Gardner, & Symonds, 2021) and the feeding behaviours of fish (Prakash, 2021). These disadvantages and subsequent changes in biological behaviours have a direct correlation to anthropogenic activities. Using the existing biological behaviours of species in zoos, information can be interpreted to share conservation educations universal message of human impact on the planet.

23.5 NATIVE SPECIES

The conservation of local species should be of priority to modern zoos for no lesser reason than it lends itself wholly to their purposes; conservation, education, recreation, and research, with the added benefits of locale and being able to include learners. It can include the community in agency

Figure 23.4 Engaging the young. Conservation education sessions "in situ" (left) in Mandalay, Burma and "ex-situ" in Hanwell Zoo, London, UK (right). Equitable practices, such as ensuring all learners have access to the equipment and fully funded placements are available to ensure participation. In these sessions, learners are engaging with local native wildlife completing tasks they set themselves.

driven action and satiate areas of behavioural change that often lack tangible links, boosting the feeling of collective action (Fraser & Switzer, 2021). From pollination of local produce, providing homes for local species or encouraging locally extinct species back, in situ conservation programmes run as conservation education programmes can be more meaningful and create relatable relationships with conservation and the natural world (see Chapter 24).

Allowing learners to choose, explore, and discover calls to action on their very own doorstep will encourage longer lasting relationships and help to establish empathy and agency. As with other programmes, in situ kinaesthetic conservation education should always be developed with learners, incorporate their needs, and be made equitably (see Figure 23.4).

There is a plethora of enriching activities for a range of ages and abilities that accomplish the aims and ideals of conservation education. In lieu of "touching" animals, these activities involve a high volume of physicality with equipment and devices used to conserve and preserve. Association-wide campaigns, such as "Grab That Gap" (BIAZA, 2020) maintains the desired principles and agency, with the added boost of collective action.

While covering areas of local curriculums (lifecycles, food webs, and identification of species etc.) in situ conservation programmes provide any locale with an abundance of opportunities. Synurbic species can be concentrated on in most organisations, and their behavioural adaptations are highlighted as a direct cause of their adaptations to human activity. A solid understanding of the behavioural biology of native species is required and should be the basis of these education sessions to engage and inspire future involvement, including possible career paths.

23.6 CONCLUSION

By incorporating extensive behavioural biology in conservation education, aquariums and zoos will be better positioned to serve their animals and learners. Behavioural biology enables areas in education, empathy, engagement, enjoyment, equity, and ethics, all key components in the future of conservation education.

In creating a diverse and sustainable future, today's aquariums and zoos will play a critical role. To achieve this, greater cooperation is required. Existing disparities between peers, departments, and organisations worldwide weaken the positive force zoos can attain together. Better communication needs to be facilitated. Educators need to be present for horticultural and animal space design activities. Keepers need to have input on conservation education, sharing their expertise and knowledge. The equitable ethos needs to be represented in the restaurant, the gift shop, and on social media accounts. Exploring biological behaviours, of all species inclusive of humans, will help to deliver this. As the ecoholistic approach runs through the organisation, a strong mind should be directed towards practices and behaviours, and these addressed appropriately. Contradictions such as the availability of single-use plastics in the gift shop, types of food served in catering outlets, representation of species, language used by staff and in an online content are all areas that should have parity.

An emphasis on ethics, exceeding current practices, opinions, and laws must support the existence of zoos as primary caretakers of conservation, thus being the backbone of conservation education. The animals that share our spaces should have their value raised in all areas of the organisation, ensuring individuals and species thrive, not just survive, and that their care careers into welfare.

More research, and better availability of current research in biological behaviours, is needed to facilitate these aims. Emphasis on beauty, positivity, and love in content and pedagogy will better serve learners, but censorship avoided. Working with the 90% of the population not currently reached by aquariums and zoos around the world will see an increase of learners through gates, bringing their own expertise and taking away all that a good modern aquarium or zoo can give.

It is in the equitable teaching of animal behaviour that zoos will meaningfully increase engagement with education and beneficial nature-centric experiences for learners. Ultimately, equipped with conservation capital, a learner can develop a lifelong affinity with the natural world, creating personal connections, their own sustainable behaviours, and a sense of place with a determination to protect nature locally and globally.

REFERENCES

Archer, L., Godec, S., Barton, A.C., Dawson, E., Mau, A., & Patel, U. (2020). Changing the field: A Bourdieusian analysis of educational practices that support equitable outcomes among minoritized youth on two informal science learning programs. *Science Education*, *105*(1), 166–203.

AZA. (2021). Code of professional ethics. https://www.aza.org/code-of-ethics?locale=en

BIAZA. (2020). Grab that gap. https://biaza.org.uk/campaigns/detail/grab-that-gap-2

Bourdieu, P. (1979). Symbolic power. *Critique of Anthropology*, *4*(13–14), 77–85.

Dawson, E. (2014). "Not designed for us": How science museums and science centers socially exclude low-income, minority ethnic groups. *Science Education*, *98*(6), 981–1008.

Dawson, E. (2019). *Equity, exclusion and everyday science learning: The experiences of minoritised groups*. (1st edition). Routledge, London, UK.

Fraser, J., & Switzer, T. (2021). *The social value of zoos*. (1st edition). Cambridge University Press, Cambridge, UK.

Kagan, R., Allard, S., & Carter, S. (2018). What is the future for zoos and aquariums? *Journal of Applied Animal Welfare Science*, *21*(1), 59–70.

Kidd, A.H., Kidd, R.M., & Zasloff, R.L. (1995). Developmental factors in positive attitudes toward zoo animals. *Psychological Reports*, *76*, 71–81.

Kimmerer, R.W. (2015). *Braiding sweetgrass: Indigenous wisdom, scientific knowledge, and the teachings of plants*. Milkweed Editions, Minneapolis, USA.

NZP Delhi. (2022). List of free ranging animals of NZP. https://nzpnewdelhi.gov.in/en/page/list_of_free_ranging_animal

Prakash, S. (2021). Impact of climate change on aquatic ecosystem and its biodiversity: An overview. *International Journal of Biological Innovations*, *3*(2), 312–317.

Rabb, G., & Saunders, C. (2005). The future of zoos and aquariums: Conservation and caring. *International Zoo Yearbook*, *39*(1), 1–26.

Rose, P.E. (2018). The relevance of captive flamingos to meeting the four aims of the modern zoo. *Flamingo*, *e1*, 23–33.

Ryding, S., Klaassen, M., Tattersall, G.J., Gardner, J.L., & Symonds, M.R. (2021). Shape-shifting: Changing animal morphologies as a response to climatic warning. *Trends in Ecology & Evolution*, *36*(11), 1036–1048.

Salas, M., & Manteca, X. (2017). Visitor effect on zoo animals. *Zoo Animal Welfare Fact Sheet*, *5*. https://www.zawec.org/en/fact-sheets/111-visitor-effect-on-zoo-animals

Space for Life. (2022). Space for life app. https://espacepourlavie.ca/en/space-life-app

Thomas, S. (2016). Future perspectives in conservation. *International Zoo Yearbook*, *50*, 9–15.

Thomas, S. (2020). Social change for conservation. https://www.waza.org/wp-content/uploads/2020/10/10.06_WZACES_spreads_20mbFINAL.pdf

WAZA. (2020). WAZA guidelines for animal-visitor interactions. https://www.waza.org/wp-content/uploads/2020/05/ENG_WAZA-Guidelines-for-AVI_FINAL_-April-2020.pdf

WAZA. (2021). WAZA homepage. https://www.waza.org/

YESTEM Project Team. (2021). YESTEM insight 1. The equity compass: A tool for supporting socially just practice. http://yestem.org/wp-content/uploads/2021/09/2021-YESTEM-Insight-1-Equity-Compass-for-ISL-updated-Sept-2021.pdf

24

Behavioural biology and the zoo as a nature reserve

JAMES E. BRERETON
University Centre Sparsholt, Winchester, UK

24.1 INTRODUCTION

Whether a city zoo or a country safari park, animal collections always have native species in their locale. Whether the local fauna comes in the form of ubiquitous feral pigeons (*Columba livia domestica*) or relatively rare pine martens (*Martes martes*) will be dependent upon the zoo's geographic region. Native species can play a valuable role in the wider function of a zoological collection, providing another facet to the term *conservation*. It is important to encourage native species into zoo sites, particularly those species which play a role in ecosystem services or are of conservation concern and in occur locally. In embracing their native species, some zoological collections have transcended the role of "ex-situ conservation institutions" to feature also as nature reserves.

Pollinators, for example, are fundamental to the health of grassland ecosystems and are essential for the production of many crops. Zoos may encourage these threatened and keystone species onto their sites by using aspects of behavioural biology. More specifically, zoos can use their knowledge of the natural history and behaviour of selected species to develop attractive, useful habitats for them. Developing the zoo as a nature reserve satisfies two of its key roles: education and conservation.

The conservation benefits of native species in zoos are clear. Zoos support *in situ* conservation, and for native species, the zoo may well form part of the species' original range. The environmental conditions within the zoo, such as temperature, humidity, and day length, are the same as those that the animal evolved to cope with. In this respect, many of the challenges associated with captive animal management disappear when native species are managed in "the nature reserve of the zoo."

One of the main roles of the modern zoo is in education (Consorte-McCrea et al., 2017), a subject further expanded upon in Chapter 23. Zoos have a duty to inform their visitors about current conservation challenges, and then encourage them to engage in pro-conservation behaviour change (Clayton et al., 2011). However, some of the conservation issues raised during a zoo visit may be

DOI: 10.1201/9781003208471-27

Figure 24.1 (A) A bug hotel, used to provide a habitat for a range of native invertebrate species, while also providing education and awareness for the public. Picture taken at Birdworld, Farnham (2021). (B) Sparrow nests, built for conservation at Parc de Sigean, France. (C) Conservation education dedicated to the plight of the house sparrow (*Passer domesticus*). Photo credit Paul Rose.

difficult for visitors to relate to. For example, visitors learning about the ivory trade or bushmeat may not always know how to address these issues.

Zoos, as providers of education, can engage in meaningful discussions surrounding native species. This could be through challenging preconceptions of local fauna, or by demonstrating how to create environments for native species (Figure 24.1). Some zoological collections go further still, providing workshops for visitors on the construction of environments for native species (e.g., bird boxes and bee hotels). These actions may encourage visitors to engage in behaviour change. In doing so, new environments for native species may be developed.

Clearly, some native species are important for conservation purposes. The zoo is tasked with the challenge of identifying native species in need of conservation, and then identifying strategies to make these animals at home in the collection. However, the zoo must also identify which species are already on site. This may become a challenge, especially with speciose taxa such as bees (Order Hymenoptera), and ants (Family Formicidae).

The notion of the zoo as a nature reserve is not a novel concept. Many zoological collections actively attract native species onto their sites for conservation purposes. An excellent example is the Wildfowl and Wetland Trust (WWT) Llanelli, which provides expansive wetlands dedicated to wild birds within the zoological collection (Figure 24.2) (WWT, 2021). This mix of reserve and zoo allows the collection to provide soft release reintroductions for local species. For example, the sister site WWT Slimbridge have successfully reintroduced the common crane (*Grus grus*) into their reserves: this allows staff to monitor the health of the reintroduced animals (O'Brien et al., 2017). The mix of wild and captive birds also allows visitors to develop a much wider understanding of the challenges of avian conservation.

The purpose of this chapter is to demonstrate how behavioural biology can be used to encourage native species into zoological collections, and their potential value on site. The chapter also explores some of the challenges associated with assessing animal diversity within the confines of the zoo, and strategies to enhance the value of zoos as nature reserves.

24.2 NATIVE SPECIES IN THE ZOO

Native species vary by locale. In the United Kingdom, common species found in zoo grounds include opportunistic mallards (*Anas platyrhynchos*) and woodpigeons (*Columba palombus*), though rarer species may also be found, such as goldcrests (*Regulus regulus*). Species often vary by habitat, with watervoles (*Arvicola amphibious*), great crested newts (*Triturus cristatus*) and common

Figure 24.2 WWT Llanelli's Millennium Wetlands, which provide habitats separate from the captive collection. Here, visitors may view wild birds in the shelter of hides (right) with minimal disturbance to the birds.

kingfishers (*Alcedo atthis*) found in zoos containing wetlands. Even small zoos can boast great diversity: across a three-year window, Basel Zoo staff identified 3,110 species on their city zoo site!

Native species that are a common sight in zoos vary geographically: tropical zoos for example may have snakes, parrots, and mongooses as regular visitors. Some zoos even have threatened species on site: the Delhi Zoo, for example, have been noted for their wild colony of painted storks *Mycteria leucocephala*), which bred on the zoo's site (Meganathan & Urfi, 2009).

Some species found in zoo sites may be locally or globally threatened. Many zoological collections provide a valuable refuge for wild animals, in part due to the relative rarity of free-ranging predators on their sites. Supplemental feeding, either on purpose by zoo staff or inadvertently through dropped food by visitors or pilfered animal feed, may also provide extra resources for animals. Additionally, zoos may represent a more wildlife-friendly environment than the surrounding environment, particularly in big cities where trees and shrubs are scarce. For example, the house sparrow (*Passer domesticus*), a species with a decreasing population trend, has been frequently sighted at the Zoological Society of London (ZSL) London Zoo. Such species may use the zoo as a refuge, venturing outside the grounds in search of food, but using the site to reduce predation risks.

Some species use the zoo either as part of their migratory route, or as their destination for the summer or winter. For example, the Bewick's swan (*Cygnus columbianus bewickii*) is an annual visitor to WWT Slimbridge, where it spends its winter (Rees, 1982). The swans then migrate back to arctic Russia for their breeding season. The provision of a safe, resource-rich environment during the winter has provided the swans with a consistent site to return to, and a valuable opportunity for researchers to investigate swan behavioural ecology (Wood, Newth, Hilton, & Rees, 2018).

24.3 ENCOURAGING WILDLIFE IN THE ZOO

So far, we have considered the role of the zoo as simply a refuge for wild animals. However, with some attention to detail, zoos can focus on encouraging wild animals onto their sites. However, there is no "silver bullet" that is effective in attracting all animals to the site. Instead, a thorough understanding of the behavioural biology of the target species, along with some information on its dispersal methods and prevalence, is necessary. It is important to identify the "right" species for the zoo and its conservation plans.

The barberry carpet moth (*Pareulype berberata*) is an example. This small, brown-white moth does not look especially extravagant, but its remaining British populations are threatened with extinction. As a result of its decreasing population trend, the moth is a priority species in the UK Biodiversity Action plan: a programme developed to provide conservation support for the locally threatened species. The moth is also a Red Data Book species: this highlights the need for continued conservation support for the moth. The caterpillars of this moth feed exclusively on common barberry (*Berberis vulgaris*): a shrub that was historically common in hedgerows. Unfortunately, common barberry shrubs were destroyed throughout the country because of the discovery that the plant was a host for wheat rust fungus, an agricultural pest (Barnes, Saunders, & Williamson, 2020).

Encouraging the barberry carpet moth into the zoo is not quite so simple as planting a hedgerow. The hedgerow must contain common barberry, and the plants must be sufficiently mature that they can withstand the appetites of the caterpillars. Additionally, the common barberry plants must be managed conservatively, with any cutting of bushes taking place late into autumn to ensure caterpillars have emerged. Finally, the zoo must also consider whether barberry carpet moths are present in the local area so that dispersal into the newly developed habitat can take place.

Not all species have such specific habitat requirements as the barberry carpet moth: some management strategies may encourage a plethora of different species. Flowers, for example, may encourage a wide range of species to use the zoo, both as a resource and a habitat. As an example, the "Grab that Gap" campaign was initiated in 2015 by the British and Irish Association of Zoos and Aquariums (BIAZA) to maximise the number of wildflowers found on their zoo member sites. The campaign encouraged zoological collections to set aside a portion of land or "Gap" for planting wild, native flowers. Through a partnership with Kew's "Grow Wild Initiative," member collections were provided with wildflower seeds to help make the most of these areas. Overall, in 2020, a total of 46 collections signed up to the campaign, making use of areas including standoff barriers, wheelbarrows, and old exhibits for planting (BIAZA, 2021). These areas in turn allowed the zoos to encourage pollinators such as butterflies, hoverflies, and bees onto their site. Carefully considered placement of signage also allowed zoos to enhance their conservation education output.

Whenever considering the zoo as a potential nature reserve, careful attention needs to be paid to the species being encouraged on site (Baur, 2011). While many species have great conservation and education value, there are also potential risks from wildlife. The risk of pests is ever-present in the zoo, especially where animal feed and cafes are concerned. However, further risks may present themselves, especially when wild and captive animals come into close proximity.

By creating exhibits that allow their captive animals to thrive, zoos also create environments that are intrinsically valuable to local, wild animals. Attracted by an abundance of food, coupled with the relative safety from the challenges of urban and wildlife, many additional animal species may be found in zoological collections. Some of the species that enter zoos are unattractive and potentially harmful to zoo animals and public health and safety. Examples of zoo pests include house mice (*Mus musculus*), brown rats (*Rattus norvegicus*), and cockroaches (*Periplaneta americana*), who may pose a disease risk, or provide an extra protein source for some animals. Other potentially problematic species include the grey heron (*Ardea cinerea*), that can be a common site around penguin and pelican enclosures where it steals fish meant for the zoo's inhabitants.

In many countries, zoos are surrounded by a perimeter fence. This fence acts as a barrier, not only to prevent zoo animals from escaping from the facility, but to prevent wild animals from entering the site. These fences may be effective, particularly in preventing large terrestrial animals from entering the site. However, the perimeter fence is less effective for animals which are small enough to squeeze through mesh, burrow below the fence foundations, climb, or fly over. The "partially permeable" zoo perimeter can allow some species to thrive. For example, birds may be less prone to predator attack if nesting in zoo grounds, as the majority of terrestrial predators of eggs and chicks will be unable to enter the zoo's grounds.

Disease is a particularly great risk, especially where closely related taxa come into contact. For example, wild birds may be asymptomatic carriers of diseases such as avian influenza. Zoo birds, after coming into contact with wild birds or their droppings, may be much more susceptible to disease than their wild counterparts. Previous studies have shown that wild birds have passed avian influenza into zoos, with infections seen in species ranging from emus (*Dromaius novaehollandiae*) to golden pheasants (*Chrysolophus pictus*) (Hassan et al., 2020). On the other hand, diseases may be passed from captive stock to wild animals. Disease must therefore be carefully considered, and strategies should be put in place to minimise risks.

Hybridisation with captive animals is another potential risk caused by the presence of native species in the animal enclosures at the zoo. Wild ducks, particularly mallards, may enter the zoo and hybridise with birds of greater conservation value. In order to avoid this issue, many collections provide alternative habitat locations for wild birds (such as mallards) and keep conservation-sensitive species in exhibits where hybridisation does not occur. This could be, for example, through the use of netted aviaries or indoor exhibits (Figure 24.3).

In some cases, it is therefore important to consider not only how native species can be invited into the zoo, but also how wild animals can be kept separate from captive stock. In the avian disease example, the solution may be the development of a netted aviary, or in presenting the food in such a way that only zoo animals have access. Providing a separate area for wild birds, such as a lake, may provide value for native species.

Figure 24.3 Separating wild from captive. The netted aviary (left) at WWT Llanelli houses a breeding pair of yellow-billed ducks (Anas undulata) (right). To reduce risks of hybridisation with mallards (to which they are closely related) and disease risk, the ducks are maintained in a netted aviary.

24.4 MANAGEMENT OF NATIVE SPECIES PROGRAMMES

There are occasions where native species have been taken into ex-situ management with a view to reintroduction. Rather than act as a nature reserve in these cases, the zoo instead takes the role of a research facility and breeding centre. For many species, the behavioural biology of the animal may remain understudied: the zoo therefore provides an opportunity to understand the habitat requirements and behaviour of the species.

One example is the pine hoverfly (*Blera fallax*), a species that is restricted to the highlands of Scotland and listed as endangered in the UK Red Data Book (Rotheray, Goulson, & Bussiere, 2016). The life history of the pine hoverfly is intriguing and is linked to their decline. The larva of this hoverfly feed exclusively on the microbes found in rotten stumps of the Scots pine (*Pinus sylvestris*). Loss of larval habitat has resulted in rapid declines for this insect.

Knowledge of the natural history of the pine hoverfly is key to successful management. In order to safeguard the pine hoverfly, conservationists have developed artificial rot hole sites in the wild. However, with the level of threat that the pine hoverfly is facing, there is also a need for captive management. Scientists have developed a captive population of pine hoverflies and breed their stock for reintroduction purposes. Conservation breeding has revealed unexplored facets of pine hoverfly biology, such as the fact that between 2 and 20% of pine hoverfly larva do not pupate until their second year (Rotheray et al., 2016). This extended lifespan, it is believed, allows the hoverfly population to overcome issues with poor breeding years.

The Royal Zoological Society of Scotland (RZSS, 2021) is involved in an ambitious conservation breeding project for this threatened species. With specialist facilities for conservation breeding (the Pine Hoverfly Shed), the RZSS has built up a captive population with a view to reintroduction attempts. Collaboration with forestry and conservation organisations means that pine forests in Caledonia are being prepared for the eventual reintroduction attempt.

The fen raft spider (*Dolomedes plantarius*) is another species which has been the focus of conservation breeding and reintroduction. One of the largest spider species in the UK, the fen raft spider was historically common in wetland habitats (Leroy et al., 2013). Changes in land management practices resulted in the localised extinctions of spider populations, with only three small populations in the UK remaining. To best protect the spiders, conservationists took fen raft spider specimens from the wild. A collaborative effort across several BIAZA zoos took place, in which hundreds of spiderlings were raised in captivity. In the wild, fen raft spiderlings have a very low chance of surviving their first few months after hatching, so head-starting the animals in captivity was a viable strategy. Once spiderlings had matured after a few months, the animals were introduced to several historic and novel wetland sites as part of the reintroduction strategy.

Zoos and aquariums have been involved in several successful reintroduction projects. Some further examples may be found in Table 24.1.

24.5 MONITORING AND SURVEILLANCE IN THE ZOO

It is important to determine which native species are present in the zoo. For many species, this is much more challenging than might be initially thought. For example, many moths are nocturnal and well camouflaged. These animals spend their days inert, hidden among tree bark and lichen and almost invisible to the observer. Even if, by lucky chance an observer managed to spot a moth, there are over 2,500 moth species in the United Kingdom alone! Clearly, strategies are needed if the variety of native species is to be successfully quantified. How can the variety of moths on site be measured?

Zoos make use of surveys to investigate their on-site biodiversity. Surveys may be informal or formalised. One formal survey method is known as the BioBlitz. During a BioBlitz, the zoo engages in a broad-spectrum survey of on-site biodiversity. The BioBlitz is normally time-limited: the entire event may take place within 24 hours, or over the course of a single weekend. The zoo makes use of volunteers to help it record the animals spotted: this could be in the form of citizen science: with the help of visitors, or taxonomic specialists.

Back to the challenge of the enigmatic moth. Moth sampling may be a part of the wider BioBlitz event. However, how can moths be caught and identified? Here, knowledge of the behavioural biology of the moth becomes essential. Most moth species are nocturnal and are attracted to bright lights. Specialist equipment can be used to take advantage

Table 24.1 Examples of support for native species from credible zoos.

Species	**How the zoo as a nature reserve has aided native species conservation**	**Reference**
Bewick's swan (*Cygnus olor*)	Wetlands are protected at WWT Slimbridge to allow wild Bewick's swans to return to a safe location following their annual migrations. The organisation has also conducted longitudinal research into the behaviour and ecology of the species, along with records of migration habits, that provide useful data to support conservation action.	Rees (1982) Wood et al. (2018)
Common crane (*Grus grus*)	WWT Slimbridge and other conservation organisations were involved with the captive rearing of common cranes for the purpose of reintroduction. During the breeding and rearing process, staff investigated the blood biochemistry and disease status of birds to determine whether birds were ready for release. WWT's own wetland reserves have also served as a location for reintroduced individuals to become re-established.	Bridge (2014). O'Brien et al. (2017)
Red-billed chough (*Pyrrhocorax pyrrhocorax*)	Founder populations of chough were sourced from several UK zoos for the purpose of a multi-partner reintroduction attempt back into Jersey, where the species had previously been extirpated. Zoos were involved in selecting the sites for reintroduction, along with post-release monitoring of the birds.	Corry, Jones, Hales, and Young (2021)
Wood mouse (*Apodemus sylvaticus*)	Dudley Zoo provides patches of woodland and grassland on-site, dedicated to native species conservation. On-site census revealed a large population of wood mice, suggesting that the zoo is functioning as a habitat for native species.	Elwell, Leeson, and Vaglio (2021)
Hazel dormouse (*Muscardinus avellanarius*)	Several UK zoos were involved in the initial reintroduction of the hazel dormouse back into several of its historic sites across the UK. More recently, zoos have been involved in funding post-release monitoring and promoting conservation education for this species.	Gubert et al. (2021)
Red squirrel (*Sciurus vulgaris*)	Welsh Mountain Zoo (among others) operates as a reservoir for the breeding of red squirrels, which are native to the UK. To prevent inbreeding and monitor population demographics, a studbook has been set up. Zoo scientists have also been involved in assessing the disease risk posed by the invasive grey squirrel (*Sciurus carolinensis*), with a view to advising on reintroduction attempts.	Ogden, Shuttleworth, McEwing, and Cesarini (2005). Sainsbury et al. (2020)
Sand lizard (*Lacerta agilis*)	Captive breeding of sand lizards at Marwell Zoo resulted in lizards being released into 26 different sites across the UK. Zoos also provided research into site use, along with consistent funding over a 25-year period to sustain reintroductions and post-release monitoring.	Woodfine et al. (2017)

of these traits. A bright light, mounted above a box filled with egg boxes, can be used to capture moths during the night. The moths are then removed from the moth trap and identified by those with a specialist knowledge in the taxa.

The next challenge is moth identification. It's easy for a non-specialist to confuse the winter moth (*Operophtera brumata*) with the V-moth (*Macaria wauaria*): a UK threatened species. Here, taxonomic specialists are invaluable: they can help identify species or could signpost to other specialists where necessary. However, new strategies are also becoming increasingly available that may reduce the challenges associated with species identification.

Developments in modern technology provide new opportunities in the realm of citizen science and data collection on the presence of biodiversity. The public is now better technologically furnished than ever, to not only identify native species during their time in the wild, but also share their results more widely. Smartphone apps, such as *iNaturalist*, *Seek*, and *iRecord* allow individuals to share pictures of animals that they have seen. These pictures are then identified, either by crowdsourcing of individuals with expert knowledge, or using machine learning to identify the species in question. This allows huge amounts of information to be collected on which species call the zoo home. With a better understanding of rare species on site, zoos will be in a better position to appraise threatened native species and put in place management strategies that match their habitat requirements.

24.6 CONCLUSION

Encouraging native species to make the zoo their home can aid animal collections in meeting their wider strategic plans in conservation. Native species not only allow the zoo to provide context to their educational provisions, but also allow them to provide local *in situ* conservation. Native species behavioural biology should underpin the management strategies used to support animals on site. There are many examples where native species have been successfully managed on site. However, for some taxa, there remain gaps in the knowledge of natural history and behaviour. Further research into the behavioural biology of a range of native species would therefore have value for the zoo community.

Monitoring and surveillance also allow the zoo to quantify its success. It is therefore important that collections continue to assess, using standardised and consistent methods, the types of native species found on their sites. There is great value in the use of modern technology, particularly animal identification apps such as *iNaturalist*. There is also some potential in citizen science and incentivising the public to search for and identify native species during their visit to the zoo. Focusing on native species in addition to exotic animal conservation should provide visitors with a more holistic overview of the types of conservation, while also arming them with knowledge on how to create habitats for native species in their own back gardens.

REFERENCES

Barnes, G., Saunders, D.G., & Williamson, T. (2020). Banishing barberry: The history of *Berberis vulgaris* prevalence and wheat stem rust incidence across Britain. *Plant Pathology*, 69(7), 1193–1202.

Baur, B. (2011). Basel Zoo and its native biodiversity between the enclosures: A new strategy of cooperation with academic institutions. *International Zoo Yearbook*, 45(1), 48–54.

BIAZA (2021). Grap that Gap campaign. https://biaza.org.uk/campaigns/detail/grab-that-gap-2.

Bridge, D. (2014). The UK common crane reintroduction 2010–2014. In *Proceedings of the VIII European Crane Conference*, Gallocanta, Spain (pp. 10–14).

Clayton, S., Fraser, J., & Burgess, C. (2011). The role of zoos in fostering environmental identity. *Ecopsychology*, 3(2), 87–96.

Consorte-McCrea, A., Bainbridge, A., Fernandez, A., Nigbur, D., McDonnell, S., Morin, A., & Grente, O. (2017). Understanding attitudes towards native wildlife and biodiversity in the UK: The role of zoos. In *Sustainable development research at universities in the United Kingdom* (pp. 295–311). Springer, Cham.

Corry, E., Jones, C.G., Hales, A., & Young, G. (2021). Reintroduction of the red-billed chough in Jersey, British Channel Islands. In *Global conservation translocation perspectives: 2021. Case studies from around the globe* (p. 103

Elwell, E., Leeson, C., & Vaglio, S. (2021). The effects of a zoo environment on free-living, native small mammal species. *Zoo Biology*, 40(4), 263–272.

Gubert, L., McDonald, R.A., Wilson, R.J., Chanin, P., Bennie, J.J., & Mathews, F. (2021). The elusive winter engineers: Structure and materials of hazel dormouse hibernation nests. *Journal of Zoology*, 316(2), 81–91.

Hassan, M.M., El Zowalaty, M.E., Islam, A., Rahman, M.M., Chowdhury, M.N., Nine, H.S., & Hoque, M.A. (2020). Serological evidence of avian influenza in captive wild birds in a zoo and two safari parks in Bangladesh. *Veterinary Sciences*, 7(3), 122.

Leroy, B., Paschetta, M., Canard, A., Bakkenes, M., Isaia, M., & Ysnel, F. (2013). First assessment of effects of global change on threatened spiders: Potential impacts on *Dolomedes plantarius* (Clerck) and its conservation plans. *Biological Conservation*, 161, 155–163.

Meganathan, T., & Urfi, A.J. (2009). Inter-colony variations in nesting ecology of Painted Stork (*Mycteria leucocephala*) in the Delhi Zoo (North India). *Waterbirds*, 352–356.

O'Brien, M.F., Beckmann, K.M., Jarrett, N.S., Hilton, G.M., Cromie, R.L., & Carmichael, N.G. (2017). Blood biochemistry and haematology values of juvenile Eurasian cranes (*Grus grus*) raised in captivity for reintroduction. *Journal of Zoo and Aquarium Research*, 5(1), 38–47.

Ogden, R., Shuttleworth, C., McEwing, R., & Cesarini, S. (2005). Genetic management of the red squirrel, *Sciurus vulgaris*: A practical approach to regional conservation. *Conservation Genetics*, 6(4), 511–525.

Rees, E.C. (1982). The effect of photoperiod on the timing of spring migration in the Bewick's swan. *Wildfowl*, 33(33), 119–132.

Rotheray, E.L., Goulson, D., & Bussiere, L.F. (2016). Growth, development, and life-history strategies in an unpredictable environment: Case study of a rare hoverfly *Blera fallax* (Diptera, Syrphidae). *Ecological Entomology*, 41(1), 85–95.

RZSS. (2021). Pine hoverfly restoration. https://www.rzss.org.uk/conservation/our-projects/project-search/field-work/pine-hoverfly-restoration/

Sainsbury, A.W., Chantrey, J., Ewen, J.G., Gurnell, J., Hudson, P., Karesh, W.B., & Tompkins, D.M. (2020). Implications of squirrelpox virus for successful red squirrel translocations within mainland UK. *Conservation Science and Practice*, 2(6), 1–4.

Wildfowl and Wetlands Trust (WWT) (2021). Wetlands for life. https://www.wwt.org.uk/

Wood, K.A., Newth, J.L., Hilton, G.M., & Rees, E.C. (2018). Has winter body condition varied with population size in a long-distance migrant, the Bewick's Swan (*Cygnus columbianus bewickii*)?. *European Journal of Wildlife Research*, 64(4), 1–8.

Woodfine, T., Wilkie, M., Gardner, R., Edgar, P., Moulton, N., & Riordan, P. (2017). Outcomes and lessons from a quarter of a century of Sand lizard *Lacerta agilis* reintroductions in southern England. *International Zoo Yearbook*, 51(1), 87–96.

25

Behavioural biology for the evidence-based keeper

CHRISTOPHER J. MICHAELS
Zoological Society of London, London, UK

LOUISE JAKOBSEN
Browse Poster (Registered Charity in England & Wales no. 1178456), London, UK

ZOE NEWNHAM
Marwell Wildlife, Hampshire, UK

25.1 ZOO-KEEPING AND BEHAVIOURAL BIOLOGY

Since the Zoological Society of London opened the world's first scientific zoo in 1828, research and observation have been key to understanding the ecology and behavioural biology of wild animals. Animal husbandry, including how behaviour is managed, has historically been founded on tradition, received wisdom and so-called folklore husbandry (Arbuckle, 2013). Whilst folklore husbandry may keep captive animals alive and breeding, it is, by definition, prone to or even destined for stagnation, as new approaches are generally disregarded in favour of the old ways. In modern zoos, evidence-based husbandry is the watchword, and this approach is gradually permeating all areas of husbandry including those that cater to the behaviour of captive animals.

Evidence-based husbandry uses the scientific principle to inform how animals are maintained, combining data on captive and wild biology from existing scientific evidence and novel research with the aim of consistently improving animal care. This approach is the perfect example of why undertaking zoo science matters, in practical terms, to animal welfare and zoo operations. The impact of this approach can be substantial. Identification of

DOI: 10.1201/9781003208471-28

Figure 25.1 Comparison of the old (2012; a) and new (2021; b) Galapagos giant tortoise facilities at ZSL London Zoo; the latter was designed incorporating behavioural data from wild and captive giant tortoises. The enclosure is designed to facilitate the behavioural biology of the animals, including duplicate resources (shelter, ponds, basking sites), line-of-sight breaks, and wild-informed environmental control to stimulate natural behavioural patterns. Evaluation of the exhibit confirmed resolution of historic behavioural concerns.

behavioural needs of animals may result in evidence-based phasing out of a species via collection planning (Michaels, Gini, & Clifforde, 2020). Alternatively, it can drive investment in new exhibits that meet institution-wide goals for animals and public alike. Figure 25.1 shows old (opened 2012) and new (2021) facilities for housing Galapagos giant tortoises (*Chelonoidis* sp.) at ZSL London Zoo. The design of the new facility was instigated, and heavily guided by, behavioural data from the animals in question. Keepers conducted in-house research to identify and address behavioural and welfare needs in the old facility (Freeland et al., 2020); these data were used to leverage the decision-making process to secure funds for a new facility. The new exhibit is appropriate in size, resource provision, environmental control, human-animal-interaction management, and layout to meet the behavioural requirements of the tortoises and ongoing monitoring is used to quantify the impact of environmental changes on behaviour.

25.2 BEHAVIOURAL BIOLOGY AND ZOO ANIMAL DIETS AS AN EXAMPLE

Much attention is often given to the formulation of captive diets that replicate the nutritional profiles of wild diets. However, the presentation and physical structure of dietary items are also important as they have direct implications for the behaviour of captive animals. Evidence-based approaches to this are key to appropriate dietary management and this topic is a perfect illustration of the evidence-based approach to behaviour in zoo animals. Animals have evolved many different strategies for diet selection, and finding and processing food, but in almost all feeding is a major contributor to the behavioural repertoire and activity budgets of both captive and wild animals. In captivity, appropriately selected and evidenced food provision strategies can help stimulate the animals' cognitive abilities to problem-solve, prolong feeding time, maintain muscle mass and agility, and benefit the overall welfare, psychological and physical health of the animals (Kistler, Hegglin, Rebel, & Nig, 2009). Knowledge of wild feeding behaviour is also relevant when it comes to identifying and reducing abnormal repetitive behaviours and evidence-based husbandry is important to tackling such welfare challenges. Giraffes (*Giraffa camelopardalis*), alongside many other browsing ungulates, are prone to oral stereotypic behaviours, but these can be mitigated by providing enrichment devices and browse to promote natural feeding behaviour (Fernandez et al., 2008).

The evidence-based approach to feeding captive animals draws on evidence from wild conspecifics to inform dietary composition and food presentation strategies, and to evaluate its impact on behaviour and welfare. Identification of specialist feeding behaviours from wild animals; for example, the three callitrichid genera (*Mico*, *Callithrix*, and *Cebuella*) that have evolved dental adaptations for tree gouging to obtain sap and gum from underneath the bark (Ruivo & Stevenson, 2017), is crucial to understanding how to promote natural behaviour in the zoo. Similarly, field data can provide information on how foraging features in the activity budgets of wild animals. While replicating the wild is usually the goal of evidence-based feeding strategies, ethical trade-offs exist and must be considered. For example, the ability to hunt live prey is impossible to provide to captive large mammalian carnivores, despite the fact that the home range size and hunt chase distance of individual species, as well as the social benefits derived from pack hunting, can have a direct behavioural impact on how we manage those animals in captivity (Clubb & Mason, 2007; Kroshko et al., 2016).

Indirect effects on health due to wild behaviours not being properly promoted can also be encountered (Kapoor, Antonelli, Parkinson, & Hartstone-Rose, 2016). Once a feeding presentation is developed, evidence can be used to appraise it and to inform future practice, not only based on welfare but also on logistics. Januszczak et al. (2016), for example, developed a feeding device for zoo-housed tree-runner lizards (*Plica plica*) based on field data indicating that they often fed on ants emerging from holes in tree bark. When this was compared with simple and less resource-heavy scatter feeding, however, the latter promoted more even enclosure use and wild behaviour and extended feeding time. Zookeepers can extend this approach to diet provision and apply it to all areas of animal husbandry; using behavioural data with comparison against wild or other standards to design, assess and re-shape husbandry practices.

25.3 MISCONCEPTIONS, RESOURCES, AND CULTURE: CURRENT BARRIERS TO EVIDENCE-BASED HUSBANDRY IN ZOOS, AND SUGGESTED SOLUTIONS

Nevertheless, evidence-based husbandry is still not a ubiquitous way of thinking for zoo husbandry staff, and there is often a disconnect between the overall intention to adopt this approach, and the way that animals are managed in practice. One critical misconception that must be tackled in order for evidence-based husbandry to fully penetrate zoo-keeping practice, is the notion that doing the science needed to underpin evidence-based husbandry is the domain of career scientists and students, and that the keeper's role is simply to "shovel poo." In reality, keepers spend the most time working with animals out of anyone in a zoo environment, often have the scientific knowledge and certainly have the competence to, if properly managed and resourced, engage in behavioural science and its application to husbandry practices.

So how can zoos develop their keeping teams to fully engage in evidence-based husbandry using behavioural biology? To facilitate the integration of behavioural science into a keeping role, it is important to ensure that keepers are hired, trained, managed and resourced to do so. Recruitment of staff with ethological training into roles designed to incorporate behavioural biology can be a key step towards the integration of science and husbandry in animal-keeping teams. Appropriate management and Continuous Professional Development (CPD) investment can facilitate keepers to research the species in their care, evaluate their behavioural needs and strategies to meet them, and implement, evaluate, and disseminate their findings (see Box 25.1). A robust welfare auditing process (Wolfensohn et al., 2018), among other tools, can be used to prioritise species for behavioural research attention to ensure a structured and collaborative approach to research within a living collection. Although the provision of high welfare standards is a key aim of the modern zoo (see previous chapters), there is a deficiency of robust welfare auditing processes specifically designed for the assessment of zoo-housed animals. This can limit the appropriate allocation of research resources and therefore the overall impact of behavioural research. One evidenced-based solution of several available is the implementation of the animal welfare assessment grid (AWAG) (e.g., Justice et al., 2017). Measuring four key parameters, physical, psychological, environmental, and procedural, by scoring individual factors of each parameter on a daily basis, a lifetime welfare record is achieved. Changes in the scores over time can be used to identify factors affecting welfare either positively or negatively and these factors can be addressed, such as the complexity of the enclosure and how closely it resembles their natural habitat. If low scores in this area are identified, keepers can use this evidence to plan enclosure modifications to address these scores. Although this process can identify several factors impacting welfare, it relies on the keeping team to have the time to complete the assessments, the species knowledge to do so

BOX 25.1: Behavioural biology for the evidence-based keeper

- Evidence-based approaches to zoo animal behaviour are fundamental to husbandry practice in a modern zoo.
- Keeper involvement in these approaches is critical to best understand the behaviour of animals in zoos and ensure the implementation of evidence-based husbandry.
- Current zoo practice often handicaps zookeeper involvement in behavioural science, and this can be addressed by:
    - **Recruitment.** To deliver behavioural biology outputs, managers should aim to represent behavioural and other relevant scientific skill sets in their animal teams through recruitment. Scientifically trained keepers as well as dedicated research office style roles, provided that they work closely with keeping teams, are both viable routes to this end. Internal training (see below) is also crucial but is not realistically capable of replacing the four to seven years of training provided by under- and post-graduate study – the importation of expertise provides the nucleus needed to grow scientific capacity within a team. This is not a recommendation to avoid hiring non-scientists as keepers, but rather to view this skill set as important to include within animal teams. Critically, there is no *a priori* reason why a candidate with scientific training should be less good at the practical aspects of the keeper role; appropriate interview techniques will identify candidates who are suited to both.
    - **Training and management.** Incorporation of behavioural biology, and other scientific outputs into the expected outcomes of teams via their managers, alongside operational and conservation targets, ensures that this sort of work is prioritised appropriately. In turn, managers should work with external and internal expertise to train staff in behavioural biology and evidence-based practice, so that an evidence-based culture of identifying and assessing problems and solutions and disseminating findings through peer-reviewed literature develops. Every keeper should understand this process and see how they and their work fit into it to prevent single-points-of-failure and obstructive mindsets. The basics of experimental design (controls, sample size etc.) are key to incorporate into training, to ensure robust results and to avoid the disheartening situation of personnel discovering that they have wasted their time collecting useless data. Recognising scientific outputs through internal communications and encouraging curiosity in a structured way is important in building a research-valuing institutional culture.
    - **Building confidence.** For a team with little experience in behavioural biology, starting simple with small, manageable projects requiring little investment, generating results over a short timescale, and addressing key practical issues are an ideal way to start. For example, simple before/after assessment of a training programme, or the documentation of novel behaviours. As confidence and expertise build, more challenging projects can be developed. This approach can lead to exponential growth in outputs as more staff members develop the skill sets needed to lead their own projects and support the training of further staff members.
    - **Collaboration.** External researchers and students represent important sources of experience, knowledge, expertise and time and may contribute financially. Relationships can be developed via direct contact, or via zoo associations such as EAZA, BIAZA and AZA. It is critical that such relationships are true collaborations and that memoranda of understanding and other agreements recognise the knowledge and skill sets contributed by keeping teams and incorporate equal opportunity to develop and be recognised for publications and similar deliverables. To cater to this approach, camera footage can be collected and stored for use in student projects later.

- **Resourcing.** Managers should require proposals for research including resources required and expected outputs and should be supportive of good proposals. Allowing staff a modicum of time to conduct well-designed research is probably the most facilitating thing a manager can do.
- **Collection planning.** Rationalising collections frees up resources to be spent using the collection rather than just caring for it.
- Time-saving data collection, including camera traps and remote technology, simple manual observation, use of existing record-keeping and designing data collection to fit into husbandry routines are useful tools in making keeper-led research happen.

correctly and the ability to address factors of concern. However, using an evidence-based approach fits in with the evolution of the modern zoo and is a method all keepers should adapt to. Such approaches to quantify welfare not only benefit the management of focal animals but allow further behavioural research to focus on the animals most in need of evidence-based welfare intervention.

Unlike some other forms of research, where substantial laboratory bench time is required, the collection of behavioural data is readily included in the busy, time-bound routines of keepers by, for instance, designing data collection sheets and a rapid protocol that can be easily replicated across multiple observers, or by using camera trap footage to be reviewed in short sessions alongside husbandry routines. For example, Carter, Hicks, Kane, Tapley, and Michaels (2021) is an entirely keeper-conceived, led, implemented, and executed behavioural study of ontogenetic shifts in resource and enclosure use by Chinese crocodile lizards (*Shinisaurus crocodilurus*), showing that lizards adjust their behaviour as they age in a manner that reflects similar changes in niche use by wild conspecifics. This information directly informs enclosure design.

Similarly, Boultwood et al. (2021) and Carter et al. (2021) used the same approach to enable behavioural data collection by keepers and vets from amphibians and reptiles, respectively, across COVID-19 lockdown phases; evidence that can be used to inform enclosure design and species selection. The involvement of students through engagement with teaching institutions can also be used to support keeping staff in doing behavioural research and can produce effective results at PhD (Passos, Garcia, & Young, 2021), Master's (e.g., Januszczak et al., 2016), and Undergraduate (Rose, Evans, Coffin, Miller, & Nash, 2014) level. The extraction of behavioural data from existing records is another means of compiling insightful data sets with little additional time required; for example, centralised systems such as the Zoological Information Management System (ZIMS; Species360) (e.g., Hosey et al., 2016), or keeper records from feeding or enrichment (e.g., Waterman et al., 2021).

Probably the biggest barriers to the engagement of keepers in behavioural biology in zoos are institutional cultural attitudes to the roles of keepers and the relevance of science to zookeeping, which creates research-resistant recruitment and management strategies, and the lack of funds required to access and publish in the scientific literature. While Open Access means that the literature is increasingly accessible to keepers, this is accompanied by large article processing charges, which are outside the budget of the majority of zookeeping teams and waivers for which zoos outside of developing countries are not typically eligible (Wood, Newth, & Hilton, 2021).

A further barrier for keepers wanting to increase the use of evidence-based husbandry is getting around the daily working hours of most keepers, and how this impacts the ease of in situ data collection. While keepers are only on site for a limited proportion of the animals' daily time budget, animals remain at the zoo 24 hours a day. Even for diurnal species, as keepers our knowledge of how the animals under our care spend their time when we are not around is lacking, and data collected after keeper working hours can provide a wide range of evidence of how to improve animal husbandry and promote welfare.

This is especially important for nocturnal, crepuscular, or cathemeral species when traditional data collection times available to keepers often does not take their natural biology into account. For example, Burger, Hartig, and Dierkes (2020) used

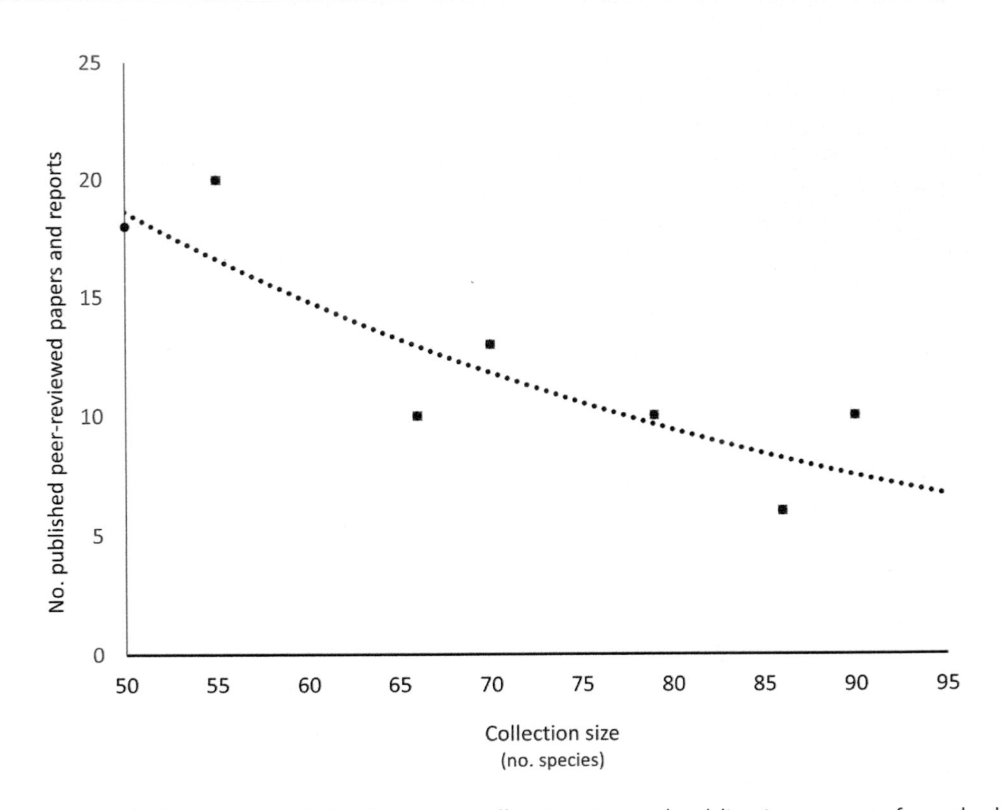

Figure 25.2 Graph showing correlation between collection size and publication outputs from the herpetology section at ZSL London Zoo. The animal collection underwent a systematic rationalisation process between 2015 and 2021, which freed up time to invest in keeper-led research.

infrared-sensitive cameras to show that nocturnal rest-activity rhythms are age-dependent in giraffe, with nightly activity increasing with age. Evidence such as this could be used by keepers looking after this species to adapt husbandry routines to provide more feeding and enrichment devices overnight, especially for herds with individuals of varying ages – this technology can be applied to almost any zoo animal. Results like this show how important it is for keepers to collect data outside of traditional working hours and where possible use this evidence to improve the welfare of the species under their care. To achieve this moving forward, zoos need to focus on an approach that allows keepers not only the time to undertake such research, but also the funding for research technology, and the resources needed to implement husbandry changes based on the results. However, both these factors are particularly problematic in the modern zoo, where a zookeeper's time is limited, and funds are being stretched even further. Detailed collection planning aimed to avoid the accumulation of animals without clearly defined roles can free up funds and time for staff to focus on outputs including research (Figure 25.2). Moving forward, the challenge for keepers in the future will be to continue to utilise evidenced based husbandry, incorporate research into their busy daily schedule and continue to push for more opportunities to provide evidence on which to base husbandry decisions. The management of captive wild animals is constantly evolving, and keepers should be striving to be at the forefront of this.

25.4 EXPORTING AND IMPORTING ZOO BEHAVIOURAL BIOLOGY TO AND FROM THE FIELD

Evidence-based husbandry is not just about using data collected by keepers in the zoo, but also about how keepers can use data collected in-situ to inform husbandry. For many zoo-housed species, there is a vast number of studies published about their wild ecology, and this information can be used to evaluate captive husbandry – for example, by comparing wild and captive behavioural activity budgets (Howell & Cheyne, 2019). However, for others this information is lacking, making applying the evidenced-based approach difficult in this respect. For example, for the okapi (*Okapia johnstoni*) most captive husbandry decisions until recently have been based on only three studies undertaken in the wild (see Hart, 1992) due to the species natural habitat, behaviour and political unrest in the country making undertaking research difficult, providing little evidence for keepers to use. For such species, the absence of wild data to act as baseline may be overcome through the careful use of analogue species, although ecological differences between taxa should be borne in mind (e.g., Michaels et al., 2016). Zoo-housed animals can even provide information about the behaviour of species that is difficult or impossible to observe in the wild; for example, the complex and unique parental care mode used by mountain chicken frogs (*Leptodactylus fallax*) is only known due to careful observations of captive animals (Gibson & Buley, 2004). Furthermore, behavioural research on captive animals can inform field conservation, for example, identifying behavioural patterns that may affect field survey success (Edwards, Bungard, & Griffiths, 2021), or training captive-bred animals to avoid predators that otherwise limit translocation success (Crane & Mathis, 2011). Keepers can contribute to this valuable work. All this demonstrates the importance of zoos facilitating keeper familiarity with (and access to) scientific literature to continue to inform husbandry.

25.5 CHAPTER SUMMARY

Zoo and aquarium keepers play a pivotal role not only in day-to-day animal care but also in driving and directing the development of their field and fostering the use of evidence-based approaches, including using behavioural biology data. Staff recruitment and management, professional development support, and resourcing are all key to enabling the evidence-based approach. If these factors are properly aligned, a zoo's impact on science, conservation, animal welfare and public education is substantially bolstered.

REFERENCES

Arbuckle, K. (2013). Folklore husbandry and a philosophical model for the design of captive management regimes. *Herpetological Review, 44*, 448–452.

Boultwood, J., O'Brien, M., & Rose, P. (2021). Bold frogs or shy toads? How did the COVID-19 closure of zoological organisations affect amphibian activity?. *Animals*, 11(7), 1982.

Burger, A.L., Hartig, J., & Dierkes, P.W. (2020). Shedding light into the dark: Age and light shape nocturnal activity and sleep behaviour of giraffe. *Applied Animal Behaviour Science, 229*, 105012.

Carter, K.C., Hicks, J.J., Kane, D., Tapley, B., & Michaels, C.J. (2021). Age-dependent enclosure use in juvenile Chinese crocodile lizards, *Shinisaurus crocodilurus crocodilurus. Journal of Zoological and Botanical Gardens, 2*, 406–415.

Carter, K.C., Keane, I.A.T., Clifforde, L.M., Rowden, L.J., Fieschi-Méric, L., & Michaels, C.J. (2021). The effect of visitors on zoo reptile behaviour during the COVID-19 pandemic. *Journal of Zoological and Botanical Gardens, 2*, 664–676.

Clubb, R., & Mason, G.J. (2007). Natural behavioural biology as a risk factor in carnivore welfare: How analysing species differences could help zoos improve enclosures. *Applied Animal Behaviour Science, 102*, 303–328.

Crane, A.L., & Mathis, A. (2011). Predator-recognition training: A conservation strategy to increase post-release survival of hellbenders in head-starting programs. *Zoo Biology, 30*, 611–622.

Edwards, W.M., Bungard, M.J., & Griffiths, R.A. (2021). Daily activity profile of the golden mantella in the "Froggotron"- A replicated behavioral monitoring system for amphibians. *Zoo Biology*. 10.1002/zoo.21650

Fernandez, L.T., Bashaw, M.J., Sartor, R.L., Bouwens, N.R., & Maki, T.S. (2008). Tongue twisters: feeding enrichment to reduce oral stereotypy in giraffe. *Zoo Biology*, 27(3), 200–212.

Freeland, L., Ellis, C., & Michaels, C.J. (2020). Documenting aggression, dominance and the impacts of visitor interaction on Galápagos tortoises (*Chelonoidis nigra*) in a zoo setting. *Animals*, 10(4), 699.

Gibson, R.C., & Buley, K.R. (2004). Maternal care and obligatory oophagy in *Leptodactylus fallax*: A new reproductive mode in frogs. *Copeia, 2004*, 128–135.

Hart, J.A. (1992). Forage selection, forage availability and use of space by okapi (Okapia johnstoni) a rainforest giraffe in Zaire. In F. Spitz, G. Janeau, G. Gonzalez, & S. Aulagnier (eds.), *Proceedings International Symposium "Ongulés/Ungulates 91"*, Toulouse, France, p. 18.

Hosey, G., Melfi, V., Formella, I., Ward, S.J., Tokarski, M., Brunger, D., Brice, S., & Hill, S.P. (2016). Is wounding aggression in zoo-housed chimpanzees and ring-tailed lemurs related to zoo visitor numbers? *Zoo Biology, 35*, 205–209.

Howell, C.P., & Cheyne, S.M. (2019). Complexities of using wild versus captive activity budget comparisons for assessing captive primate welfare. *Journal of Applied Animal Welfare Science*, 22: 78–96.

Januszczak, I.S., Bryant, Z., Tapley, B., Gill, I., Harding, L., & Michaels, C.J. (2016). Is behavioural enrichment always a success? Comparing food presentation strategies in an insectivorous lizard (*Plica plica*). *Applied Animal Behaviour Science, 183*, 95–103.

Justice, W.S.M., O'Brien, M.F., Szyszka, O., Shotton, J., Gilmour, J.E.M., Riordan, P., & Wolfensohn, S. (2017). Adaptation of the animal welfare assessment grid (AWAG) for monitoring animal welfare in zoological collections. *Veterinary Record, 181*, 143–143.

Kapoor, V., Antonelli, T., Parkinson, J.A., & Hartstone-Rose, A. (2016). Oral health correlates of captivity. *Research in Veterinary Science, 107*, 213–219.

Kistler, C., Hegglin, D., Rebel, H., & Nig, B. (2009). Feeding enrichment in an opportunistic carnivore: The red fox. *Applied Animal Behaviour Science, 116*: 260–265.

Kroshko, J., Clubb, R., Harper, L., Mellor, E., Moehrenschlager, A., & Mason, G. (2016). Stereotypic route tracing in captive Carnivora is predicted by species-typical home range sizes and hunting styles. *Animal Behaviour, 117*, 197–209.

Michaels, C.J., Fahrbach, M., Harding, L., Bryant, Z., Capon-Doyle, J.S., Grant, S., Gill, I., & Tapley, B. (2016). Relating natural climate and phenology to captive husbandry in two midwife toads (*Alytes obstetricans* and *A. cisternasii*) from different climatic zones. *Alytes, 33*, 2–11.

Michaels, C.J., Gini, B.F., & Clifforde, L. (2020). A persistent abnormal repetitive behaviour in a false water cobra (*Hydrodynastes gigas*). *Animal Welfare, 29*, 371–378.

Passos, L., Garcia, G., & Young, R. (2021). Do captive golden mantella frogs recognise wild conspecifics calls? Responses to the playback of captive and wild calls. *Journal of Zoo and Aquarium Research, 9*, 49–54.

Rose, P., Evans, C., Coffin, R., Miller, R., & Nash, S. (2014). Using student-centred research to evidence-base exhibition of reptiles and amphibians: Three species-specific case studies. *Journal of Zoo and Aquarium Research, 2*, 25–32.

Ruivo, E.B., & Stevenson, M. (2017). EAZA Best Practice Guidelines for Callitrichidae – 3.1 Edition. Available at: https://www.eaza.net/assets/Uploads/CCC/2017-Callitrichidae-EAZA-Best-Practice-Guidelines-Approved.pdf.Accessed 23/12/2021.

Waterman, J.O., McNally, R., Harrold, D., Cook, M., Garcia, G., Fidgett, A.L., & Holmes, L. (2021). Evaluating environmental enrichment methods in three zoo-housed Varanidae lizard species. *Journal of Zoological and Botanical Gardens, 2*, 716–727.

Wolfensohn, S., Shotton, J., Bowley, H., Davies, S., Thompson, S., & Justice, W.S. (2018). Assessment of welfare in zoo animals: Towards optimum quality of life. *Animals, 8*, 110.

Wood, K.A., Newth, J.L., & Hilton, G.M. (2021). For NGOs, article-processing charges sap conservation funds. *Nature, 599*, 32–32.

26

Behavioural biology and the future zoo: Overall conclusions

PAUL ROSE
University of Exeter, Exeter, UK
Slimbridge Wetland Centre, Slimbridge, UK

BRENT A. HUFFMAN
Toronto Zoo, Toronto, ON, Canada

26.1 WHAT TO TAKE HOME FROM THIS BOOK?

Knowledge of the natural behaviour of species housed under human care in zoos and aquariums is essential to the optimal care and management of the animals, to the smooth operations and functioning of the organisation, and to the attainment of the existing four core aims of the modern zoo (conservation, education, research, and recreation). Further consideration of animal welfare suggests that these aims should be updated to include well-being as the fifth aim of the modern zoo. The taxonomic-focussed chapters of this book highlight numerous examples of how appropriate husbandry, based on evidence from the wild and from the evidence of good practice across zoos, improves the ease of keeping of a species as well as the value of the species to the zoo's aims. Animals that are displayed in an ecologically relevant manner will attain positive welfare states more readily and display a more diverse range of natural or normal behaviour patterns, providing added value to their scientific and educational contributions. It is therefore in the best interest of the zoo to ensure that welfare is one of its key aims and a top priority.

The introductory chapters show that collection of behavioural data and use of appropriate evidence is fundamental to building strong support for species management plans. They also demonstrate how research questions can be designed to fill holes in existing knowledge and can yield

DOI: 10.1201/9781003208471-29

valuable information on what zoo-housed species need. Numerous papers discuss the importance of positive welfare states to overall wellbeing in the zoo, and while it is essential that animals survive and thrive, grow, and develop in the captive environment, providing an environment that is too comfortable and removed from natural challenges may ultimately damage lifetime welfare experience. Short-term challenges, such as through the use of environmental enrichment, enclosure complexity and the creation of heterogenous inside and outside spaces, provide opportunities for animals to add range and variation to activity patterns. Such plasticity enables problem solving and cognitive development as behaviours become more nuanced, dextrous, and refined in their performance. Such positive challenge shapes behaviour to be more wild-type in its performance and increases the individual's behavioural repertoire that enhances the usefulness of zoo-housed species to conservation and research aims and outcomes. Functional ecology information, i.e., an understanding of how and why species possess specific traits that convey adaptive advantages, is being collected for an ever-wider range of taxonomic groups (Tobias et al., 2022) and such data need to be embedded more frequently into zoo animal care regimes. Extending the use of functional trait data into zoo animal management further supports the implementation of One Plan Approaches (integrated conservation strategies) that include species populations housed in ex-situ facilities (Rose, 2021).

Keeping this in mind, it is essential that evidence, such as that presented in this book, is used first and foremost to design, implement, and evaluate husbandry and housing systems in the zoo before welfare assessments take place. Figure 26.1 shows the ideal and suboptimal scenarios for evaluating the state of species husbandry and welfare in the zoo. Running welfare assessments first, before biology and husbandry have been reviewed, is a sticking plaster approach to fixing clear deficiencies in the context of existing husbandry practices. The more ideal situation is to research a species' needs first and develop husbandry regimes accordingly, with adjustments made as further evidence becomes available and as welfare assessments are performed. Such methodology is the best way of using the findings of welfare assessments to improve the (already evidence-based) husbandry for each individual animal.

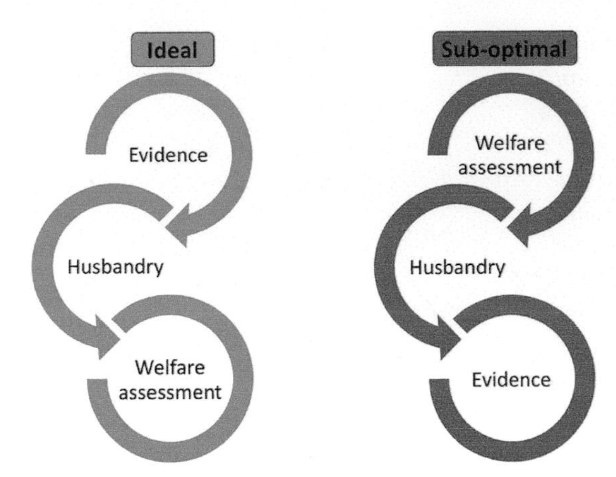

Figure 26.1 Building evidence (natural history, evolutionary ecology, behavioural ecology) into husbandry allows such husbandry to be further improved using information from welfare assessment (green diagram). Evidence-based husbandry at the species-specific level is augmented with data on individual welfare assessments, and tailors the zoo's management approach to the individuals in its living collection. Attempting to perform a welfare assessment when husbandry and housing are not species-specific or relevant to the animals being kept will have limited value as an individual will not be able to reach higher welfare states if care standards are fundamentally inappropriate to start with (orange diagram).

26.2 A NEW MODEL FOR ZOO ANIMAL WELFARE?

The overarching aim of this book is to demonstrate how understanding animal behaviour is a core component in providing and reviewing husbandry and management for zoo- and aquarium-housed species. Ultimately, the welfare state experienced by individual zoo animals will be determined by the environment and care regime provided for it. Building on the idea that welfare is the state of the individual as it attempts to cope with its environment (Broom, 1986), zoo animal welfare definitions and measurements need to include all key husbandry and management factors and how these interact with the individual's perception of its environment, and how extraneous factors (those beyond the animal's control) impact on this perception of its current welfare state.

It is important to consider what is meant by coping. In human terms, coping is defined by Oxford Languages (2022) as "the ability to deal effectively with something difficult." Human and non-human animals must use psychological, behavioural, and physiological strategies to recognise, respond to, and remedy challenges from the environment around them. While coping may seem a very human term – how do we know what animals believe to be coping or not? – it still allows us a way of judging the "normality" of behaviour as well as inferring what the behaviour might mean to the individual as a response to wider environmental conditions. Figure 26.2 defines the inherent characteristics of a zoo species (its wild ancestry and evolutionary ecology), the things that it is provided with in the zoo, how it may or not may not respond to this provision and factors that levy such responses based on the animal's wider perception of other anthropogenic and life experience factors around it.

More detailed examination of Figure 26.2 shows that:

1) Knowledge of behavioural ecology, evolutionary history, and physiological adaptations, together with information on a species' biome and habitat characteristics plus climatic conditions and ecosystem structure, should be the foundation for relevant housing and husbandry, population management guidelines and principles, choice of species in collection plans, and relevant factors to consider for species-specific welfare assessment tools.
2) Husbandry and housing, consideration of species' inclusion in collection plans, as well as population management (and attainment of these goals) and use of zoo-housed animals for conservation purposes, are all influenced by behavioural and ecological information.
3) The points listed in (2) will be affected by species' responses to their captive environment,

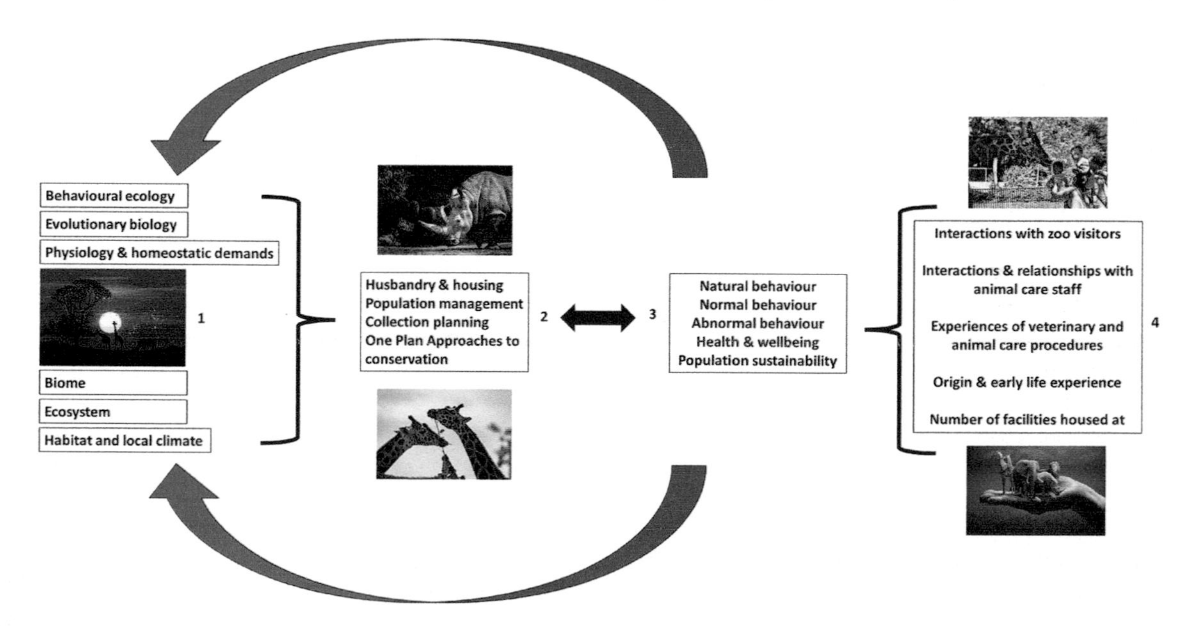

Figure 26.2 Fundamental species information (1) should support management decisions and guidelines for captive care (2). This information in turn influences animal behaviour and welfare (3), which also feeds back for review and reconsideration of (2). An animal's state – be it not developing, surviving, or thriving in the zoo – will be determined by interactions with humans, individual animal behaviours and personality traits, and overall species characteristics (4). Green arrows show how outputs of the animals in the zoo can be judged as relevant to that species (by comparing back to ecology, behaviour, and evolution). This can include the normality of behaviour patterns as well as what may be abnormal (and why this is being performed); indications for how natural behaviours can be promoted; valid criteria for health and wellbeing assessments; and factors to consider for viable breeding to be instigated (e.g., mate choice and mating system information, sexual selection, and traits used for courtship display).

i.e., what behaviour they display, how they remain healthy and in a good state of welfare, whether or not they are able to breed and produce viable young, and if they develop an ecologically relevant behavioural repertoire over the course of their lifetimes. Similarly, consideration of behavioural needs and assessment of highly motivated behaviours (Learmonth, 2019) and what animals do if they cannot perform these due to restricted behavioural choice (Browning & Veit, 2021) will require changes to housing, husbandry, conservation roles etc. listed in (3).

4) All of the behaviours performed by an animal in the zoo, as well as how it breeds and how the population it is part of is managed, and its overall health and wellbeing, will be influenced by the presence of humans (visitors, keepers, vets, managers, and so on). The animal's own experiences of management at its current zoo and of other zoos that it has resided at will affect behaviour patterns and welfare states, as well as early life experiences and rearing conditions. Consideration of the individual animal's personality and pedigree is required when judging responses and reactions to the complete environment of the zoo (people, other animals, conspecifics, enclosures, and routines).

26.2.1 The autonomous zoo animal

Behavioural biology is essential to how current husbandry standards evolve into the best practice guidelines of the future. If we do not understand the fundamental elements of a species' natural history and what it has evolved "to do" in its wild habitat, we will be unable to provide a captive environment where the species is thriving and developing normally. Part of this natural development includes the provision of challenges, albeit in a more controlled fashion than in the wild; the removal of all hardship may be detrimental to behavioural diversity and attainment of positive welfare. Considering beneficial, short-term stressors (eustress) and providing enrichment as a routine (but not routine in nature or design), developing feeding regimes to give a degree of unpredictability and enabling contrafreeloading (working for a reward) where it is relevant to a specific species can all add complexity into the other static nature of the zoo, its enclosures and care regimes. Providing ecologically relevant challenges in captive environments evokes responses in zoo animals that will be similar to those experienced by wild counterparts (such as acute stress, dissatisfaction, or frustration at not being able to instantly access a valued resource). These are likely to be needed by our zoo animals to develop fully in their behavioural responses and capacities to process, store, and recall information that can be used to solve a problem (Meehan & Mench, 2007). In such a situation, the animal's ability to experience choice and control over its environment and become autonomous (self-governing over what it can or cannot do) is enhanced, and the attainment of longer term positive affective states is possible. Figure 26.3 outlines how such an approach sits around the original Broom (1986) definition of animal welfare.

The practical elements for using behavioural biology evidence to the advancement of zoo animal management practices and animal welfare assessments are based on how such data/information/knowledge support a holistic approach to measurement, observation and evaluation of current husbandry practices and animal responses (Figure 26.3). We can adjust the degree of autonomy experienced by animals housed in zoo enclosures if we understand their motivations for specific behaviours and their behavioural needs and choices. Highly motivated actions are likely to be based on evolutionary traits and functional ecology so providing for these in the zoo will allow animals to experience control and choice over their current situation, as well as enabling challenges (and positive experiences of eustress) to be built into husbandry schedules. As outlined in several chapters of this book, the complexity and design of environmental enrichment and the planning and layout of enclosures can contribute to experiences of eustress, which enhances the animal's perception of control over its situation. Enrichment activities for marine and freshwater fish, the planting, and considerations of microhabitats available in finch aviaries, the intricacies of enrichment devices for primates, and the fundamental need of browsing species to have forage and leafy materials as daily nutrition show the range of opportunities that animal care staff have for making the zoo environment a more positively challenging one, regardless of the taxon being cared for. Any device or resource provided to give challenge and provide

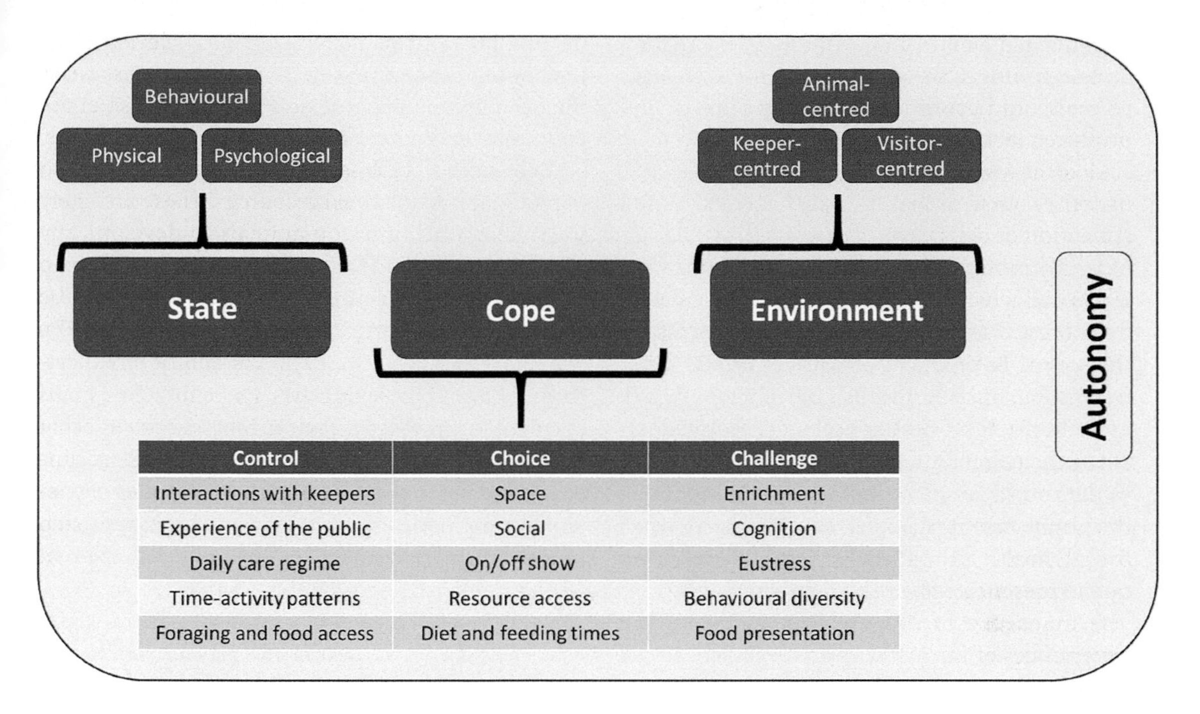

Figure 26.3 The ultimate aim of providing good welfare in the zoo and using behavioural evidence to infer animal welfare states is to instil a sense of autonomy in each individual animal. This can be done by first identifying the state of the animal, which takes many forms as behavioural, physical, and psychological outputs. Then by determining how the individual is coping with the presence of environmental factors and collecting information on opportunities for control (over what the animal can do), choice (to make decisions that lead to autonomy) and challenge (biologically relevant stimuli that cause a diversity in animal's responses and decision making). Finally, evaluating the different environmental variables that influence animal behaviour (those elements of the environment specifically created for the animal itself, those tailored towards zoo visitors and those designed for zoo staff). The autonomy of the individual animal can then be judged partly based on behavioural responses, normality of time-activity patterns, degree of behavioural flexibility, and range of behavioural choices available.

a beneficial stressor has to be considered on the species-specific level and be ecologically relevant; it also has to be within the animal's abilities to solve or remedy, or else it will become a negative chronic stressor. This would have an opposite effect of a lack of choice and control: an erosion of feelings of autonomy and therefore a poor state of welfare. The importance of this species-specific and individual animal-specific approach is expanded upon in Chapter 21.

26.2.2 Measuring behaviour and measuring other factors too

There are a multitude of other factors (inputs that the animal will react to and outputs as a result of such responses) that should be measured alongside behaviour patterns. Examples of this wider review of the animal's behaviour in the context of its zoo environment are suggested below:

- Evaluation of diets to ensure that zoo foods are providing the relevant plane of nutrition and not just "empty calories." Behaviour patterns will alter when diets are inappropriate and therefore incorrect feeding can affect opportunities for behavioural diversity. The importance of dietary quality, variation, and seasonality is a running theme across many of the taxonomic chapters of this book.
- Seasonal and temporal hormone profiles can be measured to provide further review and consideration of the meaning behind and relevance of behaviour patterns of individual animals. Measurement of endocrine responses to external or internal stimuli are useful for the evaluation

of abnormal behaviours, reproductive activities, and responses to the presence of a mate, changes to health and condition, suitability of diets, and how such diets are being metabolised.
- Endocrine responses to acute stressors and excitement are very similar in many species. A full review of time-activity patterns as well as data collection on event behaviours will provide relevant explanation of peaks and troughs in circulating hormones and the metabolites of such hormones that could be markers of stress. Chronic stress responses may be further hampered by any degree of learned helplessness or prolonged elevated circulating hormone concentrations, which may require further investigation and deciphering based around behavioural data collected when the individual animal is placed in different locations, environments, or social groups.
- The importance of pelage or plumage colour and integrity to behaviour (e.g., mate choice, sexual signalling, and an indicator of health status or as a defence or warning mechanism) should be further investigated in the zoo. As can be seen in Chapter 17, many amphibian activity patterns are dependent on the colours that they display.

A mixed-methods approach to behavioural study in the zoo can be useful, as explained in Chapters 4 and 5, and there are a wide range of methods and tools available when observing zoo animal behaviour in a systematic and repeatable manner. Many approaches can be combined into a research project to provide the fullest review of behaviour patterns possible and therefore judgement of biological relevance and wider meaning. For example:

- Construction of time-activity budgets
- Sampling of (e.g.) faeces, urine, or saliva to assay for stress hormone metabolites and then compare to behavioural data
- Application of Social Network Analysis that uses individual behavioural characteristics as attributes to explain an animal's position in a social group
- Space occupancy equations or automated tools that collect data on an animal's position or location to estimate enclosure usage.

Methods need to be planned accordingly based on species-specific anatomy, physiology, and activity; for example, using stress hormone metabolites will correlate with behaviour patterns from a length of time previous to the collection of the hormone metabolite sample (e.g., a 12 or 24hr lag between the behavioural response and that illustrated in the endocrine response).

Behavioural biology allows us to assess what animals are adapted to do, what behaviours provide them with fitness and psychological benefits, and therefore what they require in their enclosures. We can then determine collection plans, and the resources provided to species and the relevance or suitability of such species to the zoo. A final case study on the complexities of captive ungulate management is provided to show how species-specific behavioural and ecological evidence as well as general trends and patterns across a related group of animals are both useful to review when appraising how we keep a particular "type of animal" in the zoo.

26.3 THE CASE FOR ECOLOGICAL EVIDENCE... ULTIMATE UNGULATE QUESTIONS

Humans are well-versed in the husbandry of hoofed mammals – the domestication of sheep (*Ovis aries*) more than 9,000 years ago is one of the earliest examples of animal care. After millennia of experience with more than a dozen domestic ungulate species, it is not surprising that wild ungulate care in zoos was first based on agricultural knowledge. Agricultural practices, however, are aimed at production and tend to treat animals as commodities, whereas modern zoos focus on species conservation and individual animal welfare. These differing goals require different approaches. Additionally, domestic ungulates, despite their taxonomic breadth (two orders and five families), represent only a fraction of ungulate ecology: they are overwhelmingly herd-living grazers. Acknowledging these perspectives requires modern zoos to adjust their mindsets. Otherwise, they will face the impossible task of providing biologically relevant husbandry in systems that were not designed with their goals, nor species diversity, in mind.

26.3.1 The arguments for behavioural biology – practical case studies

Ungulates provide several unfortunate examples where, in hindsight, behavioural biology should have been integrated more quickly into captive husbandry. Consider the critically endangered Sumatran rhinoceros (*Dicerorhinus sumatrensis*), a tropical forest browser. A concerted effort began in the 1980s to capture rhinos from the wild for conservation-oriented captive breeding. However, the initial husbandry protocols were based on those developed for other rhino species (savannah-dwelling grazers and browser-grazers). Fed hay and typical concentrates, three of the seven Sumatran rhinos brought to North America died within 3.5 years. A delayed review of their functional ecology provided a key piece to the husbandry puzzle: the Sumatran rhino requires large quantities of browse.

Adapting husbandry from inappropriate model species also proved catastrophic for the critically endangered saiga antelope (*Saiga tatarica*). Traumatic mortality was very high in "standard" ungulate pens as the saiga's flight response is especially strong, and since it evolved in the open steppe it does not include obstacle avoidance. Saiga reproductive biology is also highly specialised: they have high reproductive potential but very short lives. In treating saiga like generic antelope, zoos failed to provide sufficient physical space and had insufficient capacity to sustain the naturally high population turnover of this species. Even after acknowledging such challenges, success with the species was fleeting. The zoo population in North America and Europe grew to 160 individuals in 1985 but just five years later crashed to 25 animals, and in 2009 it dwindled to zero.

Zoos have always used some level of improvisation to troubleshoot optimal husbandry for exotic species. While basic husbandry has been solved for the most commonly kept zoo ungulates, mismatches between the captive environment and wild ecology abound, sacrificing optimal welfare in ways that we are only now starting to understand. Swamp-dwelling one-horned rhinoceroses (*Rhinoceros unicornis*) frequently develop foot issues. This is now linked to time spent on unnaturally hard concrete, and recent recommendations advise zoos to provide a thick substrate of wood chips for cushioning. Forest-dwellers, such as the Sumatran rhinoceros and Malayan tapir (*Tapirus indicus*), tend to develop eye issues when housed with insufficient shade: they are simply not adapted to sunny environments. As demonstrated throughout this book, understanding a species' functional ecology is the best starting point in providing evidence-based care, with optimal welfare achieved when observations are made to further refine husbandry practices.

The ever-popular giraffe (*Giraffa camelopardalis*) is one of the most ubiquitous zoo ungulates, but (as illustrated in Chapter 7) this gregarious browser still faces numerous deficiencies in human care. Some issues have been alleviated by targeting symptoms, such as training for voluntary hoof care and creating "busy board" enrichment to direct tongue use. However, to address the underlying causes of these issues, husbandry needs to better reflect behavioural biology. Improper hoof wear and arthritis are exacerbated by lack of exercise, a particular problem in cold-climate zoos where giraffes are confined in traditional box-stall barns for long periods. Such barns also restrict herd sizes and decrease the social welfare of giraffes. The most elegant (if expensive) solution for temperate zoos that want to display giraffes is to invest in expansive indoor spaces. Building an enlarged giraffe barn may not be on a zoo's immediate horizon, but other welfare challenges can be addressed through smaller, incremental changes. The abnormal oral behaviours displayed by giraffes represent a highly motivated action without an appropriate outlet. Research into existing husbandry has shown that these oral stereotypies are linked to improper diet and feeding practices, with physiological as well as behavioural consequences (Monson et al., 2018). Sourcing more forage (particularly lots of browse and free choice alfalfa) diminishes stereotypies and improves overall welfare by providing more fibre; this increases time spent ruminating (creating a more natural time budget), improves rumen physiology (thereby improving digestion and nutrient absorption), and reduces metabolic disorders (improving overall health). The behaviour of tongue use is a key component of wild giraffe foraging; providing functionally similar feeding opportunities in the zoo is an easy way for keepers to directly address the underlying problem. Potential solutions include offering smaller concentrate portions throughout the day (rather than bulk feeding) and using enrichment feeders to provide challenge and promote adaptive tongue use. Such thoughtfully considered approaches are simple husbandry changes that can have a dramatic effect on giraffe welfare.

26.3.2 The way forwards

The 250+ species of wild ungulates play a significant role in ecosystems around the globe. Their presence in zoos adds educational value, "wow" factor, and demonstrates real conservation outcomes, exemplified by early work with Arabian oryx (*Oryx leucoryx*), European bison (*Bison bonasus*), and Pere David's deer (*Elaphurus davidianus*) among others. While ungulates will continue to be a part of zoo collections in the future, their diversity precludes the one-size-fits-all approach that is still apparent in generic hoofstock pens, diets, and management approaches. Further research into ecology and natural history is needed to determine specific animal needs (Figure 26.4), which can be used to guide improvements in animal care and management.

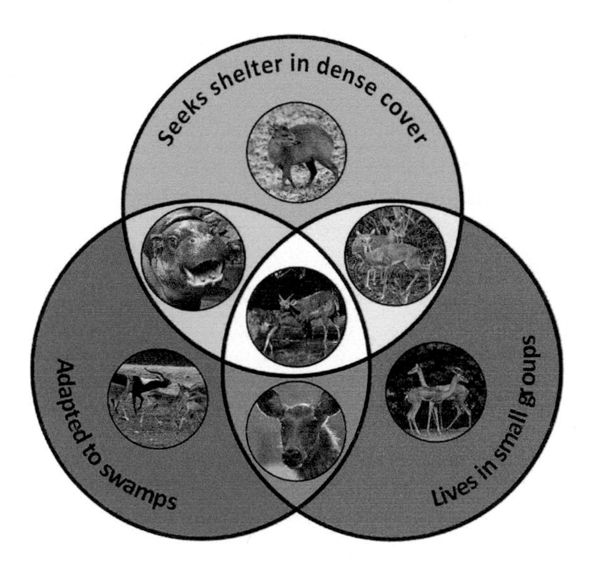

Figure 26.4 Functional ecology is the sum of a species' adaptive characteristics and their relationship with the environment. Here, three of the many adaptive variables – group size, biome, and sheltering behaviour – illustrate how seven ungulate species may overlap in certain key aspects of behavioural biology but differ significantly in others, precluding a one-size-fits-all approach. Understanding such species-specific needs enables zoos to design environments and husbandry protocols to complement a species' behavioural biology, such as by providing appropriate space, substrate, shade, visual barriers, diet, and conspecifics. A customised environment provides the proper context for natural adaptive behaviours and thereby improves welfare. Clockwise from top: red-flanked duiker (*Cephalophus rufilatus*), Guenther's dik-dik (*Madoqua guentheri*), gerenuk (*Litocranius walleri*), sambar (*Rusa unicolor*), Nile lechwe (*Kobus megaceros*), pygmy hippo (*Choeropsis liberiensis*), and [in centre] sitatunga (*Tragelaphus spekii*). Photos: B. Huffman.

Ungulate husbandry based in behavioural biology will challenge many well-accepted practices: to date, most ungulate management has prioritised human requirements, such as containment. Where herd size was once dictated by the available housing, future enclosures should be designed to accommodate a species' natural social structure. In cold climates, zoos intending to keep tropical ungulates (which need to be kept indoors for long periods) must invest in sufficiently large indoor spaces to allow for natural herd sizes, social interactions, and exercise. The environment itself should be chosen or customised with appropriate features, including substrates, visual barriers, and shade, to reflect functional ecology. Zoo diets based on commercially available forages (grown for grazing domestics) should be re-evaluated in the light of natural diets, and replaced, where appropriate, with improved functional equivalents. Zoos are already innovating with browse provision (planting browse farms and developing preservation methods) as a first step to feeding leafy forage as a principal diet component.

Integrating behavioural biology into husbandry practices requires zoos to ask themselves a number of challenging questions. Which species have the potential to thrive in a given institution, based on space, climate, available nutrition options, and budget? Which of those species meet institutional mandates regarding conservation and guest engagement? The ungulate species most in need of conservation attention, such as the saiga, may not be "easy" to care for in established systems; to succeed, zoos need to shed traditions and assumptions and design programmes with the species' functional ecology in mind. The same is true for zoo ungulates with high public draw, such as rhinos and giraffes: if zoos are to actually educate people about these animals in an ecological context, they need to be kept in a fashion that exemplifies their actual behavioural biology.

Ungulate species once passed fleetingly through generic exhibits depending on popular trends. But as ungulate populations decline both in the wild and in captivity, this short-term approach is increasingly unsustainable. Keeping animals in accordance with their functional ecology requires greater institutional investment and long-term commitment to the species. Behavioural biology, therefore, serves both individual welfare and population sustainability, ensuring that prioritised species – particularly those of conservation importance – will thrive into the future.

26.4 BEHAVIOURAL BIOLOGY... FINAL THOUGHTS

One book is never going to cover all of the intricacies and species-specific details of modern-day zoo and aquarium animal behaviour and husbandry, but it can highlight common challenges that zoos face in providing species-appropriate care. Throughout this text, the authors have exemplified the good work of zoos and how much good work can be valued and promoted with further evolution of animal husbandry and care. From elephants to army ants and all in between, best practice care regimes based on behaviour and ecology is a must for population sustainability, good welfare, and effective completion of the zoo's missions and aims.

Moving forwards, new scientific disciplines are using big data to assess interrelated variables that can predict future events, which are helpful in developing the field of zoo animal husbandry. For example, the innovative approach of functional ecology – evaluating traits that determine a species' effect on ecological processes and its response to environmental factors (Reiss, Bridle, Montoya, & Woodward, 2009) – and the addition of data, collected from zoo specimens, to big questions could enable conservation action to be more successful. Figure 26.5 illustrates the principles of functional ecology data and how this could be relevant to the zoo. What is clear overall, however, is that without continual collection and evaluation of behavioural data, zoo animal husbandry and welfare will not fully progress to ensure that the needs of all individuals are met in ex-situ facilities. The four ungulates illustrated in Figure 26.5 all share common features but are different in their phenotype, genotype, and the functional traits they possess. Although habitat

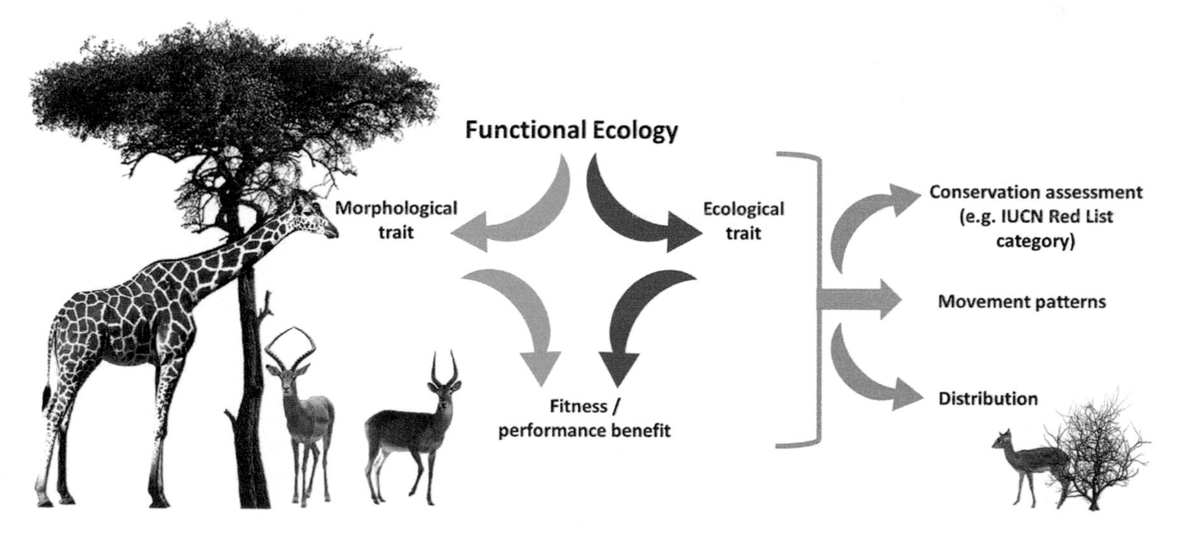

Figure 26.5 Functional ecology assesses continuous morphological characteristics (e.g., size and shape, variation in shared anatomical and physiological features) and continuous ecological characteristics (e.g., variation in range or habitat choice) to determine the adaptive benefits (promotion of reproduction and survivorship) of the traits that species possess. Information on these continuous trait patterns across species can then be used to understand how species are likely to cope with environmental changes, how they may alter or change their distribution and movement patterns to access valued resources and therefore what help they may need from the zoo community to maintain viable populations in the future. Inputs (in green and blue) can predict outputs (in orange) and therefore determine, based on inherent aspects of a species' behavioural ecology and evolutionary history, which species are likely to thrive under prevailing and future environmental conditions.

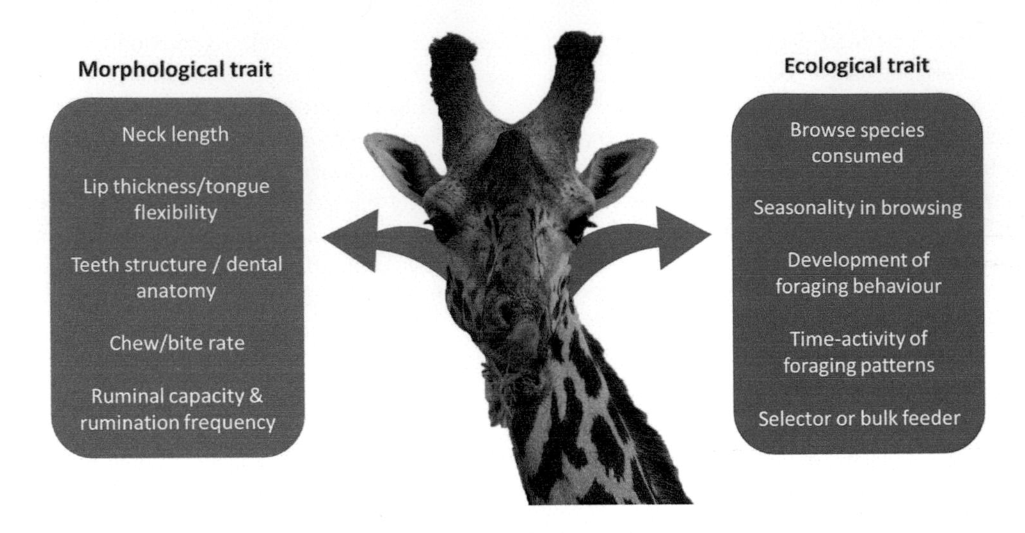

Figure 26.6 Morphological and ecological functional traits of the giraffe. Knowledge of such adaptations can guide and improve zoo animal management for the benefit of the animal. Photo: G. Kautz.

and environmental conditions experienced by these species may be similar, they will respond to changes in selection pressures in different ways, some more successfully than others. Selection pressures that are shaping the evolution of the size and shape of a species (morphology) and how its behavioural ecology enables fitness benefits to be realised continually shape the evolutionary pathway of wild species. Identification of morphological and ecological traits is beneficial to the evolution of zoo animal husbandry and management, too (Figure 26.6).

The continual embedding of such scientific information like functional traits will affect how zoos and aquariums care for their animals into the future, as well as adding more value to their living collections overall. The four aims of the modern zoo (conservation, education, research, and recreation) should be expanded to five (wellbeing), covering the welfare of the animals themselves and the enhancements to the mood and mental health of the humans that engage with the animals on different levels (including the keeper, curator, visitor, and vet). Researching behavioural biology for a given species and then implementing this knowledge into zoo husbandry is in the best interests of all. Zoos and aquariums need to look deeply inside themselves and fully critique their animal collection plans. Can a species be housed appropriately? Do we know of its needs and how to provide for them? Does the presence of this species enhance the aims of the zoo and its wider messaging? Are we aware of the challenges with animal care of this species? Are these challenges logistically and practically fixable if we create biologically relevant housing, husbandry, and management? Not all species belong in the zoo. But behavioural biology belongs in the zoo for all of its species.

REFERENCES

Broom, D.M. (1986). Indicators of poor welfare. *British Veterinary Journal, 142*(6), 524–526.

Browning, H., & Veit, W. (2021). Freedom and animal welfare. *Animals, 11*(4), 1148.

Learmonth, M.J. (2019). Dilemmas for natural living concepts of zoo animal welfare. *Animals, 9*(6), 318.

Meehan, C.L., & Mench, J.A. (2007). The challenge of challenge: can problem solving opportunities enhance animal welfare? *Applied Animal Behaviour Science, 102*(3–4), 246–261.

Monson, M.L., Dennis, P.M., Lukas, K.E., Krynak, K.L., Carrino-Kyker, S.R., Burke, D.J., & Schook, M.W. (2018). The effects of increased hay-to-grain ratio on behavior, metabolic health measures, and fecal bacterial communities in four Masai giraffe (Giraffa camelopardalis tippelskirchi) at Cleveland Metroparks Zoo. *Zoo Biology, 37*(5), 320–331.

Oxford Languages (2022). *Oxford Languages*. Oxford University Press, https://languages.oup.com/

Reiss, J., Bridle, J.R., Montoya, J.M., & Woodward, G. (2009). Emerging horizons in biodiversity and ecosystem functioning research. *Trends in Ecology & Evolution*, *24*(9), 505–514.

Rose, P.E. (2021). Evidence for aviculture: Identifying research needs to advance the role of ex situ bird populations in conservation initiatives and collection planning. *Birds*, *2*(1), 77–95.

Tobias, J.A., Sheard, C., Pigot, A.L., Devenish, A.J.M., Yang, J., Sayol, F., Neate-Clegg, M.H.C., Alioravainen, N., Weeks, T.L., & Barber, R.A. (2022). AVONET: morphological, ecological and geographical data for all birds. *Ecology Letters*, *25*(3), 581–597.

Index

Note: Page numbers in *Italics* refer to figures; **Bold** refer to table

C

F

G

H